COMPREHENSIVE CHEMICAL KINETICS

COMPREHENSIVE

CHEMICAL KINETICS

EDITED BY

C. H. BAMFORD

M.A., Ph.D., Sc.D. (Cantab.), F.R.I.C., F.R.S.
Campbell-Brown Professor of Industrial Chemistry,
University of Liverpool

AND

C. F. H. TIPPER

Ph.D. (Bristol), D.Sc. (Edinburgh)
Senior Lecturer in Physical Chemistry,
University of Liverpool

VOLUME 3

THE FORMATION AND DECAY OF EXCITED SPECIES

ELSEVIER PUBLISHING COMPANY
AMSTERDAM - LONDON - NEW YORK
1969

ELSEVIER PUBLISHING COMPANY
335 JAN VAN GALENSTRAAT
P.O. BOX 211, AMSTERDAM, THE NETHERLANDS

ELSEVIER PUBLISHING CO. LTD.
BARKING, ESSEX, ENGLAND

AMERICAN ELSEVIER PUBLISHING COMPANY, INC.
52 VANDERBILT AVENUE
NEW YORK, NEW YORK 10017

LIBRARY OF CONGRESS CARD NUMBER 68-29646
STANDARD BOOK NUMBER 444-40802-9

WITH 53 ILLUSTRATIONS AND 30 TABLES.

PRINTED IN THE NETHERLANDS

COMPREHENSIVE CHEMICAL KINETICS

Contributors to Volume 3

C. S. BURTON   Department of Chemistry,
University of Texas,
Austin, Texas, U.S.A.

A. B. CALLEAR   Physical Chemistry Laboratory,
University of Cambridge,
Cambridge, England

T. CARRINGTON   Institute for Basic Standards,
National Bureau of Standards,
Washington, D.C., U.S.A.
(*Now* Department of Chemistry,
York University,
Toronto, Ont., Canada)

D. GARVIN   Institute for Basic Standards,
National Bureau of Standards,
Washington, D.C., U.S.A.

G. HUGHES   The Donnan Laboratories,
University of Liverpool,
Liverpool, England

J. D. LAMBERT   Physical Chemistry Laboratory,
University of Oxford,
Oxford, England

W. A. NOYES, JR.   Department of Chemistry
University of Texas,
Austin, Texas, U.S.A.

Preface

The rates of chemical processes and their variation with conditions have been studied for many years, usually for the purpose of determining reaction mechanisms. Thus, the subject of chemical kinetics is a very extensive and important part of chemistry as a whole, and has acquired an enormous literature. Despite the number of books and reviews, in many cases it is by no means easy to find the required information on specific reactions or types of reaction or on more general topics in the field. It is the purpose of this series to provide a background reference work, which will enable such information to be obtained either directly, or from the original papers or reviews quoted.

The aim is to cover, in a reasonably critical way, the practice and theory of kinetics and the kinetics of inorganic and organic reactions in gaseous and condensed phases and at interfaces (excluding biochemical and electrochemical kinetics, however, unless very relevant) in more or less detail. The series will be divided into sections covering a relatively wide field; a section will consist of one or more volumes, each containing a number of articles written by experts in the various topics. Mechanisms will be thoroughly discussed and relevant non-kinetic data will be mentioned in this context. The methods of approach to the various topics will, of necessity, vary somewhat depending on the subject and the author(s) concerned.

It is obviously impossible to classify chemical reactions in a completely logical manner, and the editors have in general based their classification on types of chemical element, compound or reaction rather than on mechanisms, since views on the latter are subject to change. Some duplication is inevitable, but it is felt that this can be a help rather than a hindrance.

Volumes 1 and 2 having dealt with the practice and the theory of kinetics, we turn in Volume 3 to the effect on chemical systems of radiation of relatively low and relatively high energy, the processes giving rise to excited molecules, atoms and free radicals, and charged species, all being considered. The formation of excited products in exothermic elementary chemical reactions (the nature of the excess energy and its distribution over the degrees of freedom) is dealt with, and finally the transfer of translational, electronic, vibrational and rotational energy between chemical species (methods of investigation, results and theory) is reviewed.

The Editors wish to express once again their sincere appreciation of the advice and support so readily given by the members of the Advisory Board.

Liverpool　　　　　　　　　　　　　　　　　　　　　　　C. H. BAMFORD
October, 1969　　　　　　　　　　　　　　　　　　　　　C. F. H. TIPPER

Contents

Chapter 1

Effect of Low Energy Radiation

C. S. BURTON

AND

W. A. NOYES, Jr.

1. Introduction

Radiation may consist of particles, of electromagnetic waves, and of compressional (*i.e.* sound) waves. Thus a discussion of the chemical effects of low energy radiation could have a vast scope and would be very superficial and indeed meaningless unless limited carefully. Such a limitation must needs be arbitrary. Since the action of compressional waves is different in character from that of either moving charged particles or electromagnetic waves we will omit completely any discussion of their effects.

It is of course true that the reactivities of molecules depend on their content of vibrational energy and this in turn may be altered both by change in temperature and by absorption of relatively long wave radiation. Nevertheless, absorption of infrared radiation to give high overtones of fundamental frequencies is weak and hence high overtones could generally be excited by absorption of radiation only under experimental conditions such that vibrational energy would be rapidly equilibrated with the surroundings. Direct chemical effects other than those due to increase in temperature are rarely to be expected from absorption of microwaves and of infrared radiation. Such effects will be mentioned later. Similarly shock wave techniques lie outside the scope of this chapter.

Electronic states of molecules may be changed either by inelastic collisions with charged particles or by absorption of electromagnetic radiation. Electronic states of very little energy above the ground state do exist and they may have reactivities dependent on other characteristics than energy alone. Data on this point are meager. Thus for practical purposes only electronic states which may be formed by absorption of wavelengths shorter than the infrared have been extensively studied. Energies of 1.5 to 2.0 electron volts constitute a lower limit to the electromagnetic energy convenient for use. Charged particles of such low energies may suffer inelastic collisions with molecules and some chemical effects may result. However high intensity beams of charged particles with such low energies are not available under conditions suitable for observing chemical effects.

As energy is progressively increased by shortening wavelengths, or by increasing kinetic energies of charged particles, all molecules can be made to shift to new electronic states and ultimately to ionize. Ionization potentials extend from four or five volts for the alkali metals in the gas phase to about 25 volts for helium, *i.e.* from a wavelength of about 3200 A to about 500 A. Special techniques will permit studies to be made of the effects of electromagnetic radiation to wavelengths of a few hundred Angstrom units. Most research has been limited by the transmission of quartz and hence to wavelengths longer than 2000 A. High quality quartz of small thickness may transmit to about 1500 A. Lithium fluoride and calcium fluoride may be used to about 1200 A or less.

Nearly all substances begin to become transparent somewhere below 100 A, *i.e.* to X-rays and gamma rays. We consider this to be a high energy region and do not include it in this discussion.

The discussion in this chapter will be limited to the absorption of electromagnetic radiation of wavelengths from about 1000 A to perhaps 10000 A or from say 12 electron volts down to 1 or 2 electron volts. Mention will be made when possible of the effects of charged particles with corresponding energies, but such studies are so rare as not to constitute an important part of the chapter.

2. Survey of the primary effects of electromagnetic radiation

Absorption of electromagnetic radiation by monatomic gases leads either to production of atoms in excited states or, if the wavelength is short enough, to ionization. Absorption lines have widths which depend on the following factors[1]: (*a*) the temperature, *i.e.* there will be a Doppler effect which will broaden the line if the absorbing atoms are in motion; (*b*) an intrinsic factor dependent on the nature of the electronic state and on the extent of perturbations by other states; (*c*) the number of isotopes and hence of the magnitude of the hyperfine structure; (*d*) perturbations either from induced effects from neighboring atoms or by external fields.

Absorption becomes weaker as the change in principal quantum number becomes larger, so that absorption by a monatomic gas to give ions and electrons is usually weak at the convergence limit of a spectral series. Ions may initiate chemical reactions but chemical effects initiated by ions and electrons of low velocity have received relatively little attention because of the experimental difficulties. Allusion will be made to these effects later but they will not be discussed in detail.

At wavelengths longer than those required to produce ionization, excited or energy-rich atoms or molecules are formed. Ultimately they must either return to the ground state or disappear by reaction. One or the other of the following things can happen: (*a*) emission of radiation; (*b*) collision with transfer of all or part of the energy to other molecules; (*c*) direct chemical reaction with colliding molecules.

Reactions sensitized by monatomic gases have been very extensively studied. The majority of the cases involve mercury vapor. Direct reaction of excited mercury atoms with other gases has also received considerable attention of late.

2.1 SENSITIZATION BY MONATOMIC GASES

A monatomic gas at low pressures shows line absorption at wavelengths longer than those required to produce ions. Transitions from atoms in low energy states to higher states will be observed. If levels and their assignments are well known, one may use the Boltzmann equation

$$\frac{N_i}{N} = \frac{p_i e^{-E_i/kT}}{\sum_{ij} p_{ij} e^{-E_{ij}/kT}} \tag{1}$$

where N_i is the number of atoms in the state i, N is the total number of atoms, p_i is the *a priori* probability or statistical weight of the state i and E_i is the difference in energy between the state i and the one of lowest energy. The summation in the denominator must be performed over all states which participate statistically in the equilibrium.

The quantity p_i is the number of states into which this level will split in the presence of a foreign magnetic field. For cases which show the usual coupling between L and S one may write

$$p_i = 2J + 1 \tag{2}$$

where J is the total angular momentum quantum number. Other cases need not concern us.

Since hyperfine structure may sometimes be important we must consider in more detail absorption by atoms with nuclear moment i. If atoms with nuclear moments i have total electronic angular momenta J, the i and the J combine to give a total angular momentum which may have the values

$$J+i, J+i-1, \ldots |J-i| \tag{3}$$

The selection rules state that the total angular momentum quantum number may change by ± 1 or 0. Thus an element with several isotopes each with its own nuclear spin will present a line spectrum with a very complex and, under most experimental conditions, unresolved hyperfine structure. Nevertheless, as we shall see later, the overlap between the hyperfine components of a spectrum line is sufficiently incomplete to permit preferential excitation of one isotope in a mixture of isotopes by radiation from a lamp containing that same isotope.

For light atoms the selection rules are quite rigorously obeyed. The important rules which govern transitions when only one electron is involved are

$$\Delta L = \pm 1 \tag{4}$$
$$\Delta S = 0; \Delta J = \pm 1 \text{ or } 0 \tag{5}$$

Transitions are also forbidden if $0 \to 0$ for J. Other transitions, and particularly those which involve coincidental transitions of two electrons, may occur under special conditions. For such transitions different selection rules are followed but, at least at present, photochemists are not often concerned with such transitions.

As atomic weights increase, selection rules are less rigorously obeyed so that many transitions occur with violation of one or more of them. This is particularly true for the transitions with change in spin, so that $\Delta S = 1$ (or a change, for example, of multiplicity from singlet to triplet or *vice versa*) is often found for heavy atoms. Transitions for which the rules are obeyed always occur with higher probability than those for which one or more of the rules is disobeyed.

It must be kept in mind that the distribution of atoms in a monatomic gas among the various energy levels will follow equation (1) under equilibrium conditions. Atoms in sufficiently high concentrations in high energy states will not be in thermodynamic equilibrium with their surroundings. In the absence of energy dissipation by collision they must eventually lose energy by radiation. Thus "forbidden" emissions are sometimes observed, particularly under very low pressure conditions in comet tails, in certain stellar spectra, and in the upper atmosphere. In the presence of strong magnetic and electrostatic fields the selection rules must also be modified and "forbidden" transitions appear more often.

If the ground state of a monatomic gas is singlet, fully allowed transitions will lead in absorption to single lines (except for hyperfine structure). If the ground state is doublet, *e.g.* $^2S_{\frac{1}{2}}$, the absorption will consist of doublets. On the other hand if the lowest level is a P level there will be two levels more or less close together: $^2P_{\frac{1}{2}}$ and $^2P_{\frac{3}{2}}$. Absorption from these two levels will give at least five transitions for each change in principal quantum number. If the energy difference between these two ground states is small the ratio of statistical weights will largely determine the distribution of the atoms between them and $^2P_{\frac{3}{2}}$ will have twice the abundance of $^2P_{\frac{1}{2}}$. The five lines will be: $^2P_{\frac{1}{2}} \to {}^2S_{\frac{1}{2}}$, $^2P_{\frac{1}{2}} \to {}^2D_{\frac{3}{2}}$, $^2P_{\frac{3}{2}} \to {}^2S_{\frac{1}{2}}$, $^2P_{\frac{3}{2}} \to {}^2D_{\frac{3}{2}}$, $^2P_{\frac{3}{2}} \to {}^2D_{\frac{5}{2}}$. The sum of intensities of the last three will be approximately twice the sum of the intensities of the first two.

Generally speaking, atoms in doublet and triplet states are reactive and tend to form chemical bonds either with themselves or with other atoms or molecules. Hence the photochemist is concerned in practice with gas phase photochemical reactions of mercury, cadmium, zinc, and the noble gases whose atoms exist normally in singlet states.

2.2 PHOTOSENSITIZATION BY MERCURY VAPOR[2,3]

The ground state of mercury vapor is $6\,^1S_0$ and the transition which would not violate selection rules would be

$$^1S_0 + h\nu \rightarrow {}^1P_1 \tag{6}$$

where for the sake of brevity we omit the prefix 6 which refers to the principal quantum number (see Fig. 1). The longest wavelength which will cause such a transition lies at 1849 A. This wavelength is appreciably absorbed by oxygen of the air, while some quartz is not transparent to it. Few studies have used this wavelength exclusively[4]. The energy is 6.71 electron volts or 154.7 kcal.mole^{-1}.

The atomic weight of mercury is high so that some selection rules will not be rigorously obeyed. There is one other prominent transition at 2537 A

$$Hg(6\,^1S_0) + h\nu \rightarrow Hg(6\,^3P_1) \tag{7}$$

The transition to form the state $^3P_0$ would violate the rule which prevents $0 \rightarrow 0$ for change in J and is not observed in absorption. Under certain extreme experimental conditions the reverse transition

$$^3P_0 \rightarrow {}^1S_0 + h\nu \tag{8}$$

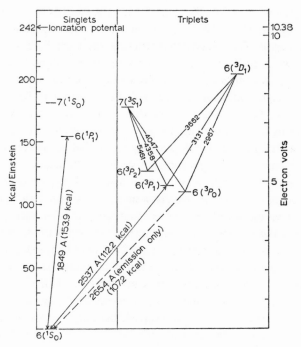

Fig. 1. Energy level diagram for Hg. From ref. 1.

is observed. 6^3P_0 mercury atoms would not be in thermodynamic equilibrium with their surroundings. If they are unable to lose energy in some other way, *viz.* by collision, they must eventually lose it by emission. The energy for (7) is 4.89 electron volts or 112.7 kcal.mole^{-1}.

Reaction (6) as an allowed transition means that absorption of the 1849 A line by mercury vapor is extraordinarily high, and with mercury vapor in equilibrium with liquid mercury at room temperature nearly all of the radiation will be absorbed in a distance of less than one millimeter from the window through which radiation enters the vessel.

In treating line absorption one may use as incident light either a continuum or a line spectrum. If mercury vapor is at room temperature the absorption line may be very sharp and indeed hard to observe without good resolution. There is by Rayleigh scattering a slight shift in wavelength. Thus the wavelengths removed by absorption tend to be supplied by scattering from the wings. The distribution of excited atoms does not, therefore, follow any simple exponential variation with distance.

If a mercury lamp is used as a light source the mercury in the lamp removes some of the radiation before it leaves the lamp. This leads to the phenomenon of reversal and only specially designed mercury lamps emit radiation which will be absorbed by cold mercury vapor. Such special lamps operate usually with a foreign gas and a low pressure of mercury vapor.

Mercury emission lines usually come from light sources in which the emitting atoms are in rapid motion. The lines are much broader than absorption lines in cold mercury vapor. Thus Rayleigh scattering for such light sources will also reintroduce slowly the absorbed components and there will be a long and gradual tailing off of the absorption as one proceeds away from the window of incidence. Because of Rayleigh scattering a "parallel" beam ceases to be strictly parallel.

The mean life of the $^1P_1$ state is very short, about 10^{-9} sec. At 1 atmosphere pressure the mean time between collisions of a given molecule is about 10^{-10} sec so that appreciable emission by $^1P_1$ mercury atoms can occur even in the presence of strongly quenching gases at pressures of several millimeters.

When a fluorescent radiation is strongly absorbed by the emitting gas there occurs the process of radiation imprisonment, *i.e.* a quantum emitted will not always escape from the vessel before it is absorbed, and after absorption it may be re-emitted. Thus the radiation will be passed from atom to atom in random directions, and with the very high absorption of the 1849 line, many such events can occur before radiation escapes from the vessel. The detailed theory has been treated by several authors[2, 5, 6]. Since each atom which interacts with a photon may, in the absence of foreign gas, have a mean life of about 10^{-9} second regardless of the previous history of the energy it has received, the apparent mean life will depend on the number of events between absorption and final escape from the vessel. Quite naturally the geometry of the incident light beam and of the vessel will determine this quantity. The observed rate of decay of fluorescence should not be simply exponential in character and a simple kinetic picture will not suffice.

Let us assume that $^1P_1$ mercury atoms (which we will designate as ^{1}Hg for convenience) collide with other molecules which remove energy to form radicals

$$^1\text{Hg} + \text{M} \rightarrow \text{R}_1\cdot + \text{R}_2\cdot + \text{Hg} \qquad (9)$$

and that this process is competitive with emission of radiation

$$^1\text{Hg} \rightarrow \text{Hg} + h\nu \qquad (10)$$

By the assumption of the steady state (6), (9), and (10) lead to

$$I_a/I_f = 1 + k_9[\text{M}]/k_{10} \qquad (11)$$

This is the well known Stern–Volmer relationship which relates the number of photons absorbed per unit volume per second (I_a) to the number emitted per unit volume per second (I_f). The quantity I_f/I_a is the emission efficiency and is often given the symbol Q.

A detailed treatment of radiation imprisonment would be beyond the scope of the present chapter. The situation would be simple if the entering radiation had exactly the same wavelength distribution as the emitted radiation and if all emitted radiation would traverse a distance to the walls exactly equal to that traversed by the incident radiation in reaching the point of absorption. Neither of these requirements is met in practice. In a general way one may say that if photons are half absorbed on the average from each point of emission and if the mean life τ is defined by the equation

$$\tau = 1/(k_{10} + k_9[\text{M}]) \qquad (12)$$

the apparent mean life for the emission of radiation would be

$$\tau = 2/(k_{10} + k_9[\text{M}]) \qquad (13)$$

With high absorption, as would be the case for the 1849 A line of mercury, the apparent mean life of the radiative process might be as much as one or two orders of magnitude greater than the true radiative life. For these and other reasons quantitative data concerning chemical effects produced by $^1P_1$ mercury atoms are meager. Nevertheless the rather extensive data relative to $^3P_1$ mercury atoms are sometimes marred by failure to remove the 1849 line completely.

The $^1P_1$ mercury atoms which possess nearly 155 kcal.mole^{-1} of energy are capable of initiating many chemical reactions. Nearly all single bond energies (probably all of them) are less than this quantity. Many double bonds have dissociation energies less than 155 kcal and indeed nearly the only bond with energy greater is that between two nitrogen atoms in the nitrogen molecule, N_2.

Let us revert to (9). The rate of production of the R_1 and R_2 radicals will be

$$d[R_1\cdot]/dt = d[R_2\cdot]/dt = (1/V)\int_0^V k_9[^1Hg][M]dV \tag{14}$$

where V is the volume of the reaction vessel and the rate is given in each case as the average rate in moles of radicals formed per unit volume per second. If either R_1 or R_2 or both can be completely scavenged by some process, first order with respect to R_1, such as

$$R_1\cdot + X \rightarrow Y \tag{15}$$

then the rate of formation of Y will be

$$d[Y]/dt = (1/V)\int_0^V k_{15}[R_1\cdot][X]dV \tag{16}$$

where the rate is once again the average rate in the entire volume V.

If it is desired to calculate relative rates of the various reactions it now becomes necessary to evaluate $[^1Hg]$. If the concentration of M can be maintained sufficiently high to prevent diffusion of radiation, *i.e.* if essentially every excited mercury atom collides effectively with a molecule M before it emits, and if the concentration of X can be kept so low that reactions between it and ^1Hg may be neglected, the average rate of formation of excited mercury atoms per unit volume will be

$$(1/V)\int_0^V I_a dV = \bar{I}_a \tag{17}$$

where I_a is the microscopic rate of absorption of quanta per unit volume and $\bar{I}_a$ is the average absorption per unit volume in volume V.

Under these simplified conditions (which may, however, be approached very closely in practice)

$$\bar{I}_a = k_9[^1Hg][M] = k_{15}[R_1\cdot][X] \tag{18}$$

The measurement of $\bar{I}_a$ for a line such as the 1849 A line of mercury will not be easy although in principle it could be determined if mercury vapor were mixed with some gas which gave a well defined product with a known quantum yield. In converting such measurements to measurements with the gas M one would perforce make some assumption about the breadth of the absorption line. If pressures and the characters of the two gases are such that no "broadening" could occur, I_a could be determined merely by measuring the total amount of product formed per second by the calibrating gas and dividing by the volume and by the known quantum yield. This procedure would be difficult to apply with the 1849 A line but has been used with some success for the 2537 A line.

If I_a has been determined and the concentration of X is known it is now possible to determine $k_{15}[R_1\cdot]$. The classical problem confronting the kineticist now arises: how can one obtain a useful quantitative measurement of the rate coefficient, k_{15}, so that it will be of value either to theoreticians or to persons interested in applying kinetic data to complex systems? It must be remembered that, if $R_1\cdot$ radicals react sufficiently rapidly with X so that they may be considered to disappear right at the place where they are formed, it might be possible to use electron spin resonance, or in certain rare cases absorption spectroscopy, to determine $[R_1\cdot]$. Suffice it to say that some measurement other than a purely kinetic one must be used if k_{15} is to be obtained with the system in question. If one value of k_{15} is known, values for other gases can be obtained since relative rate coefficients are easier to determine than absolute ones.

It would lie beyond the scope of this treatment to describe the various methods now available for determining radical concentrations. Most of them are difficult to apply to the gas phase and none of them is really useful unless (a) the intensity I_a is very high, i.e. over 10^{18} quanta.ml^{-1}.sec^{-1} and (b) experimental conditions are such that radical lifetimes are at least 10^{-4} second or more. In a way these two conditions are mutually exclusive.

In the absence of a precise measurement of $[R_1\cdot]$ it is necessary to use methods of doubtful value even though some may give the right order of magnitude. For example, scavengers may be used and for a few of them, such as oxygen and nitric oxide, the order of magnitudes of the rate coefficients for the reaction with simple radicals are known. The rate of reaction of methyl radicals with oxygen[7] is such that the activation energy is probably zero, and at pressures of 100 or more torr where third body effects are unimportant, reaction occurs at about one collision in a thousand. Unfortunately, oxygen reacts with excited mercury atoms and so would most other scavengers. Studies of absolute rates of reaction by the techniques of mercury sensitization are difficult.

One other case must be mentioned, although we reserve a more detailed discussion for another section.

Let us assume that (15) competes with a second order removal of the radicals $R_1\cdot$

$$R_1\cdot + R_1\cdot \rightarrow \text{products (P)} \tag{19}$$

The microscopic rate of (19) is

$$+d[P]/dt = k_{19}[R_1\cdot]^2 = -(\tfrac{1}{2})d[R_1\cdot]/dt \tag{20}$$

The macroscopic rate must be obtained by integration

$$-(\tfrac{1}{2})(d[R_1\cdot]/dt) = (1/V)\int_0^V k_{19}[R_1\cdot]^2 dV \tag{21}$$

If one assumes that radicals react in the same volume element in which they are formed, the steady state condition for $R_1\cdot$ leads to the following equation

$$I_a = k_{15}[R_1\cdot][X] + 2k_{19}[R_1\cdot]^2 \tag{22}$$

It must be remembered that I_a for the 1849 A line of mercury varies rapidly from one point to the other in the absorbing system. The integration of I_a to obtain $\bar{I}_a$ may be accomplished in principle using actinometry but the localized values of I_a are difficult to obtain. Let experimental conditions be such that in most of the reaction volume

$$2k_{19}[R_1\cdot]^2 \gg k_{15}[R_1\cdot][X] \tag{23}$$

Then

$$[R_1\cdot] = (I_a/2k_{19})^{\tfrac{1}{2}} \tag{24}$$

Various methods have been used to obtain k_{19} for simple radicals. Thus the rate coefficient for the reaction

$$CH_3\cdot + CH_3\cdot \rightarrow C_2H_6 \tag{25}$$

has been carefully determined[8-10].

Returning now to reaction (15), one obtains for the microscopic rate

$$+d[Y]/dt = k_{15}(I_a/2k_{19})^{\tfrac{1}{2}}[X] \tag{26}$$

The integration over the volume V can now be accomplished only if I_a is known as a function of the volume. Since the rates of reactions (19) and (15) depend on different powers of $[R_1\cdot]$ it is impossible to obtain relative rates of any great value. However, if the inequality (23) is fulfilled it still might be possible to compare rates for various molecular species X provided the integrated rate in (21) is kept constant.

Nevertheless it must be emphasized that, for any system for which the radical concentration changes rapidly from point to point in the reaction vessel, the use of comparative rates when one of them depends on a different power of the radical concentration from the other is not recommended. Thus mercury sensitization is not a particularly useful method for determining such rates and this is even more true for the 1849 A line than it is for the 2537 A line.

There may, however, be certain experimental conditions for which radical concentrations may be considered to be uniform throughout the reaction vessel. These would be mainly (a) low intensity so that the rate of reaction (19) is small and (b) either a low steric factor or a high activation energy for reaction (15). Under such experimental conditions radicals could diffuse far from points of formation. Unfortunately, these conditions will also permit numerous collisions between radicals and walls. If accommodation coefficients are high, so that wall collisions are effective in radical elimination from the system, the concentrations of radicals throughout the vessel may still be far from uniform[10-12].

We have emphasized these various points not because studies with the 1849 A line of mercury are at present very important but because they illustrate certain fundamental principles which need to be understood in studying all photochemical and even many thermal reactions.

The 2537 A line of mercury has been used far more extensively than the 1849 A line, and many significant results have been obtained from mercury sensitized reactions studied with this wavelength.

The $^3P_1$ state of mercury formed by (7) has a mean life[2] of about 1.1×10^{-7} sec, *i.e.* the rate coefficient for the reaction

$$Hg(6^3P_1) \rightarrow Hg(6^1S_0) + h\nu \qquad (27)$$

is about 0.91×10^7 sec^{-1}. Stated in this way the mean life is the so-called radiative life. Thus in a large number of isolated $^3P_1$ mercury atoms not subjected either to collisions or to the action of foreign fields, half would emit 2537 A quanta in less than 1.1×10^{-7} second and half in a time longer than this.

Apparently monochromatic resonance radiation of mercury which passes through mercury vapor at the saturated pressure at 25 °C is about half absorbed in four millimeters distance. Beer's law is not obeyed at all because the incident radiation cannot be considered to be actually monochromatic, and absorption coefficients of mercury vapor vary many times between zero and very high values in the very short space of one or two hundredths of an Angstrom unit. Moreover, absorption of mercury resonance radiation by mercury vapor is sufficiently great even at room temperature to make radiation imprisonment a very important phenomenon. If the reaction vessel has any dimension greater than a few millimeters the apparent mean life of $Hg(6^3P_1)$ may be several fold the true radiative life of 1.1×10^{-7} sec, reaction (27), because of multiple absorption and re-emission.

Several phenomena may occur when 6^3P_1 mercury atoms encounter other molecules. The simplest of these would be the transfer of electronic energy from one atom to another

$$Hg(6^3P_1)+A \rightarrow A^*+Hg(6^1S_0) \tag{28}$$

Reaction (28) must be exothermic, although this step would occur despite a slight endothermicity and activation energy. The better the resonance, and hence the smaller the amount of energy which must appear as kinetic energy, the higher the rate coefficient. The atoms A* will be formed by (28) only if certain rules are well obeyed. Thus the sum of the electron spins on the right should equal the sum of the electron spins on the left.

Reaction (28) may be followed by emission by A*

$$A^* \rightarrow A+h\nu \tag{29}$$

If the entire mechanism consists merely of steps (7), (27), (28), and (29) the following expressions based on the assumption of the steady state may be derived

$$1/Q_{Hg} = 1+k_{28}[A]/k_{27} \tag{30}$$

$$1/Q_A = 1+k_{27}/k_{28}[A] \tag{31}$$

where Q_{Hg} and Q_A are the emission efficiencies for Hg and A atoms respectively. If these efficiencies can be measured accurately, equations (30) and (31) provide means for testing this simple mechanism. Since molecular velocities at 25 °C are about 2×10^4 cm.sec^{-1} for Hg and the same for A if it is a heavy atom, the excited atoms will move less than 0.01 cm from the point of formation before they emit. This simplifies the study of the system to some extent.

Since the frequency of the radiation emitted in (27) will be different from that from (29) the relative intensities can be determined without too much difficulty. Thus one can obtain

$$Q_A/Q_{Hg} = k_{28}[A]/k_{27} \tag{32}$$

k_{27} is known and hence k_{28} can be determined from this ratio. To determine whether the mechanism is adequate one could with enough care obtain absolute values of Q_A. A plot of $1/Q_A$ vs. $1/[A]$ should be a straight line of slope k_{27}/k_{28}.

The mechanism just discussed is a simplified one which has not been tested experimentally in great detail. The thallium radiation emitted from a system of Hg and Tl vapor on excitation was studied many years ago by Compton and Turner[13]. It might be interesting to pursue these studies further with modern techniques, including the use of mercury isotopes. One difficulty would be that of obtaining other appropriate monatomic gases at reasonable pressures under feasible experimental conditions.

The transfer of electronic energy from $6\,^3P_1$ mercury atoms to di- and polyatomic molecules by steps analogous to (28) is known to occur[3]. This may occur wholly or in part by the processes

$$6\,^3P_1 + M'' \rightarrow 6\,^3P_0 + M''_x \tag{33}$$

$$\rightarrow 6\,^1S_0 + M' \tag{34}$$

where we follow the customary practice of using '' to designate a lower (in this case the ground) electronic state and ' to designate an upper electronic state. The energy difference between $6\,^3P_1$ and $6\,^3P_0$ is small, only about 0.22 electron volt or 5.07 kcal.mole^{-1}. This corresponds to transitions in the infrared. M''_x represents a vibrationally excited molecule still in the ground electronic state. Evidence for (33) exists and good resonance apparently increases the probability of energy transfer. However, nitrogen is one of the most effective gases in promoting $^3P_0$ formation and yet resonance is not efficient.

The reactions of $6\,^3P_0$ mercury atoms will be dismissed briefly. As mentioned before, the transition

$$\text{Hg}(6\,^3P_0) \rightarrow \text{Hg}(6\,^1S_0) + h\nu \tag{35}$$

not only violates the spin conservation rule but the rule which forbids $0 \rightarrow 0$ for J. Hence the radiative mean life of $6\,^3P_0$ mercury atoms should be long, and is in fact about 10^{-3} second[14]. With such a long mean life, bimolecular processes must be considered and there is some evidence for the dissociation of molecules such as nitrogen which require more energy than that available by collision with one excited mercury atom.

It must also be kept in mind that approximately one collision in 3000 at 25 °C provides sufficient energy for the reverse of (33). Thus at pressures of 1 torr, or even less, the reverse reaction may occur provided that the colliding molecules do not undergo the reaction

$$\text{Hg}(6\,^3P_0) + M''_x \rightarrow \text{Hg}(6\,^1S_0) + M \tag{36}$$

This would be true for inert compounds such as N_2, CO_2, and SF_6.

While reactions of $6\,^3P_0$ mercury atoms must also be considered as well as those of $6\,^3P_2$ atoms (which might be formed by collision from $6\,^3P_1$ mercury atoms with an activation energy of about 8 kcal.mole^{-1}) they have not been an important part of photochemistry up to the present time and we will not discuss them further.

Most mercury sensitized reactions are based on (34) where M' may be either a stable upper electronic state or a repulsive state. The literature in this field is so extensive[3] that we must limit our discussion to a few examples.

If the upper electronic state of M is either repulsive or if the energy exceeds the dissociation energy into radicals or atoms, the primary process will be the same as (9)

$$^3Hg_1 + M \rightarrow R_1\cdot + R_2\cdot + Hg \qquad (37a)$$

where we use $^3Hg_1$ as a shorthand notation for $Hg(6\,^3P_1)$. The equations (16) through (26) with appropriate changes in rate coefficients may be used just as well for $^3Hg_1$ as for ^1Hg atoms. Due to lower absorption of the 2537 A than of the 1849 A line, difficulties due to radiation imprisonment and of localization of effects to proximity to the walls will be lessened, although still serious.

One example will be discussed. The action of $^3Hg_1$ atoms on ammonia is certainly represented by

$$^3Hg_1 + NH_3 \rightarrow NH_2\cdot + H\cdot + Hg \qquad (37b)$$

One might almost generalize that hydrogen atom elimination is important whenever $^3P_1$ mercury atoms encounter a molecule with hydrogen atoms which are not bound with too great an energy.

The complete mechanism either for the mercury sensitized or for the direct photochemical decomposition of ammonia is complex and there exists reasonable doubt about some of the steps even after many years of intensive work. Under static experimental conditions hydrogen and nitrogen are virtually the only products and they are formed in the ratio of 3 : 1. Other products must, therefore, be of negligible importance under ordinary conditions.

Hydrazine may, however, be an important product and it is almost certainly formed even in static systems. If this is so it must be attacked by free radicals and atoms so rapidly that its steady state concentration is small, less than can be determined quantitatively.

Gunning and his coworkers[16, 17] have worked extensively on the ammonia reaction. They have shown that hydrazine is formed in a flow system and that the fraction of hydrazine in the products increases markedly with the flow rate. These facts would seem to lead to the following conclusions: (a) the reaction

$$NH_2\cdot + NH_2\cdot(+M) \rightarrow N_2H_4(+M) \qquad (38)$$

is reasonably fast under ordinary experimental conditions. The third body M for dissipation of energy is included because it is certainly necessary, although its presence may not be rate controlling if the pressure is 100 torr or more. (b)

Hydrazine is attacked rapidly either by H atoms or by NH_2 radicals or both. The reactivity toward hydrogen atoms has indeed been known for many years. (c) Hydrogen atoms must be removed from the system by several different processes only one of which would be by the attack on hydrazine. Two well-recognized steps are

$$H\cdot + H\cdot + M \rightarrow H_2 + M \tag{39}$$

$$H\cdot + walls \rightarrow \tfrac{1}{2}H_2 + walls \tag{40}$$

The necessity for the third body to dissipate the energy of reaction (39) is well known. (40) is also an accepted reaction whose rate will depend on the characteristics of the wall surface.

Let us for the sake of convenience write the attack on hydrazine as follows

$$H\cdot + N_2H_4 \rightarrow H_2 + N_2H_3\cdot \tag{41}$$

The fate of N_2H_3 need not concern us here. If (39) is the only hydrogen atom combination reaction and if the reaction

$$H\cdot + NH_2\cdot + M \rightarrow NH_3 + M \tag{42}$$

is at least partially responsible for the low quantum yield in this system, the steady state concentration of hydrazine would be given by

$$k_{38}[NH_2\cdot]^2[M] = k_{41}[H\cdot][N_2H_4] \tag{43}$$

At high hydrogen atom concentrations (39) may be taken as the main method of removal of hydrogen atoms. If this is so

$$I_a = k_{39}[H\cdot]^2[M] \tag{44}$$

to a rough first approximation. (43) then becomes

$$\frac{k_{38}k_{39}^{\frac{1}{2}}}{k_{41}I_a^{\frac{1}{2}}}[M]^{\frac{1}{2}}[NH_2\cdot]^2 = [N_2H_4] \tag{45}$$

The concentration of $NH_2\cdot$ would be a complex function of the various parameters including the intensity. $[NH_2\cdot]$ might be expected to vary roughly as the square root of I_a and therefore the hydrazine concentration would increase as the square root of the intensity. Its absolute concentration would depend on the rate coefficient k_{41} and the data indicate this to be large.

This simple treatment of the data with a static ammonia system is not adequate for a flow system. In a rapidly flowing mixture the concentrations of all radicals and atoms will be low and hence (40) will predominate over (39). Hydrazine once formed will be swept rapidly into a zone in which the hydrogen atom concentration is low and it will be thus free from attack by these atoms. The results of Gunning and his coworkers indicate that the reaction for ammonia disappearance approaches the stoichiometry

$$2NH_3 = N_2H_4 + H_2 \tag{46}$$

at infinite flow rate. Thus reaction (38) must be fast compared to either (39) or (42), perhaps because of the longer lifetimes of the primary collision complexes.

The complete ammonia reaction has, among other things, been hard to elucidate because little is known about the fate of N_2H_3 radicals. Some uncertainty exists even as to the step which leads to N_2 formation, although

$$2NH \rightarrow N_2 + H_2 \tag{47}$$

is the leading candidate.

During recent years real progress has been made in understanding a few of the reactions of excited mercury. If the primary step is (9) the reactions of the radicals will be the same in a given system no matter how they are formed. Mercury has seven stable isotopes and when ordinary mercury vapor is irradiated by a resonance lamp made of ordinary mercury all seven will be excited at rates proportional to their abundances and to the relative intensities of the hyperfine components from the resonance lamp. Even though some isotopes are more abundant than others and the intensities from the resonance lamp will be higher for the abundant isotopes than for others, there is no specificity about (9) and no dependence on isotopic composition would be expected.

The situation will be very different, however, if the primary process leads directly to a final product[17]. An example would be found in a system consisting of mercury vapor and an alkyl halide such as methyl chloride. Three different primary processes can be visualized

$$^3Hg_1 + CH_3Cl \rightarrow Hg + H\cdot + \cdot CH_2Cl \tag{48}$$

$$\rightarrow Hg + \cdot Cl + \cdot CH_3 \tag{49}$$

$$\rightarrow HgCl + \cdot CH_3 \tag{50}$$

Reactions (48) and (49) are of the type of (9) and each introduces two radicals into the system. (50) introduces only one radical. Reaction (48) seems to be unimportant.

Reactions started by the methyl radicals may be numerous and we will not attempt here to discuss them in detail. Four reasonable ones may be postulated

$$CH_3\cdot + CH_3Cl \rightarrow CH_4 + \cdot CH_2Cl \tag{51}$$

$$CH_3\cdot + \cdot CH_2Cl \rightarrow CH_3CH_2Cl \tag{52}$$

$$CH_3\cdot + CH_3\cdot \rightarrow C_2H_6 \tag{53}$$

$$\cdot CH_2Cl + \cdot CH_2Cl \rightarrow C_2H_4Cl_2 \tag{54}$$

Reactions (51)–(54) do not lead to calomel formation and each reaction (52), (53), and (54) causes two radicals to disappear.

The net number of radicals introduced in the system is $2I_a - N_{HgCl}$ where N_{HgCl} is the number of molecules of HgCl formed per quantum absorbed. Hence

$$2I_a - N_{HgCl} = 2N_{C_2H_5Cl} + 2N_{C_2H_6} + 2N_{C_2H_4Cl_2} \tag{55}$$

HgCl is formed by two steps, (50) and

$$Cl\cdot + Hg + M \rightarrow HgCl + M \tag{56}$$

Reaction (56) can occur between chlorine atoms and any isotope of mercury in the system while (50) will occur only with those isotopes of mercury which have been excited by the absorption of radiation.

In some of the experimental work on this system a mercury resonance lamp made with isotope 202 has been used. Mercury with the usual isotopic distribution was mixed with the methyl chloride. Let X^i be the atomic fraction of isotope 202 in the HgCl formed and $\bar{X}^i$ be the atomic fraction of isotope 202 in ordinary mercury. Then

$$N_{HgCl} \frac{(X^i - \bar{X}^i)}{(1 - \bar{X}^i)} = N_{HgCl}^{50} \tag{57}$$

where N_{HgCl}^{50} is the number of moles of HgCl formed by reaction (50).

The chlorine atoms formed by reaction (49) either appear as HgCl or as HCl, viz.

$$\cdot Cl + CH_3Cl \rightarrow HCl + \cdot CH_2Cl \tag{58}$$

Therefore

$$N_{HgCl} - N_{HgCl}^{50} = N_{HgCl}^{56} \tag{59}$$

$$(N_{HgCl}^{56} + N_{HCl})/I_a = \phi^{49} \tag{60}$$

$$N_{HgCl}^{50}/I_a = \phi^{50} \tag{61}$$

where ϕ^{49} and ϕ^{50} are the quantum yields of reactions (49) and (50) respectively.

References pp. 63–66

The quantities X^i, $\overline{X}^i$, N_{HgCl}, N_{HCl}, and I_a can all be obtained experimentally. Of these I_a may be the most difficult to determine accurately, but Gunning and his coworkers have determined the quantum yield of hydrogen formation from several hydrocarbons (e.g. cyclohexane) with quite high precision so that, barring pressure broadening and uncontrolled variations in radiation imprisonment, data can be obtained for testing these various relationships.

Many experiments have been performed with mercury lamps made of single isotopes and methyl chloride or other substances mixed with mercury with a normal isotopic composition. These experiments serve probably better than any others to illustrate the true meaning of monochromatic light from a photochemical standpoint. The emission line from a mercury resonance lamp has many hyperfine components arising from each of the seven isotopes of mercury. These lines are broadened for several reasons, including the Doppler effect. The absorption by cold mercury vapor also has a great number of hyperfine components and will cover a smaller wavelength range than the emission line because each absorption component in cold mercury vapor will be broadened less than each component in a source at higher temperature. Self-reversal will also modify the emission lines to some extent. And yet the whole absorption by ordinary mercury at room temperature covers a wavelength range of only a few hundredths of an Angstrom unit or perhaps 0.4 cm^{-1}.

In spite of this very small range of absorption there is a preferential absorption by a single isotope when a mercury resonance line from a single isotope is used. The quantum yield of the isotopically specific step (50) with methyl chloride is 0.28 and of course part of the nonspecific step (49) will also be brought about by the isotope used to make the resonance lamp.

These various studies on reactions caused by monochromatic light from specific isotopes have been of great value in elucidating the entire question of mercury sensitization. By the use of scavengers which will remove atoms and radicals it has been possible to eliminate certain secondary reactions and further to determine the true nature of the primary processes.

An interesting case is the action of excited mercury on oxygen[18, 19]. The energy required to dissociate the oxygen molecule is 5.13 electron volts or 118 kcal.mole^{-1}. $6\,^3P_1$ mercury atoms are thus unable to cause this dissociation. However, the reaction

$$^3Hg_1 + O_2 \rightarrow HgO + O(^3P) \tag{62}$$

is almost certainly exothermic although the heat of sublimation of HgO is not known and hence this statement may not be made with certainty.

There is another possible process

$$^3Hg_1 + O_2 \rightarrow Hg + O_2' \tag{63}$$

where O_2' may represent one of several electronic states with varying amounts of vibrational energy. Some of the excess energy would appear as kinetic energy. The reaction

$$O + O_2 \rightarrow O_3 \tag{64}$$

is known to occur with high probability, at least at high total pressures[19]. The reaction

$$O_2' + O_2 \rightarrow O_3 + O \tag{65}$$

may also occur provided each O_2' retains at least 40 kcal.mole^{-1} of energy after (63). O_3 reacts thermally with mercury to give mercuric oxide.

Less than two per cent of the reaction between excited mercury and oxygen follows (62) and only this part is isotopically specific. This finding is in agreement with data obtained in 1926 by Dickinson and Sherrill[18] and verified by Volman[19, 20] in 1954. In a flow system many molecules of ozone are formed per mercury atom passing through the system under experimental conditions such that radiation absorbed by the oxygen would be negligible. Thus (62) must not be a very important primary step. The formation of mercuric oxide is beyond question due to a thermal reaction between ozone and mercury, and this can occur even after condensation of the ozone together with some mercury in a cold trap.

Attention has been given at some length to sensitization by mercury vapor partly because of the very real value of this work and partly because it illustrates principles which must be used in discussing more complex systems.

2.3 ABSORPTION BY DIATOMIC GASES

Ever since the classical paper by Franck[21] in 1926 on the relationship between spectroscopic phenomena and the photochemical primary process in diatomic molecules and the equally important papers in 1924 by Henri and Teves[22] who clearly pointed out that diffuse bands in the spectrum of the sulfur molecule S_2, as shown in Fig. 2b, meant that the molecule must dissociate upon absorption, the study of photochemical reactions initiated by absorption by diatomic molecules has been on a sound basis.

The spectroscopy of diatomic molecules has been treated so extensively that it will not be necessary either to describe the phenomena in great detail or to summarize the rules governing the complex symbols for designating energy states of such molecules. Readers are referred to any of the excellent treatises on this subject[23].

It is, however, necessary for us to summarize a few of the fundamentals of the photochemical primary processes for diatomic molecules.

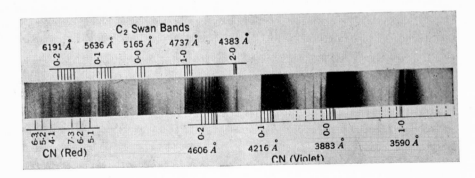

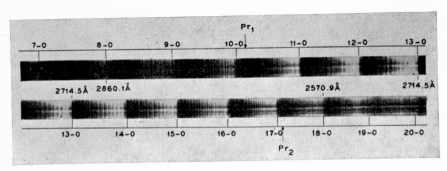

Fig. 2. (a) Band spectrum of the molecules CN and C_2. (b) Predissociation spectrum of S_2.
From ref. 23a.

(a) Electrons in diatomic molecules, as for atoms, may have spins paired or one or more of them may be unpaired. If the number of electrons is odd (e.g. NO) at least one electron cannot have its spin paired with another. The multiplicity is $2S+1$ where $S = n/2$, n being the number of unpaired electrons. Transitions between states with different multiplicities are "forbidden" but the degree of forbiddenness will depend on the masses of the atoms which make up the molecule. Molecules which are not in thermodynamic equilibrium with their surroundings must sooner or later lose energy. If collisions are not sufficiently frequent to permit equilibration, "forbidden" transitions will occur. Thus forbidden transitions are observed in the aurora borealis.

(b) The quantum numbers l of the individual electrons combine to give a quantum number L as for atoms. The projection of L on the nuclear axis is given the symbol Λ. When $\Lambda = 0$ the state is a Σ state and the wave function is symmetric about the nuclear axis. When $\Lambda = 1$ the state is a Π state, etc. The selection rule for Λ is $\Delta\Lambda = \pm 1$ or 0.

(c) The spins of the individual electrons add, as for atoms, to give a resultant S. The projection of S on the nuclear axis is given the symbol Σ (not to be confused with the Σ which stands for $\Lambda = 0$).

(*d*) There is a quantum number

$$\Omega = |\Lambda + \Sigma| \tag{66}$$

Ω is strictly analogous to J for atoms.

(*e*) If the electronic wave function is unaltered by reflection in a plane through the two nuclei it is called positive and is given the symbol $+$. If there is alteration the symbol $-$ is placed as a superscript.

The symbol for the normal hydrogen molecule is $^1\Sigma_g^+$ and means the following: (*i*) there is no net orbital angular momentum around the axis of the molecule; (*ii*) the two electron spins are paired (as they have to be if the Pauli Exclusion Principle is not to be violated); (*iii*) the sum of the l's is an even number (zero in this case) and leads to the subscript g, which means that the wave function does not change sign by inversion through the center of symmetry (otherwise the symbol u would be used).

Most simple diatomic molecules with like atoms have the symbol $^1\Sigma_g^+$ for their ground states but there are exceptions. Thus the ground state of the oxygen molecule is $^3\Sigma_g^-$.

(*f*) The selection rules which concern us are

$$+ \to + \quad \text{and} \quad - \to - \tag{67}$$

$$\Delta\Lambda = \pm 1 \text{ or } 0 \tag{68}$$

$$\Delta\Sigma = 0 \tag{69}$$

$$\Delta\Omega = \pm 1 \text{ or } 0 \tag{70}$$

$$g \to u \quad \text{and} \quad u \to g \tag{71}$$

There are many permutations and combinations of quantum numbers used to describe the states of diatomic molecules. We do not make any attempt to give them all. It must be kept in mind that although to a first approximation it is possible to give the total energy of a diatomic molecule as the sum of three separate energies, electronic, vibrational, and rotational, the wave function which describes the state of the molecule must involve all motions. While usually the energy difference between successive rotational levels is small compared to the energy difference between successive vibrational levels, and the latter in turn small compared to the difference between electronic levels, these generalizations are not necessarily true.

The rotational quantum number for a diatomic molecule (to which the symbol J is usually given) is often visualized as the number of units of angular momentum of a dumbbell rotating end over end. This picture enables us to describe some of the events which occur in reaction kinetics and is often helpful in obtaining orders of

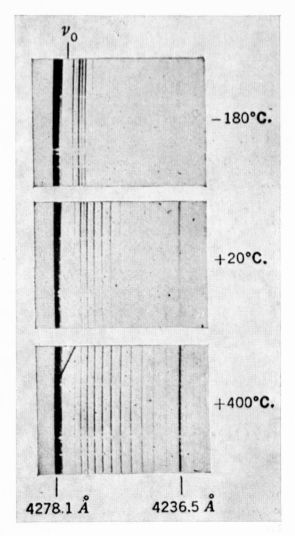

Fig. 3. Spectrum of N_2^+ ($\lambda = 4278.1$ A). Fine structure showing alternating intensities. From ref. 23a.

magnitude for the time intervals and energy differences in certain processes. Nevertheless J does not refer to a simple rotating dumbbell but is the number of units of angular momentum accorded to the molecule as a whole and depends on the complete wave function for the molecule. Symmetry considerations may dictate which values of J are possible and which are not. Such an effect may be observed in the spectrum of N_2^+ (see Fig. 3).

The classic example of restrictions in J due to symmetry is found in the hydrogen molecule. Each proton has a spin of $\frac{1}{2}$ and the two protons in the hydrogen molecule may give a total nuclear spin of either 1 or 0. The electronic wave function must be antisymmetric in the electrons, that is, it must change sign when the electrons are interchanged. With a nuclear spin of zero the antisymmetry of the electronic wave function is not affected by the rotation of the molecule as long as the number of units of rotational momentum is an even number and hence with $I = 0$, $J = 0, 2, 4, 6, \ldots$. Conversely when $I = 1$, $J = 1, 3, 5, 7, \ldots$. Since rotation of the hydrogen molecule does not vanish except when $J = 0$, *para* hydrogen for which $I = 0$ is the stable form at temperatures approaching the absolute zero. At high temperatures (room temperature may be considered almost but not quite "infinite" in this respect) ordinary hydrogen consists of three parts *ortho* and one part *para* since the statistical weights will be in the ratio 3 : 1 with $2I+1$ the contribution of the nuclear spin to the statistical weight. For large values of J the ratio of the contributions to the statistical weight for *para* to that for *ortho* hydrogen by the rotational function $(2J+1)$ will essentially be unity.

Nearly pure *para* hydrogen can be obtained at very low temperatures and upon being brought to higher temperatures it will return very slowly to the normal 3 : 1 ratio of *ortho* to *para*[24]. The return is catalyzed both by certain surfaces and in the homogeneous gas phase by paramagnetic molecules including free atoms and radicals with unpaired electrons. *Para–ortho* hydrogen conversion has been a convenient tool for determining relative concentrations of free radicals.

The photochemist may find it necessary to pay attention to the rotational terms of diatomic molecules in a few instances, especially when he uses *ortho–para* hydrogen conversion and also when he uses spectral analysis to identify molecular species. On the other hand only very rarely does he use light sufficiently monochromatic to excite a system of molecules to one and only one rotational level. We will, however, describe one such experiment and indicate the conclusions which may be drawn therefrom.

The selection rule which applies to the rotational quantum number is

$$\Delta J = \pm 1 \text{ or } 0 \tag{72}$$

where zero is not permitted for some homonuclear molecules. Highly monochromatic light may be obtained with the 5461 A line of mercury. If iodine vapor at its equilibrium pressure at 25 °C (about 0.2 mm) is exposed to this radiation, one and only one rotational level in a single vibration level ($v' = 25$) of the upper electronic state is excited[25]. The change in J may not be zero for iodine, and molecules in this unique upper level may now emit by descending to the ground electronic state with $\Delta J = \pm 1$. Since there is no formal restriction to the change of vibrational quantum number, for each change in vibrational quantum number two lines will appear as long as emission occurs only from the level initially formed by the absorption act.

Collisions will cause loss (or gain) of the various forms of energy. If foreign gas is added, the molecules originally excited will gain or lose rotational energy and emission will occur from many different rotational levels even though loss or gain of vibrational or electronic energy has not occurred.

The emission from iodine vapor excited by monochromatic light may easily be photographed since it consists of a series of doublets. With addition of even a low pressure (*e.g.* 1 torr of helium) of foreign gas, the iodine molecules become distributed among so many rotational levels that the emission becomes very difficult to resolve even though a phototube may indicate virtually no decrease in total intensity.

The data obtained in this and in other ways have shown that rotational energy is gained or lost in small increments at almost every collision. There are some restrictions which affect the probability of energy transfer. In a purely mechanical sense transfer of angular momentum from a light moving particle to a heavy one is less probable than the inverse, for example.

Loss of vibrational energy may also occur by collision with other molecules. In the absence of perturbations by neighboring electronic levels the radiative lifetime of a diatomic molecule is only moderately dependent on its vibrational energy. The probability of transfer of vibrational energy upon collision is a highly specific matter and depends very markedly on the properties of the colliding molecules. Data on this important question have been accumulated rapidly during the past few years. Among the first very significant experiments were those based on fluorescence of simple diatomic molecules such as iodine. Since the radiative lifetime of excited iodine is about 10^{-8} sec, a foreign gas pressure of about 0.01 atm provides about one collision per mean life[25-28]. Iodine has a low dissociation energy and there are also several low-lying electronic states. However, detailed interpretation of the data is difficult although recently some progress has been made.

Perhaps the first significant experiments on transfer of vibrational energy were performed in Norrish's laboratory at Cambridge[29]. When chlorine dioxide is exposed to radiation of the appropriate wavelength it dissociates

$$ClO_2 + h\nu \rightarrow ClO + O \tag{73}$$

and the oxygen atoms react with chlorine dioxide

$$O + ClO_2 \rightarrow ClO + O_2 \tag{74}$$

Absorption spectroscopy showed the oxygen molecules produced by (74) (see Fig. 4) to possess initially several quanta of vibrational energy over and above that expected at room temperature. By following the change in vibrational level in the presence of various pressures of added foreign gas, it was possible to estimate the

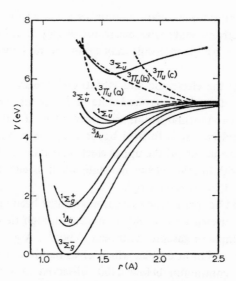

Fig. 4. Potential curves for states of O_2. From ref. 19.

probability per collision of loss of vibrational energy. Monatomic gases such as helium, neon, and argon are relatively ineffective in producing vibrational energy loss; only about one collision in 10^7 led to the loss of one quantum. On the other hand, with most polyatomic molecules, many of which have low fundamental vibration frequencies, often one quantum was lost on the average in 10 to 100 collisions.

The more detailed question as to whether multiple loss of vibrational energy quanta is probable is much more difficult to answer decisively. The view is generally accepted that stepwise loss of vibrational energy is usual. The vibrational energy is much more rapidly equilibrated for molecules with low fundamental vibration frequencies such as iodine ($\omega = 215$)[30] than it is for molecules with high frequencies such as nitrogen ($\omega = 2360$)[30].

More recent work, particularly by Broida[31] at the National Bureau of Standards and by Wasserman at the Bell Telephone Laboratories has served mainly to confirm that rotational energy is very rapidly equilibrated with the surroundings and so also is vibrational energy in most cases, although each case must be studied on its own merits.

The photochemist is little interested in rotational energy, mainly because it is an unimportant fraction of the total energy possessed by an electronically excited molecule under most experimental conditions. The vibrational energy is quite another matter, and we must now examine the possibilities which may arise after the electronic energy of a diatomic molecule is increased.

If a molecule in its ground electronic state (with electronic energy arbitrarily called zero) and with vibrational quantum number $v''(= a)$ absorbs a photon and goes to an upper electronic state with electronic energy E volts and vibrational quantum number v' $(= b)$, the molecule may suffer several fates.

Case 1. The sum of the electronic and vibrational energy is greater than D'', the energy required to dissociate a ground state molecule into two normal atoms. Dissociation to give two normal atoms can only occur if either (a) the upper electronic state is repulsive, or is formed beyond its convergence limit and dissociates into normal atoms, or (b) the upper electronic state is of such a character that it may interact with another electronic state which is either repulsive or has energy in excess of its dissociation energy.

Condition (a) would be met if some selection rule has been violated during the absorption act. Such absorptions would be weak although they are observed in certain cases, for example in gaseous hydrogen iodide[32]. A good example would be the very weak absorption in oxygen which gives bands which converge about 2400 A[19]. The weak continuum below 2400, observed at very high pressures or through very thick layers of oxygen, is due to dissociation into two normal oxygen atoms both in the ^3P state. Condition (b) is frequently found and the heavier the atoms which compose the diatomic molecule the more probable a crossover to a suitable dissociating state would be. There are many factors which govern this kind of a crossover and we must mention the subject of *predissociation*.

Predissociation was first correctly described by Henri and Teves in 1924[22]. They observed a series of sharp bands in the spectrum of the sulfur molecule S_2 followed by bands which were diffuse, that is, no rotational structure could be resolved although there were absorption maxima and minima so that remnants of the banded absorption remained. At still shorter wavelengths discrete bands were once more observed.

This phenomenon of discrete bands followed by diffuse bands and in turn followed by discrete bands may be described in simple mechanical terms as follows: transitions from the ground state to an upper electronic state occur and the upper state has a mean life sufficiently long to permit the absorbing molecule to vibrate and to rotate many times. Diffuseness is due to the molecules crossing over rapidly to a state which dissociates. Since remnants of the vibration levels persist, one must assume that the molecule after absorption remains intact for several periods of vibration, *i.e.* longer than say 10^{-12} second. However, since the rotational structure has disappeared dissociation has presumably taken place before many rotations can occur, *i.e.* in less than about 10^{-8} second. As the absorbed energy further increases, the probability of crossover to the dissociating state diminishes and rotational structure is again observed.

This qualitative description of the phenomenon of "predissociation" is adequate in many ways but it is insufficient for a quantitative calculation of lifetimes and dissociation probabilities[33].

The wave function for a molecule must include contributions from many different states and it is incorrect to label the electrons with precise quantum numbers even though such a designation may serve the useful purpose of predicting gross behavior. Contributions to the wave function by states which differ markedly in total energy from the state in question will be small, and the contribution will also be small if symmetry properties are different. Thus g states will contribute little to u states and $+$ states little to $-$ states.

The number of units of vibrational energy does not affect the symmetry of the wave function for a diatomic molecule unless at some point through near identity of energy the contribution to the wave function by a state of different symmetry becomes important. When this happens it is incorrect to designate the state by a set of quantum numbers belonging to either electronic state and the correct wave function is a complex hybrid in which simple symmetry rules no longer apply. In one sense one can describe the molecule as residing part of the time in one state and part in the other. If dissociation can happen rapidly the spectrum is no longer truly discrete. The detailed calculation of dissociation probability from such considerations is difficult except for the very simplest molecules with a total of less than about four electrons.

If one adopts the correct point of view that the complete wave function of any state of a diatomic molecule has contributions from all other states of that molecule, one can understand that all degrees of perturbation and hence probabilities of crossover may be met in practice. If the perturbation by the repulsive or dissociating state is very small, the mean life of the excited molecule before dissociation may be sufficiently long to permit the absorption spectrum to be truly discrete. Dissociation may nevertheless occur before the mean radiative lifetime has been reached so that fluorescence will not be observed. Predissociation spectra may therefore show all gradations from continua through those with remnants of vibrational transitions to discrete spectra difficult to distinguish from those with no predissociation. In a certain sense photochemical data may contribute markedly to the interpretation of spectra.

The complete wave function of a molecule will often change in the presence of foreign fields. Foreign fields, either electrostatic or electromagnetic, may under certain conditions change wave functions sufficiently to permit perturbations which would otherwise be unimportant. Perturbations may also be induced by collisions with other molecules, particularly with molecules which are themselves paramagnetic. These effects give rise to what is often called collision-induced predissociation.

Absorption under Case 1 leads with high probability to dissociation, and can only be observed if the energy is sufficient to cause dissociation. Small deficiencies of energy, however, may be supplied from thermal energy either if the absorbing molecules are in vibrational levels of the ground state above the lowest or if the excited molecules may acquire vibrational energy by collision. It is possible, therefore, for the primary photochemical dissociation of a diatomic molecule to appear to have an activation energy.

Case 2. The sum of electronic and vibrational energy is less than D''. Dissociation after absorption of radiation is now impossible without acquisition of further energy and this aspect of the problem has been covered for Case 1. Under Case 2 we are interested in the behavior of molecules in active states. If the energy is greater than D'' (the energy required to produce two normal atoms) but less than D' (the energy required to produce one normal and one electronically excited atom) the molecule may (*a*) revert to the ground state by emission of radiation; (*b*) revert to the ground state by energy loss through collisions; (*c*) react as an entity with colliding molecules. Reaction (65) has already been cited as an example of (*c*) although O_2' in that reaction would be a molecule with low electronic energy difficult to produce in the laboratory by direct absorption of radiation.

The formation of ozone after absorption of ultraviolet radiation by oxygen can occur even though the energy is unsufficient to dissociate the molecule into one normal and one excited atom. The energy required to dissociate the oxygen molecule into two normal atoms from the lowest vibration level of the ground state is 5.084 electron volts (117.2 kcal.mole^{-1}). The Schumann–Runge band system of oxygen in absorption at room temperature starts about 1950 A and reaches a convergence limit (beginning of continuous absorption) at 1759 A, corresponding to 7.05 electron volts or 162.5 kcal.mole^{-1} (see Fig. 4). One may obtain the following energy relationships at 1759 A[19,34]

$$O_2(^3\Sigma_g^-)+h\nu_{1759} \rightarrow O(^3P)+O(^1P) \tag{75}$$

$$O(^1P) \rightarrow O(^3P)+1.95 \text{ eV} \tag{76}$$

$$O_2(^3\Sigma_g^-) \rightarrow 2O(^3P)-5.084 \text{ eV} \tag{77}$$

Absorption of radiation at wavelengths below 1759 A will dissociate oxygen into one normal ^3P atom and one excited ^1P atom. Between 1759 A and the long wave limit of this band system excited molecules with more than enough energy to dissociate into two normal atoms will be formed. Perturbations leading to induced predissociation could cause formation of two normal atoms, or direct reaction of an excited with a normal oxygen molecule might occur by the following sequence of steps

$$O_2(^3\Sigma_g^-)+h\nu \rightarrow O_2(^3\Sigma_u^-) \tag{78}$$

$$O_2(^3\Sigma_u^-)+O_2(^3\Sigma_g^-) \rightarrow O_3+O(^3P) \tag{79}$$

Reaction (79) is quite exothermic. Experimental data using several wavelengths above and below 1759 A show two molecules of ozone to be formed per quantum absorbed. It is evident that the reaction between oxygen atoms and oxygen molecules to give ozone occurs quantitatively whether or not the oxygen atoms are normal or excited, but the details of the latter process are not known. Reaction (79) followed by (64) would give a quantum yield of 2 for ozone formation. One must not exclude the possibility of a collision induced predissociation to give two normal atoms

$$O_2(^3\Sigma_u^-) + M \rightarrow M + 2O(^3P) \tag{80}$$

A brief summary of some aspects of predissociation may be of service because we will have occasion to use many of them in dealing with polyatomic molecules.

(*i*) Spectrum appearance is not an absolutely safe guide to the presence or absence of predissociation, *viz*. band diffuseness is a positive indication of predissociation but there may be predissociation without observable diffuseness.

(*ii*) Predissociation may be highly restricted to certain vibration levels of the upper electronic state as shown by the fact that a very limited number of bands in a progression may appear to be diffuse.

(*iii*) If, for example, predissociation occurs with high probability at $v' = $ a, then excitation to $v' > $ a will lead to dissociation provided vibrational energy is lost in small increments and vibrational relaxation is slow relative to dissociation. The latter will necessarily be true if the spectrum is diffuse when $v' = $ a.

(*iv*) Collision and foreign field induced predissociation may not be clearly indicated by spectrum appearance even under conditions of high pressure and strong field strength but more data should be sought on this point.

(*v*) An increase in temperature will give absorption from higher and higher vibrational levels of the ground state and probably, but not certainly, increase the wavelength at which predissociation will occur.

(*vi*) Since predissociation occurs at $v' = $ a it may be found when $v' < $ a with an activation energy provided the radiative lifetime from the excited molecule initially formed is sufficiently long. To obtain a rough estimate of the probability of such an event let us assume that an activated molecule is formed with a frequency deficient by 1000 cm^{-1} or 2.9 kcal.mole^{-1}. If a molecule at atmospheric pressure undergoes 10^{10} collisions per second, about one collision in a hundred at 298 °K or 10^8 per second will provide the necessary energy to raise the molecule to $v' = $ a. The steric factor, that is, the probability that the energy will be acquired when the energy is available, is certainly below unity and may be about 10^{-1} to 10^{-2}. With a steric factor of unity about half the molecules would dissociate if the radiative lifetime is 10^{-8} second, but with a steric factor of 10^{-2} about one per cent would dissociate. However, this percentage would become four times as large at 100 °C. If the deficiency were only 200 cm^{-1} the probabilities of dissociation would be about 40 times as large as for 1000 cm^{-1}. Thus both the separation between energy levels and the temperature will markedly affect the range over which predissociation can occur even if the spectrum shows only a limited range of diffuseness. These facts should be kept in mind when we discuss polyatomic molecules.

(*vii*) Predissociation may occur over a long spectral range provided internuclear distances for the two perturbing states are about the same and have about the same dependence on vibrational quantum numbers. This is in fact a way of saying that each case must be studied in detail on its own merits.

A detailed discussion of examples of predissociation would require more space than could be given to it. However, a few cases are considered below.

Bromine. The heat of dissociation of bromine[35] is about 45 kcal.mole^{-1} and corresponds to a wavelength of 6350 A. A progression of Br_2 bands based on $v'' = 0$ converges at about 5100 A corresponding to 56.04 kcal.mole^{-1}. The difference means that absorption by bromine vapor at wavelengths below 5100 A will (provided $v'' = 0$) lead to the following process

$$Br_2(v'' = 0) + h\nu \rightarrow Br(^2P_{\frac{1}{2}}) + Br(^2P_{\frac{3}{2}}) \tag{81}$$

The $^2P_{\frac{1}{2}}$ atoms will have an electronic excitation energy of $56.04 - 45.00 = 11.04$ kcal.gram atom^{-1} or about 0.48 electron volt. The continuum usually used by photochemists in studying the photochemical reactions of bromine lies below 5100 A and the primary process is (81). It has been known for many years that there is no sudden change in quantum yields of bromination reactions as one crosses the threshold at 5100. Since wavelengths out to 6350 A provide enough energy to dissociate Br_2 molecules with $v'' = 0$, and slightly longer wavelengths can dissociate Br_2 molecules when $v'' > 0$, predissociation must occur by the process

$$Br_2 + h\nu \rightarrow 2Br(^2P_{\frac{3}{2}}) \tag{82}$$

Many attempts have been made in the past to detect differences in reactivity of halogen atoms in $^2P_{\frac{1}{2}}$ and $^2P_{\frac{3}{2}}$ states but without much success. Indeed, until recently little was known about rates of transformation of one atomic state into another but some beautiful work by Donovan and Husain[36] has provided some information on this point. Emission of infrared radiation as well as collisions may cause the reaction

$$I(^2P_{\frac{1}{2}}) \rightarrow I(^2P_{\frac{3}{2}}) \tag{83}$$

where the energy difference is about 0.9 electron volt.

Nitric oxide[37]. Nitric oxide shows several band systems in the region 1500–2300 A and there is little or no evidence from the appearance of the spectra that predissociation occurs. Nitric oxide decomposes in this spectral region. There is now also conclusive evidence that excited nitric oxide molecules react with other molecules, but we will not discuss this point in detail. Nitric oxide must dissociate by the process

$$NO(^2\Pi) \rightarrow N(^4S) + O(^3P) \tag{84}$$

Fluorescence is not emitted above certain values of v' in some of the band systems. The abrupt cessation of emission when a definite v' is reached probably means that predissociation has started, but other explanations are possible. Since the ground state of nitric oxide is doublet (*i.e.* there is one unpaired electron) the excited states formed by direct absorption will also be doublet. Excited quadruplet states (with three unpaired electrons) also are known and these lie at lower energies than the doublet states of the same electronic distributions. Perturbations which would lead to crossovers from doublet to quadruplet states would be very weak for molecules made of such light atoms as nitrogen and oxygen except in the presence of external fields. Thus collisions might be effective in inducing crossovers, the more so because nitric oxide is paramagnetic. A detailed study of the effectiveness of pressure in diminishing fluorescence in nitric oxide at high values of v' in some of these band systems would be of interest.

This relatively brief discussion of the photochemistry of diatomic molecules may be summarized as follows.

Under some relatively infrequently encountered conditions excited diatomic molecules may react with other molecules directly. Examples cited were excited oxygen plus oxygen to give ozone and the reactions of excited nitric oxide with several molecules.

Photochemical dissociation of diatomic molecules may occur either (*a*) by absorption in a spectral region showing a continuum or (*b*) by absorption in a region with bands which have more or less discrete appearance followed by crossover from one upper electronic state to another, which results in dissociation. The latter phenomenon is called predissociation and has been discussed at some length.

When a diatomic molecule dissociates photochemically it yields atoms either in their ground or in excited states. When the optical transition leading to dissociation is not forbidden, and does not arise from predissociation, one of the two atoms formed will not be in its ground state.

The reactions initiated by atoms do not belong in the realm of photochemistry since under a given set of experimental conditions these reactions will be the same no matter how the atoms are formed. The classical work on reactions initiated by chlorine after absorption of light has shown that chain reactions may be started[38]. The thermal and photochemical hydrogen–bromine reactions lead to the same products with the same mechanism, the one distinction being that bromine atoms are produced by the thermal dissociation of Br_2 in the one case and by the photochemical dissociation in the other[39].

An extended discussion of the reactions initiated by atoms formed photochemically would not be possible in a few pages. It might be interesting to deal with differences in behavior dependent on the state of excitation of the atoms. In this respect photochemical reactions may differ from thermal reactions. Data in this very interesting field are only now being obtained and the future will certainly be very bright for this type of study.

We will mention only one series of studies, mainly due to Gunning and his coworkers[40] and summarized in a review by Gunning and Strausz[41]. These relate to sulfur atoms.

Sulfur atoms in their ground state are $^3P_2$ but the $^3P_1$ and $^3P_0$ states lie only 1.136 and 1.639 kcal.gram atom^{-1} above the lowest state, respectively. The two lowest lying singlet states are $^1D_2$ at 2.640 kcal and $^1S_0$ at 6.339 kcal above the ground state.

Many of the investigations of Gunning and his coworkers have involved the molecule COS[41] as a source of sulfur atoms, viz.

$$COS + h\nu \rightarrow CO(^1\Sigma^+) + S(^1D) \tag{85}$$

at 2288 A. The sulfur atoms react with COS (carbonyl sulfide) to form S_2, viz.

$$S(^1D) + COS \rightarrow S_2 + CO \tag{86}$$

The quantum yield of carbon monoxide with pure carbonyl sulfide never reaches the theoretical value of two and a full explanation of this fact is not available. However, olefins may be used to scavenge the sulfur atoms formed in the primary process and under these conditions only one molecule of carbon monoxide is formed per photon.

It is possible to form triplet sulfur atoms in either of two ways. The first is by a mercury sensitized decomposition of COS[41]

$$Hg(6^3P_1) + COS \rightarrow CO + S(^3P) + Hg(6^1S_0) \tag{87}$$

and the second is by direct photolysis in the presence of carbon dioxide[41] which seems to be effective in converting singlet to triplet sulfur atoms

$$S(^1D) + CO_2 \rightarrow S(^3P) + CO_2 \tag{88}$$

It has been known for some years that singlet methylene (CH_2) radicals will insert in carbon–hydrogen bonds. Thus methylene plus ethane will give propane. This matter will be discussed later. Oxygen atoms are isoelectronic with methylene radicals; sulfur atoms may be expected to undergo reactions similar to those of oxygen atoms and thus to be analogous to methylene radicals.

The work of Gunning et al.[40] has indeed shown that insertion in alkanes does occur, for example

$$S(^1D) + C_2H_6 \rightarrow CH_3CH_2SH \tag{89}$$

Concurrent deactivation of sulfur atoms to the ground state also takes place and the latter either react with each other or with carbonyl sulfide to give S_2.

With olefins two reactions occur: (a) insertion in one of the carbon–hydrogen bonds, and (b) reaction at the double bond to give an episulfide. The insertion reaction is suppressed by addition of carbon dioxide as would be expected. In the case of singlet methylenes, reactions with the double bond are stereospecific while those of the triplet methylenes are not. Here Gunning and his coworkers found the sulfur atoms to behave differently and both singlet and triplet atoms reacted stereospecifically with double bonds in olefins.

We leave this discussion of differences in behavior of atoms in different electronic states with the statement that such work should be continued, particularly when it may apply to atoms formed by the primary photochemical dissociation of diatomic molecules.

2.4 ABSORPTION BY POLYATOMIC GASES

In starting the discussion of polyatomic gases it may be well to state concisely some respects in which they may be expected to differ from monatomic and diatomic gases.

Monatomic gases. If one neglects absorption in the far and intermediate infrared as well as of microwaves, all of which might lead to minor changes in electronic state such as one would expect merely from a change in J or a change in L, absorption by monatomic gases leads to electronic transitions of a few electron volts and, at short wavelengths, convergence of series corresponds to ionization. Absorption does not and cannot lead to a chemical reaction and the latter arises from secondary processes either by transfer of energy from excited atoms to other molecules or by direct reaction of excited atoms with colliding molecules.

Diatomic gases. Absorption by diatomic gases at long wavelengths will lead to increase in rotational, or in vibrational energy, or both, and neither of these alone will produce chemical reactions. As for atoms, absorption at shorter wavelengths leads to electronic excitation but with a very great difference: the binding energy between the two atoms in a diatomic molecule will vary from one electronic state to another. While the symmetry of a diatomic molecule will not *per se* be affected by the vibrational energy, it will vary from one electronic state to another. Absorption of radiation which causes changes in electronic state may lead to dissociation into atoms, a process naturally not found with monatomic gases, and also may, if the energy is great enough, lead to ionization either to form molecular ions or ionization of one of the two atoms as the molecule dissociates.

Polyatomic gases. The symmetry of a polyatomic molecule may be affected by vibration. For example, carbon dioxide is a linear molecule in its ground state but one possible mode of vibration would be a bending motion during which the molecule would not be linear most of the time. The benzene molecule is a regular hexagon in its ground state but several of the possible modes of vibration will distort the hexagon, even though it is still planar, and others will make it cease to be planar. Thus the vibrational contribution to the wave function of a polyatomic molecule may play an important part in the electronic transitions which are permitted. A polyatomic molecule may be electronically excited and ultimately ionize if sufficient energy is provided. It may dissociate to give atoms or free radicals just as do diatomic molecules. But there is an additional type of process: internal rearrangement of atoms to give isomers or even to give dissociation products which are not atoms or free radicals. These latter processes were largely ignored by photochemists until the past few years but they now constitute some of the most interesting parts of photochemistry.

A detailed description of the symbolism used to designate electronic states of polyatomic molecules would of itself require an entire book. We shall limit ourselves to a discussion of the factors which govern the appearance of polyatomic spectra and of some of the rules which generally determine intensities and behavior[42].

Electrons in these molecules may be designated as bonding, nonbonding and antibonding. As the term implies, the bonding electrons play an important part in holding the molecule together. The removal of a bonding electron will weaken one or more of the bonds in a molecule. Nonbonding electrons play little part in holding the molecule together and the removal of one of them will affect very little the strengths of the bonds and the fundamental vibrational frequencies of the molecule in question. Antibonding electrons by their presence tend to weaken some of the bonds within the molecule.

During electronic transitions, a single electron changes from one quantum state to another and in the process it may change from bonding to nonbonding or to antibonding. Changing from nonbonding to antibonding also occurs. It is unusual, although not impossible, for nonbonding and antibonding electrons to become bonding electrons.

It has been customary to designate certain transitions as $n \rightarrow \pi^*$ and $\pi \rightarrow \pi^*$[43,44]. It may be wise to try to describe briefly what is meant by these terms.

An isolated chlorine atom has the following electron distribution: $(1s)^2(2s)^2$ $(2p)^6(3s)^2(3p)^5$. The outer seven electrons are the ones which are concerned in forming compounds of chlorine and one must consider that the $3d$ shell is also not yet occupied. The $1s$, $2s$, and $2p$ electrons we will ignore as not participating effectively in chemical bonding.

If a chlorine atom is joined to a hydrogen atom there are now eight electrons in the valence shell and also the hydrogen chloride molecule has acquired an axial symmetry which the chlorine atom did not have.

p electrons have nodes and since there will be six of them they may be divided into pairs, two along the x, two along the y and two along the z axis. Since $l = 1$ for each p electron one might say that the l vectors are divided two along each axis. By the Pauli Principle each pair of electrons must have spins antiparallel.

The l vectors may be considered either to be parallel or perpendicular to the axis of the HCl molecule. Two will be perpendicular and four will be parallel. The two former are designated $p\sigma$ electrons and the four latter $p\pi$ electrons. The latter are often called merely π electrons for short and they are the ones which will be either bonding or antibonding.

The normal state of the HCl molecule is a Σ state because there is no net projection of l vectors, that is the π electrons are all paired both as to spins and as to orbital angular momenta. When a π electron is raised to a higher quantum state the Pauli Principle may be obeyed even if spins are unpaired or the directions of the l vectors are changed. Thus one may have both singlet and triplet states as well as Σ, Π, Δ states.

If one of the electrons which does not participate in the chemical bond is raised to a higher level, more specifically if a $p\sigma$ electron has become a $p\pi$ electron, the transition is labeled $n \to \pi^*$. If one of the others is raised the transition is labeled $\pi \to \pi^*$.

The designation of electronic transitions as $n \to \pi^*$ and $\pi \to \pi^*$ for polyatomic molecules is often done very qualitatively and unfortunately often very loosely. While the language is no longer very rigorous, bonding electrons are often called π electrons, even though in so doing one is in reality referring to the axis of a single bond rather than to the symmetry of the molecule as a whole. The electrons in benzene which participate in electronic transitions are called π electrons and the transition which occurs in the general region of 2500 A is called a $\pi \to \pi^*$ transition. Pyridine, on the other hand has electrons on the nitrogen atom which are essentially nonbonding and it probably has two transitions in the same region, one $n \to \pi^*$ and the other $\pi \to \pi^*$. The former lies at a slightly longer wavelength than the latter although the detailed description of these spectra leaves something to be desired[43, 45].

This brief allusion to types of transition will serve merely to give the reader an idea that the detailed description of electronic transitions in polyatomic molecules is complicated and for the moment not a realm of universal agreement. Zimmerman[46] has extended the mechanistic ideas currently used by physico-organic chemists to organic photochemistry. It is possible in this way to give systematic descriptions of many organic photochemical phenomena. A systematization of this type will ultimately be necessary merely because of the multitude of facts which must be made useful.

It is difficult to subdivide the photochemistry of polyatomic molecules in any rational way. The following discussion will doubtless appear too arbitrary to many.

2.4.1. Dissociation into atoms and radicals

The distinction between a truly continuous absorption spectrum and a banded absorption spectrum for diatomic molecules may be made by instruments of relatively low resolving power. Even though individual rotational lines are not resolved, a discrete spectrum will have sharp band heads and the appearance will in no way resemble the appearance of a continuum.

As we have seen in discussing the behavior of diatomic molecules, there is often great difficulty in identifying a predissociation spectrum since the appearance may at one extreme be essentially that of a truly discrete spectrum and at the other that of a true continuum. The facts of photochemistry may be of great use to the spectroscopist in distinguishing between predissociation and truly discrete spectra.

For many polyatomic molecules the problem is far more difficult. A few of the simple polyatomic molecules such as formaldehyde, nitrogen dioxide, carbon dioxide, and ammonia have spectra which can be adequately resolved, and treated in rigorous fashion by the theoreticians. For most of the others the treatment is much less satisfactory.

A linear polyatomic molecule may be treated by the same relatively simple rules as are used for diatomic molecules provided it is not necessary to treat those vibrations which tend to make the molecule nonlinear. Since the symmetry is changed when such vibrations are excited in the upper state, transitions to such levels will, in general, be weak.

The two linear molecules of most concern to the photochemist are carbon dioxide[37, 47] and nitrous oxide[37]. Carbon dioxide absorbs only in the relatively far ultraviolet. The primary dissociation is almost certainly

$$CO_2 + h\nu \rightarrow CO + O \tag{90}$$

If the wavelength is longer than 1200 A the carbon monoxide molecule must be formed in its ground ($^1\Sigma^+$) state. The state of the oxygen atom probably depends on the wavelength. The ^1D level lies lower than the ^1S level and may well be formed toward the long wavelength end of the absorption region, *i.e.* at wavelengths greater than about 1600 A. The detailed interpretation of the photochemistry of CO_2 is not in a very satisfactory state.

Nitrous oxide is a linear nonsymmetrical molecule, NNO. Its photochemistry has been investigated many times[37]. The known products of the photochemical decomposition of N_2O are N_2, O_2, and NO. Two primary processes may be visualized

$$N_2O + h\nu \rightarrow N_2 + O \tag{91}$$

$$\rightarrow NO + N \tag{92}$$

If (91) is the primary process, secondary reactions might be

$$O + N_2O \rightarrow 2NO \tag{93}$$

$$\rightarrow N_2 + O_2 \tag{94}$$

If (92) is the primary process the secondary process is

$$N + N_2O \rightarrow NO + N_2 \tag{95}$$

Atom combination reactions to give N_2 and O_2 can also occur. Since O_2 is a principal product and the overall yield is nearly 2, it is necessary to conclude that (91) is important.

Nitrous oxide shows many regions of absorption with the longest extending to wavelengths somewhat above 2100 A. The chemical products are undoubtedly the same no matter what region of absorption is used but the electronic states of the primary fragments undoubtedly vary from one region to another. At all wavelengths (91) is the main primary process but (92) may take place concurrently, particularly at short wavelengths.

Nitrous oxide and carbon dioxide have the same number of electrons and their linearity might be expected to make resolution of the spectrum and its interpretation possible. Nevertheless there are still uncertainties although much progress has been made.

There are $(3n-6)$ modes of vibration in a polyatomic molecule except that for a linear molecule one mode is doubly degenerate so that effectively the number is $(3n-5)$; n is the number of atoms in the molecule. Symmetry rules may restrict the kinds of transitions which can occur during an electronic transition but often the number of possible transitions is very large.

Let us consider the simple hypothetical case of a molecule with four atoms whose absorption spectrum is being observed at a sufficiently low temperature so that all of the absorbing molecules are in the lowest vibrational level of the ground state. Furthermore let us say that absorption begins at 3000 A and that the six fundamental vibration frequencies in the upper electronic state are 200, 350, 450, 550, 700, and 1000 cm^{-1} respectively. 3000 A corresponds to 33333 cm^{-1}, 2800 A to 35714 cm^{-1}. If there are no restrictions on the vibrational levels which can be excited in the upper electronic state, the number of bands in this interval of 200 A will be well in excess of 100 which implies a spacing of less than 2 A between bands. Since each band will cover a finite wavelength range there will be overlapping of bands even with high resolution spectrographs. Probably a careful study would show irregularities and successive maxima and minima but these would not be easily discernible with a low resolution instrument.

Fortunately the Franck–Condon principle, which states that an electronic transition occurs so rapidly that there is a high probability of forming the upper state with the same bond distances and angles as the lower state, means that many bands will be of very low intensity or absent. Nevertheless the resolution of poly-atomic spectra for molecules of low symmetry is usually impossible. It is useless to try to determine the onset of predissociation, or even of direct dissociation, from the appearance of such absorption spectra. Unfortunately the photochemical literature is replete with mis-statements based on the appearance of absorption spectra.

Quite evidently a bond may not be broken unless sufficient energy is available to produce that effect. Each absorbing molecule in equilibrium with its surround-ings already possesses rotational and vibrational energy, and for polyatomic molecules this may amount to several thousand calories per mole. The availabil-ity of this energy to aid in bond dissociation may not even be estimated in most instances. In thermal reactions, however, the activation energy is usually less than the bond dissociation energy for such molecules.

It is, of course, not correct to treat the wave function of a polyatomic molecule as localized in the chromophoric group considered responsible for the optical absorption. The carbonyl group in aldehydes and ketones gives rise to absorption which extends from about 3450 A to about 2200 A (as well as to absorption at shorter wavelengths). Nevertheless the carbon–oxygen bond is never broken by absorption at these wavelengths. Frequently an adjacent bond is broken but often more complex processes occur. It is sometimes possible to describe these pro-cesses in terms of quantum mechanics but some of them should not be treated as direct dissociations.

Dissociation of polyatomic molecules into free radicals and free atoms may be described in much the same terms as one uses for diatomic molecules, but a more elaborate picture is necessary.

Whenever a molecule possesses *in toto* an amount of energy sufficient to cause dissociation of one bond there will exist a probability, however small, that the energy will be so localized that dissociation will occur. As the energy in excess of the bond energy increases, the probability of dissociation increases. This is the basis for the Rice–Ramsperger–Kassel theory[48, 49] of unimolecular reactions. The extension of this theory to photochemical processes is perhaps even more difficult than its rigorous application to thermal reactions. The most that can be done at present is to provide a reasonable picture of what can happen and possibly a calculation of the correct order of magnitude of certain probabilities.

Nevertheless, predissociation in polyatomic molecules can be visualized as qualitatively similar to a unimolecular reaction in the Rice–Ramsperger–Kassel sense. With diatomic molecules possessing only one degree of vibrational freedom, the crossover to the dissociating state can follow only a limited number of paths. With polyatomic molecules each energy state must be described as a polydimensional surface and the chances of intersections and crossovers are numerous. There is little one can do at present with the theory of such processes, although one can give a rational explanation of almost any data which are provided by the experimentalists.

Polyatomic molecules do, under some circumstances, show spectra which give definite progressions and, at short wavelengths, Rydberg series. The progressions at their convergence limits correspond to a dissociation, and the Rydberg series converge to an ionization potential.

Thus in some instances it is possible to state when direct dissociation after absorption of a photon will occur, but predissociation is a very common process and its identification usually difficult. The experimentalist is also confronted with the great difficulty of measuring the quantum yield of the primary dissociation process. This may be illustrated first by a hypothetical and then by an actual case.

Suppose a molecule $R_1 R_2$ absorbs a photon and dissociates

$$R_1 R_2 + h\nu \rightarrow R_1\cdot + R_2\cdot \tag{96}$$

The radicals $R_1\cdot$ and $R_2\cdot$ may now undergo reactions to give final products which regenerate these same radicals but ultimately they combine with each other to stop the chain of events. The combination reactions would be

$$R_1\cdot + R_2\cdot + M \rightarrow R_1 R_2 + M \tag{97}$$

$$R_1\cdot + R_1\cdot + M \rightarrow (R_1)_2 + M \tag{98}$$

$$R_2\cdot + R_2\cdot + M \rightarrow (R_2)_2 + M \tag{99}$$

If $R_1\cdot$ and $R_2\cdot$ had exactly the same mass and all three reactions had the same steric factors, the rate of (97) would be twice the rate of either (98) or (99). If the masses are not the same the reduced masses must be considered and the relative rates would be proportional to $2\{(M_1 + M_2)/M_2 M_2\}^{\frac{1}{2}}$, $(2/M_1)^{\frac{1}{2}}$, and $(2/M_2)^{\frac{1}{2}}$ respectively. Actually this theoretical relationship is not obeyed and the factor 2 in the first rate expression is found on the average in many cases to be about 1.7. However, since this factor has no sound foundation it must be used with care.

Now to return to the measurement of the quantum yield of reaction (96). Each primary dissociation introduces two radicals into the system and (97), (98), and (99) each removes two radicals from the system. Hence the primary quantum yield may be found from the relationship

$$(R_{97} + R_{98} + R_{99})/I_a = \phi_{96} \tag{100}$$

where R_{97}, R_{98}, and R_{99} are the rates of the respective reactions either in molecules per unit volume per second or in moles per unit volume per second, and I_a is the number of photons (or einsteins) absorbed per unit volume per second. In practice the total amounts of products of (97), (98), and (99) can be divided by the integrated number of photons absorbed. We use the small ϕ to indicate the quantum yield of a primary process and the capital Φ to indicate an overall or net quantum yield. In this particular case

$$\phi_{96} = \Phi_{97} + \Phi_{98} + \Phi_{99} \tag{101}$$

It will be immediately apparent that the difficulty lies with reaction (97) which re-forms the original starting material. The yields of reactions (98) and of (99) can be determined and it might under some circumstances be possible from them to deduce the rate of (97). If this is impossible or subject to too many uncertainties some independent device must be used to obtain the rate of (97). Isotopic labeling might permit conclusions to be drawn as will be indicated below.

A specific example will now be given. The photochemistry of acetone has been the subject of a great many investigations[50]. The work to 1957 has been summarized and the book by Calvert and Pitts[51] has brought the summary up through 1965. More recent work has generally served to confirm the accepted mechanism, but has provided better values for some of the rate coefficients and activation energies.

The photochemistry of acetone has been studied chiefly over the spectral range 2537 to 3130 A. The temperature has been varied from 0 to several hundred °C. The primary process in this wavelength region is almost solely

$$CH_3COCH_3 + h\nu \rightarrow CH_3\cdot + CH_3\dot{C}O \tag{102}$$

The acetyl radical will dissociate

$$CH_3\dot{C}O \rightarrow CH_3\cdot + CO \tag{103}$$

partially as a result of vibrational or kinetic energy it retains after the primary dissociation of acetone and partially as a result of a thermal reaction with an activation energy. The activation energy for (103) has been the subject of some controversy but the value now generally accepted[50] is about 13 kcal.mole^{-1}.

Many secondary reactions initiated by the methyl and acetyl radicals may occur. Suffice it to say that at temperatures up to one or two hundred degrees, chain reactions are unimportant and the radical reactions we must consider are mainly the following

$$CH_3 \cdot + CH_3 \cdot (+M) \rightarrow C_2H_6(+M) \tag{104}$$

$$CH_3\dot{C}O + CH_3\dot{C}O \rightarrow (CH_3CO)_2 \tag{105}$$

$$CH_3 \cdot + CH_3\dot{C}O \rightarrow CH_3COCH_3 \tag{106}$$

$$CH_3 \cdot + CH_3COCH_3 \rightarrow CH_4 + \dot{C}H_2COCH_3 \tag{107}$$

$$CH_3 + \dot{C}H_2COCH_3 \rightarrow CH_3CH_2COCH_3 \tag{108}$$

$$CH_3 \cdot + CH_3\dot{C}O \rightarrow CH_4 + CH_2CO \tag{109}$$

The primary process (102) introduces two radicals into the system and this is true even if it is followed by (103) almost instantaneously. The reactions which remove radicals from the system are (104), (105), (106), (108), and (109). Of these all but (106) lead to products for which in principle analytical techniques can be found.

The mechanism is considerably simplified at temperatures above about 120 °C unless the intensity is very high indeed. Under such conditions reaction (103) proceeds to completion. Since no secondary reaction under these experimental conditions leads to the formation of carbon monoxide

$$\phi_{102} = \Phi_{CO} \tag{111}$$

Experimental data show that, at temperatures from about 100 to about 200 °C, $\Phi_{CO} \sim 1$ at wavelengths from 2500 to 3200 A[50, 51]. The value cannot be exactly unity toward the long wave end of this region because there is a small fluorescence of acetone even at these temperatures[52]. The exact point at which fluorescence ceases to be excited is indeterminate and should be slightly temperature dependent. There is no fluorescence at 2537 A. The quantum yield of fluorescence seems never to have been determined with high precision, partly because it is low and the intensity very weak. The value is certainly less than 0.01 under the experimental conditions cited. Hence within the experimental error one may state

$$\Phi_{CO} = 1 \ (120\text{--}200 \ ^\circ\text{C}) \tag{112}$$

and indeed the photochemical decomposition of acetone proves to be a very satisfactory actinometer for many purposes[53]. At higher temperatures quantum yields increase slightly, probably due to the thermal decomposition of the acetonyl radical

$$\dot{C}H_2COCH_3 \rightarrow CH_2CO + CH_3 \tag{113}$$

If the carbon monoxide quantum yield is indeed unity and the primary yield is essentially unity, the following processes may be neglected unless the intensity is extremely high: (105), (106), (109). Since (107) does not decrease the total number of radicals in the system, (104) and (108) are the radical removing steps and

$$\phi_{102} = \Phi_{C_2H_6} + \Phi_{CH_3COCH_2CH_3} = \Phi_{CO} \tag{114}$$

The yield of 2-butanone is low under most experimental conditions [the yield of $(\dot{C}H_2COCH_3)_2$ would be even smaller] and there are enough complicating features to make a rigorous test of (114) difficult. It is certainly obeyed within experimental error.

At temperatures below 100 °C none of the reactions (102)–(109) can be neglected and indeed further steps, including some minor radical reactions, must be added[51, 54]. Since these details illustrate some principles which have proved to be of fundamental importance in photochemistry we must consider them at some length.

As already mentioned the emission yield from acetone is wavelength and temperature dependent. At all temperatures it is zero at 2537 A and low but temperature dependent at 3130 A. At some probably indefinite point inbetween it becomes immeasurable. It was early shown by Daniels and his coworkers[55–57] that the emission consists of two parts, one mainly in the green and the other (much weaker) in the blue. They also showed that even traces of oxygen suppress the green emission[55]. Later Matheson and Zabor[58] showed that this green emission was due to biacetyl, one of the products of the reaction, and not to the acetone itself.

Whenever emission can be observed during a photochemical reaction it is an extremely useful aid to determining the behavior of some of the excited states and less frequently of some of the radicals which must be present.

For the sake of brevity let us designate an acetone molecule by A and a biacetyl molecule by B. Since there is emission, at least part of which is characteristic of the acetone molecule itself, the primary process at all wavelengths may not be represented by (102). At 3130 A at room temperature, where the carbon monoxide yield is intensity dependent but is about 0.1, the primary act of absorption should be written

$$A + h\nu \rightarrow {}^1A_n \tag{115}$$

where A is a normal ground state molecule and ${}^1A_n$ is an acetone molecule in an excited singlet state in the n'th vibrational level. The upper state is designated as singlet because the ground state is singlet and a change in multiplicity would lead to an extremely weak absorption. Attempts to find the singlet–triplet transition with long path lengths either in the vapor phase[59] or in liquid acetone[60] have failed.

Most of the photochemistry of acetone has been studied with pressures of several hundred torr because at 3130 A the absorption is weak. The radiative lifetime of singlet acetone has not been measured but the calculated value is about 10^{-6} second[60]. Since the emission yield is very low the true lifetime of the excited singlet state is probably determined mainly by processes other than emission. As far as measurements are available the wavelength of the emission is independent of the exciting wavelength under those conditions where the emission can be studied. This means that emission does not occur from the initially excited vibrational level but from lower levels with vibrational energy equilibrated with the surroundings. A step of energy equilibration must be included

$$^1A_n + M \rightarrow {}^1A_0 + M \tag{116}$$

where $^1A_0$ is an excited singlet acetone molecule with vibrational energy equilibrated with the surroundings. Part of the emission is due then to

$$^1A_0 \rightarrow A + h\nu \tag{117}$$

That emission by (117) is not the only emission is proved by the following facts. (a) Part but not all of the emission is suppressed by oxygen and the part which remains has a different wavelength distribution from the total even with a rapidly flowing system where biacetyl is not allowed to accumulate in the emission zone. (b) Part of the emission disappears as the temperature is raised and this disappearance goes hand in hand with an increase in quantum yields[61]. The interpretation of this result is very complex because the rate of reaction (103) increases with temperature, but there must be an emitting state of acetone whose importance diminishes with increase in temperature.

Other processes must be included

$$^1A_n(+M) \rightarrow {}^3A_m(+M) \tag{118}$$

$$^1A_0(+M) \rightarrow {}^3A_0(+M) \tag{119}$$

$$^3A_m + M \rightarrow {}^3A_0 + M \tag{120}$$

$$^3A_p + M \rightarrow {}^3A_0 + M \tag{121}$$

$$^3A_0 \rightarrow A + h\nu \tag{122}$$

$$\rightarrow A \tag{123}$$

$$^3A_0 + B \rightarrow {}^3B_n + A \tag{124}$$

$$^3B_n + M \rightarrow {}^3B_0 + M \tag{125}$$

$$^3B_0 \rightarrow B + h\nu \text{ (green)} \tag{126}$$

$$\rightarrow B \tag{127}$$

All of these steps are necessary and indeed the true situation must be even more complicated. For example the loss of vibrational energy by (116) must occur in steps and each level must have a step (118). That crossover must occur from several different levels is proved by the following fact. The ratio of the intensity of the emission from (122) to that from (117) is not independent of the exciting wavelength and of the pressure as it would be if (117) and (119) were the only sources of triplet state molecules[†].

Changes in multiplicity as in (118) and (119) might be expected to be slow but many lines of evidence, too numerous to discuss in detail here, indicate that these steps compete very successfully with radiative steps such as (117).

The emission step (117) does not involve a change in multiplicity whereas (122) does. The lifetime of the triplet state of acetone should be longer than that of the singlet state and indeed it is 2×10^{-4} sec as determined by Duncan and Kaskan[60]. The lifetime of the triplet state of biacetyl (126) is even longer, about 2×10^{-3} sec[60,62,63]. However in both cases as well as for (117) the molecules in question take part in several concurrent reactions and the experimentally determined lifetimes must be shorter than the true radiative lifetimes.

Since the photochemical quantum yields for acetone (as well as for biacetyl) increase with increase in temperature it is logical to say that the triplet state because of its long lifetime is subject to a thermal dissociation with an activation energy. Due to the complexity of the mechanism an unambiguous determination of this activation energy is difficult. The value for biacetyl is about 16 ± 2 kcal[50] and the most recent determination for acetone indicates a probable similar value[64].

This discussion of acetone is closed without going into detail concerning the various radical reactions at room temperature. Many of them must occur and the overall mechanism is believed to be well understood. Any mechanism which necessarily involves so many steps probably never could be proved to be correct in its entirety. The points we wish to emphasize are mainly that dissociation can occur in more than one way, sometimes from one electronic state and sometimes from another. The relative importances of the various steps are difficult to determine but studies of emission decay, of intermediate concentrations by flash photolysis, and of the yields of all of the products by modern methods such as chromatography and nuclear magnetic resonance permit reasonable mechanisms to be proposed.

[†] The terms "fluorescence" and "phosphorescence" do not have internationally agreed definitions. We follow the definitions proposed by G. N. Lewis and favored by Pitts, Wilkinson and Hammond[44] that fluorescence is an emission during which the multiplicity of the emitting molecule does not change and phosphorescence is an emission during which the multiplicity of the emitting molecule does change. Thus (117) would be called fluorescence and (122) and (126) would be phosphorescence.

2.4.2 Dissociation into complete molecules

The classical work of Norrish and his coworkers[65-69] in the middle 1930's demonstrated beyond reasonable doubt that certain ketones (such as 2-hexanone) decompose photochemically to give methyl ketones (in that case acetone) and an olefin (in that case propylene). Until that time the attention of photochemists had been directed mainly to atom and free radical reactions and the belief was fairly generally held that unless the absorbing molecules dissociated into atoms and/or free radicals no net reaction would occur.

During the past few years the number of photochemical reactions known not to occur *via* free radical intermediates has grown a very great deal until one might almost feel that they are the rule rather than the exception for many complex molecules and even for some simple ones. The detailed proof of mechanism often proves to be extremely difficult.

One of the simplest molecules studied in the early days, also by Norrish and his coworkers[70,71], is that of formaldehyde. Two primary processes may be visualized

$$HCHO + h\nu \rightarrow H\cdot + H\dot{C}O \tag{128}$$

$$\rightarrow H_2 + CO \tag{129}$$

The exact energy required for reaction (128) is a matter of debate but probably quanta at wavelengths longer than 2900 or 3000 A would not provide sufficient energy for it to happen. Nevertheless there is no sharp change in quantum yield of hydrogen formation near this wavelength.

Reaction (128) should be followed by several radical and atom steps including the formation of glyoxal and the production of H_2 both by atom recombination, reaction of $H\cdot$ with $H\dot{C}O$ and (with an activation energy) abstraction from HCHO.

Dideuteroformaldehyde should undergo similar reactions

$$DCDO + h\nu \rightarrow D\cdot + D\dot{C}O \tag{130}$$

$$\rightarrow D_2 + CO \tag{131}$$

In a mixture of D_2CO and H_2CO the hydrogen should be solely H_2 and D_2 if the primary processes are (129) and (131) but HD will be a product if there is dissociation into radicals[71]. It is also possible to add scavengers such as oxygen and iodine to remove the radicals. These various methods indicate that both primary processes occur and that at 3130 A they are about of equal importance.

A second compound studied in a similar way is methanol[72,73]. Several primary dissociations are possible

$$CH_3OH + h\nu \rightarrow CH_3O \cdot + H \cdot \tag{132}$$

$$\rightarrow \cdot CH_2OH + H \cdot \tag{133}$$

$$\rightarrow H_2 + HCHO \tag{134}$$

$$\rightarrow CH_2 + HOH \tag{135}$$

Reaction (135) does not occur, at least at wavelengths longer than about 1800 A.

Two deuterated methanols were used: CD_3OH and CH_3OD[72]. The radiation was not strictly monochromatic but lay around 1800 A. From these two deuterated methanols a large fraction of the hydrogen was HD. This fact of itself does not show that the primary process is any one of the first three rather than a mixture.

Glycol, $(CH_2OH)_2$, is a product of the photolysis of methanol vapor. This, however, does not distinguish between (132) and (133) because one of the two radicals $CH_3O \cdot$ and $\cdot CH_2OH$ will be more stable than the other. If it is assumed that (132) is the primary process this could be followed by

$$CH_3O \cdot + CH_3OH \rightarrow \cdot CH_2OH + CH_3OH \tag{136}$$

$$H \cdot + CH_3OH \rightarrow H_2 + \cdot CH_2OH \tag{137}$$

No evidence has been found that two methoxy radicals ever combine to give dimethyl peroxide, but they do disproportionate to give methanol and formaldehyde

$$2CH_3O \cdot \rightarrow HCHO + CH_3OH \tag{138}$$

By use of the deuterated methanols it was shown that the proportions of H_2, D_2, and HD were time dependent thus indicating that one or more of the products took part in secondary processes.

When all of the data were analyzed it was concluded that near 1800 A the majority of the primary process is (134) and all or nearly all of the remainder is (132). At short wavelengths emission by the $\cdot OH$ radical indicates[73] that the primary process is at least partially

$$CH_3OH + h\nu \rightarrow CH_3 \cdot + \cdot OH(^2\Sigma^+) \tag{139}$$

No conclusive evidence for (133) has been found.

The two examples of molecular hydrogen formation could be duplicated with other molecules but the two atoms which form the H_2 molecule must come from adjacent atoms in the molecular skeleton. In the far ultraviolet, below about 1500 A, at least one of the primary processes is that of molecular hydrogen formation. Thus the main process at these wavelengths in ethane is[37]

$$C_2H_6 + h\nu \rightarrow C_2H_4 + H_2 \tag{140}$$

Since (129) is an important primary process with formaldehyde one might ask whether or not the somewhat similar step with acetone

$$CH_3COCH_3 + hv \rightarrow C_2H_6 + CO \tag{141}$$

occurs. Certainly (141) has often been postulated, but the use of scavengers such as iodine vapor, oxygen, and nitric oxide (which rapidly remove methyl radicals) indicates that it is of negligible importance. Perhaps one is not able to say that it does not occur at all. However, there seem to be sound steric reasons why two methyl groups even in an excited acetone molecule would have great difficulty in joining to form a carbon–carbon bond. Repulsion between the hydrogens would tend to make (141) improbable.

Other more complex molecular elimination reactions are known to occur. One of the most interesting is the elimination of methane from acetaldehyde[74,75]. Several primary steps are possible

$$CH_3CHO + hv \rightarrow CH_3\cdot + \cdot CHO \tag{142}$$

$$\rightarrow H\cdot + CH_3\dot{C}O \tag{143}$$

$$\rightarrow CH_4 + CO \tag{144}$$

At 3130 A reaction (142) is the major process and indeed Ramsay and Herzberg[76,77] have used the flash photolysis of acetaldehyde as a means of photographing the absorption spectrum of the formyl radical. However methane is always a product and at wavelengths below about 2700 A it is formed in good yield when scavengers are present. The evidence indicates that singlet acetaldehyde molecules decompose according to (144) and that at longer wavelengths such as 3130 A both singlet and triplet molecules must be considered, as they are for acetone, in establishing a complete mechanism[78]. Reactions (142) and (144) are of a type very important for aliphatic aldehydes. Reaction (143) must occur, particularly at wavelengths toward or below the limit of transmission of quartz, since hydrogen is a product.

One of the most interesting molecular elimination reactions was first discovered by Norrish and Appleyard[65] in 1934 and studied further by Bamford and Norrish[66–69] in papers appearing in 1935 and 1938. These authors found that, on photolysis, aliphatic ketones with hydrogen atoms on carbons in the gamma position to the carbonyl yielded olefins and a methyl ketone. An early example was found in 2-hexanone, *viz.*

$$CH_3COCH_2CH_2CH_2CH_3 + hv \rightarrow CH_3COCH_3 + CH_3CHCH_2 \tag{145}$$

It was shown[79] that radicals were not involved in this reaction, that the yields were essentially the same in the vapor phase and in paraffin solution and that they were practically independent of temperature over the wide range of 0 to 300 °C. (145) competes with the generally accepted split into radicals (often called the Norrish Type I reaction) exemplified by (102).

The original proposal was that energy absorbed by the carbonyl group "flowed" to another part of the molecule and that scission occurred between the alpha and beta carbon atoms. The method of moving hydrogen atoms to give complete molecules was largely ignored in this picture.

The evidence is now practically conclusive that the intramolecular process occurs *via* hydrogen bonding between the gamma carbon atom and the oxygen, and that scission then occurs to give the enol form of a methyl ketone and the olefin in a single step. A very brief summary of the evidence which leads to this conclusion is as follows.

(*a*) The reaction is a general one. For esters it will occur in either of two ways:

$$RCH_2CH_2CH_2CO_2R' + h\nu \rightarrow RCHCH_2 + CH_3CO_2R' \tag{146}$$

$$RCO_2CH_2CH_2R' + h\nu \rightarrow RCO_2H + CH_2CHR' \tag{147}$$

Type II reactions also occur with aldehydes and possibly with certain amines[80].

(*b*) Norrish Type II reactions are found with complicated cyclic compounds[81] (*e.g.* camphor) and also with certain polymers such as polymethylvinyl ketone[82, 83].

(*c*) The reaction is often inhibited by hydrogen bonding solvents such as alcohols.

(*d*) By deuterium substitution on the gamma carbon it has been shown[84] that the olefin always contains the right amount of deuterium. The methyl ketone is, however, formed in the enol form and the latter is active in exchange reactions. It will exchange either with H_2O or with D_2O (probably on the walls). The amount of deuterium in the methyl ketone is such as to indicate strongly that an intermediate enol form had been present.

(*e*) By the use of a long infrared path, McMillan *et al.*[85] have shown beyond reasonable doubt that the enol form of acetone is present during the photolysis of 2-pentanone.

The six membered hydrogen bonded ring might be thought to be present in unexcited carbonyl compounds. The absorption of radiation by the fraction of the molecules in this form might lead to the Norrish Type II reaction. It is almost inconceivable that such a ring could be formed with zero change in entropy. If a change in entropy is postulated the mole fraction in the ring form would change (probably rapidly) with temperature. For some compounds such as 2-hexanone the Type II yield is essentially independent of temperature and for most of the others the change is small. By this reasoning the ring must be formed in the excited state, more rapidly than the competing Type I process. At a first glance this might appear to be improbable but there is accumulating evidence that many molecular rearrangements occur rapidly, *i.e.* in periods shorter than radiative lifetimes and even in periods shorter than for relaxation of vibrational energy even in the liquid phase.

The Type I process and the Type II process are not the only ones which occur concurrently. Thus Ausloos and Rebbert[86] have shown that with 2-pentanone some methylcyclobutanol is formed, also by a non-free radical process

$$CH_3COCH_2CH_2CH_3 + h\nu \rightarrow CH_3\overset{\overset{\displaystyle H}{|}\overset{\displaystyle O}{|}}{C}\text{---}CH_2 \atop CH_2\text{---}CH_2 \tag{148}$$

Since the excited state is involved in the formation of the hydrogen bonded ring it is probably useless to speculate in detail about the influences of substituent groups using data from the customary reactions of organic chemistry. Photochemistry is the chemistry of excited and not of normal molecules.

Considerable attention has been devoted to proving whether the Type II process occurs through the singlet or through the triplet state. With 2-hexanone little or no inhibition of the Type II process occurs even with large pressures of added oxygen[87], a paramagnetic substance which destroys triplet states in at least some cases. With 2-pentanone it was shown that the same state which by energy transfer excited biacetyl to emission also was responsible for the Type II reaction. Since several torr of biacetyl were necessary to cause maximum sensitized emission it was inferred that probably the singlet state was involved[88]. However, Ausloos and Rebbert[89] have shown that the triplet state is certainly the one responsible for the Type II process with this ketone. More recent evidence[90] from Hammond's laboratory indicates quite strongly that Type II reactions may arise in the liquid phase from either singlet or triplet states and that the relative amounts from each state will vary from one molecule to another.

Some words of caution are necessary, however. In the first place the sum of the various primary process yields and the reasonable estimates of fluorescence and phosphorescence yields is not unity. The best evidence is perhaps from studies with 2-hexanone where the Type II yield is about 0.40 ± 0.05 and independent both of temperature and of added oxygen[79,88]. No fluorescence or phosphorescence from this molecule has been observed. The Type I yield is very low at room temperature but increases rapidly with temperature. The accurate Type I quantum yield is difficult to determine because there are many radical reactions, including chain propagating steps, as the temperature is increased. As judged by analogy with other molecules for which the evidence is better, there is probably a triplet state which dissociates by the Type I process[91] with an activation energy[79]. However, if the Type II process competes with crossover to the triplet state and the latter is followed immediately by dissociation, the sequence of events would be difficult to distinguish from Type I and Type II processes competing in the decomposition of the triplet state. If emission occurs from one state or the other, or better still from both, and if the two emissions can be studied separately, it is possible to change several variables and with some certainty relate certain yields to particular states. However geometry and modes of vibration must play important parts in complicated reactions such as we are considering so that in fact we are dealing with a great many, rather than just two, excited states. Far too little is known as yet about the rapidity with which one mode of vibration is changed into another by collision to permit intimate details of molecular motions to be related to specific primary processes. Only now are photochemists becoming aware of the data which must be obtained to answer many important questions. The experiments will not be easy because highly monochromatic light must be used and it must have sufficient intensity so that measurable amounts of products are formed.

The Type II process does not occur when there are unsaturations in the molecule adjacent either to the carbonyl or to the gamma carbon atom[92]. An interesting reaction is

$$
H_3C-\overset{\overset{O}{\|}}{\underset{\underset{CH_2-CH_2}{|}}{C}}\quad \overset{\overset{CH_2}{\|}}{\underset{|}{CH}} + h\nu \longrightarrow H_3C-\overset{\overset{O}{|}}{\underset{\underset{CH_2-CH_2}{|}}{C}}\overset{CH_2}{\underset{|}{---}}\overset{|}{CH} \tag{149}
$$

The propensity to form ring compounds, even under conditions such that a large amount of strain would be expected in the product is of itself sufficient proof that ideas which apply to normal molecules may not be carried over without change to photochemical processes.

It should be mentioned in passing that both mercury and benzene sensitized Norrish Type II reactions have been shown to occur. Under the proper conditions the triplet states of the substrate molecules would be preferentially excited. This is further proof that Type II reactions may, with certain compounds, occur through the triplet state.

2.4.3 Rearrangements

Photochemical rearrangements of complex molecules have long been known. The classical researches of Ciamician and Silber[93] and of Berthelot and Gaudechon[94] showed that such rearrangements occurred, although much of the work was with condensed phases and little could be said at that time about the free radical or non-free radical character of the reactions. The number of cases in which rings are closed to form compounds not easily obtained by classical techniques is very large indeed, and growing almost day by day.

The detailed accounting of the number of quanta absorbed is one of the major problems encountered in the elucidation of the primary process[95]. The First Law of Thermodynamics must, of course, be obeyed for photochemical reactions as it is for all other systems. Thus the energy absorbed must appear in one or more of several well recognized ways, *viz.*

$$\text{energy absorbed} = \text{energy re-emitted} + \text{energy as heat} +$$
$$\text{chemical energy change} \quad (150)$$

From the heats of formation of starting materials and of products one can calculate the chemical energy change (*i.e.* ΔH for thermal processes). The absolute measurement of fluorescence and of phosphorescence is difficult, particularly for gas phase reactions, but in principle this quantity can be obtained. The "energy as heat" is difficult to measure if the incident intensity is low and in fact it is very rarely determined.

If light initiates an exothermic chain reaction the determination of the yield of the primary process is difficult. There are cases, however, for which the data are reasonably satisfactory and which indicate a large unknown quantity in balancing the energy terms. This means either that unknown products are formed or that a large fraction of the absorbed energy is degraded to heat by processes unknown.

An example of the difficulties we are citing is found with biacetyl photolysis at 4358 A at room temperature. The following figures have been obtained by various authors

Fluorescence efficiency	0.0025[96,97]
Phosphorescence efficiency	0.15 [98]
Primary dissociation yield	~0 [99]
Fraction of absorbed quanta accounted for	0.153

Thus about 85 per cent of the energy of the photons absorbed does not reappear in recognized forms. It is true that the quantum yield of products increases with the first power of the intensity (*i.e.* the rate is proportional to the square of the intensity) but it is close to zero at intensities generally used by photochemists.

Another example is found with benzene vapor at room temperature. The emission yield (fluorescence only) is about 0.20 ± 0.04 and essentially independent of pressure[100,101]. There are products formed, although there is some disagreement on this subject and the yields in the gas phase have not been determined. Also there is a crossover from the singlet to the triplet state so that all or nearly all of the primarily excited singlet state molecules either fluoresce or cross over to the triplet state. Nevertheless, in the vapor phase at room temperature, about 80 per cent of the molecules which absorb neither react chemically nor emit radiation[100,101].

Much useful information can be gathered about the behavior of excited molecules by studying them in glassy matrices at very low temperatures. At these temperatures (4 to 77 °K) molecules after absorption of radiation must lose vibrational energy very rapidly. Lifetimes and emission yields must be those of excited state molecules possessing only zero point vibrational energy. Benzene in glassy matrices has been studied at 4 and at 77 °K. In this temperature range both fluorescent and phosphorescent emission occur with yields of about 0.25[102]. The absorption spectrum of triplet benzene in glassy matrices has been observed by Godfrey and Porter[103]. Also it has been shown that there is chemical reaction between excited benzene and added molecules[104]. Leach and Migirdicyan[105] suggest that these reactions imply the formation of a hexatriene intermediate.

Benzene derivatives in solution and also in the gas phase undergo rearrangements[106-110]. Thus m-xylene rearranges to give, among other things, o- and p-xylene without any indication of radical formation. Kaplan et al.[109] have shown moreover that the ring carbon atoms to which isobutyl groups are attached migrate along with the isobutyl groups to the new positions in the ring. A logical explanation based on intermediate isomers with transannular bonds has been put forward by these authors.

The calculated radiative lifetime of benzene vapor based on absorption coefficients[111] is in excellent agreement with the mean lifetime measured by Donovan and Duncan[112] and is about 5×10^{-7} sec. And yet this agreement must be partly fortuitous, since the emission yield is only about 0.2 and hence there must be processes which compete with radiative emission[100, 101]. It is true that, if twenty per cent of the molecules emit, these competitive processes must have rate coefficients not greatly different from the rate coefficient of the process[100]

$$C_6H_6(^1B_{2u}) \rightarrow C_6H_6(^1A_{1g}) + h\nu \qquad (151)$$

The efficiency of emission from benzene molecules with equilibrated vibrational energy will be given by the equation

$$Q_f = k_{151} / \sum k_i \qquad (152)$$

where the summation in the denominator must be taken over all competing processes including (if necessary) any which are second order. Q_f is the fluorescent efficiency. The summation in the denominator must be about five times k_{151}. The main competing process (but not the only one) is the crossover from the singlet to the triplet state[101].

The mean lifetime of triplet benzene in the glassy matrix is not certain, but all authors agree that it is large and it is probably about 20 sec[113]. And yet the available data on decay of the triplet state in the gas phase (there is no phosphorescence) indicate a mean lifetime of the order of magnitude of 10^{-4} to 10^{-5}

second. There would seem to be no *a priori* reason for this large difference unless there are competing processes which have rates dependent on the vibrational level of excited molecules either in the singlet or in the triplet state. A complete explanation of all the facts is not possible but these competing processes apparently have yields which vary with wavelength (possibly even in the liquid phase but more data are necessary on this point) and hence are rapid compared to equilibration of vibrational energy even at pressures of 50 to 100 torr. If this is true the unimolecular rate coefficients must be greater than 10^{10} sec^{-1}. If the pre-exponential factors are 10^{13} sec^{-1}, activation energies must be not above a few kcal.mole^{-1} for the conversion of vibrationally and electronically excited molecules to isomers.

Let us now return to equation (150). An excited but metastable atom such as (Hg 6^3P_0) must either radiate or lose energy by collision to return to the ground state. The same is true of a diatomic molecule. In both instances the term "collision" is broadly used and might include the formation of an intermediate complex which would ultimately dissociate to ground state molecules.

Processes occur with polyatomic molecules which could not occur with atoms and diatomic molecules. Let us postulate the following steps

$$M + hv \rightarrow {}^1M \tag{153}$$

$$^1M(+M) \rightarrow I(+M) \tag{154}$$

$$^1M \rightarrow {}^3M \tag{155}$$

$$^1M \rightarrow M + hv \tag{156}$$

$$M(+M) \rightarrow I(+M) \tag{157}$$

$$I(+M) \rightarrow 2M \tag{158}$$

Other steps could be added. I is a molecule of isomer and it may revert to a normal molecule either by a first or second order homogeneous reaction or at the walls. The fluorescence efficiency would be given by the equation

$$Q_f = k_{156}/(k_{154} + k_{155} + k_{156})$$

The concentration attained by the isomer in its steady state would depend on its rate of formation and on the coefficient k_{158}.

For benzene several isomers are possible, most of them suggested in the last century as possible structures for benzene before the Kekule model was proposed and accepted. Prismane, Dewar benzene, and benzvalene (see Fig. 5) all might be formed from excited benzene molecules provided proper modes of vibration were

Fig. 5. Isomers of benzene.

present. Of course the heats of formation of these various forms of benzene are not known but they may be estimated by standard procedures based on bond energies, etc. Most of them would be formed endothermically from benzene with the absorption of 60 to 70 kcal.mole^{-1} or 2.5 to 3 electron volts.

For atoms and even for diatomic molecules there seems often to be no alternative but to convert relatively large amounts of electronic energy into kinetic or vibrational energy in order to reduce excited molecules to the ground state. Theoretically, processes of this sort should be rather slow, unless the colliding particles which receive the energy are easily dissociated or have many rather low vibrational frequencies. The better the resonance, *i.e.* the less the amount of energy which must be converted to kinetic energy, the more pleasing the situation is.

With polyatomic molecules the mechanism embodied in (153)–(158) provides a means of dissipating energy without doing violence to preconceived ideas. The isomer I could last long enough to revert to normal molecules on the walls.

Isomerization is an accepted photochemical process and we will not discuss further examples in detail. A few types are cited below.

(*a*) *Cis–trans* isomerization. Simple olefins such as cis-2-butene are converted to isomers, *viz.* trans-2-butene, by radiation below about 2000 A[114]. Presumably an intermediate triplet state with free rotation is formed and the molecule reverts to the ground state in either the *cis* or the *trans* form[115]. A photochemical stationary state dependent on wavelength results. One of the earliest classical examples was the maleic–fumaric acid system[116].

(*b*) Shift of a single atom from one location to another. One or two examples have been mentioned. A classical case is that of *ortho*nitrobenzaldehyde to *ortho*-nitrosobenzoic acid[117, 118].

(*c*) Shift of an atom accompanied by opening of a ring. The conversion of ethylene oxide to acetaldehyde is an example although there are concurrent radical processes[119, 120]. Another example is cyclopentanone[121] which undergoes several reactions as follows

$$\text{cyclo } C_4H_8 + CO \tag{159}$$
$$2 C_2H_4 + CO \tag{160}$$
$$CH_2{=}CH{-}CH_2CH_2CHO \tag{161}$$

Srinivasan[122] has shown that the hydrogen atom transferred to the carbonyl carbon comes from one of the two carbon atoms in the 2 positions. Formation of the pentenal appears to be favored at low vibrational energies since the addition of foreign gas improves the yield which is, nevertheless, always small. Reactions (159) and (160) are the classical steps proposed by Norrish and his coworkers and they are found also with the perfluorocycloketones[123].

(d) Concurrent shift of two atoms. Perhaps details of this process are not too well understood. Butadiene[124] is an example

$$CH_2=CH-CH=CH_2 + h\nu \rightarrow CH_3CH_2-C\equiv CH \tag{162}$$

Other less drastic reactions lead to methyl allene and cyclobutene.

(e) Ring formation. This is found especially with conjugated compounds. Cyclooctatetraene[125] bridges across the ring to give a six membered and a four membered ring, viz.

$$\tag{163a}$$

Reactions of benzene have already been mentioned. They have been studied mainly with benzene derivatives and we will not here give the complete story but will cite one pair of reactions both of which are reversible photochemically[126]

$$\tag{163b}$$

Reactions such as (163a) and (163b) have now been shown to be very common photochemical processes.

(f) Ring closure reactions. These are invariably accompanied by transfer of an atom from one location in the molecule to another. (148) is an example of this type of process and so would be the formation of cyclobutene from butadiene.

For a more extended list of reaction types longer treatises should be consulted[127].

2.4.4 Summary of basic problems

Absorption of radiation by polyatomic molecules composed of light atoms follows, as regards change in multiplicity, the same rules as those for atoms and diatomic molecules. Since most polyatomic molecules have ground singlet states the vast majority of absorption processes form upper singlet states. In the presence of paramagnetic substances (and also probably in the presence of sufficiently strong magnetic fields) singlet–triplet transitions can be observed. Thus benzene which shows a singlet–singlet transition beginning about 2680 A shows a further absorption around 3400 A in the presence of high pressures of oxygen[128].

Upper repulsive states exist for polyatomic molecules just as they do for diatomic molecules. The alkyl halides all show absorption which corresponds to dissociation into alkyl radicals and halogen atoms. Saturated hydrocarbons absorb only below 1800 A unless rings with considerable strain are present. The first absorption region for the simple alkanes is a continuum and dissociation must immediately follow the absorption act. The simple alcohols also show continua.

It must be re-emphasized, however, that a true continuum in a spectrum is more difficult to identify than for diatomic molecules, and a primary yield of unity independent of total pressure and of scavengers is better proof of the presence of a continuum than is the appearance of the spectrum.

When unsaturation is present the situation is usually much more complex and more interesting. If the molecule in its ground state is highly symmetrical, e.g. benzene, there are rigorous selection rules which govern transition probabilities. For example the first electronic transition of benzene centered around 2500 A is forbidden for reasons of symmetry. Since the complete wave function must include participation of vibrations, transitions to certain vibrational levels with the proper symmetry will be permitted. The most pronounced progression in the absorption spectrum of benzene vapor is due to the production of excited singlet molecules with one unit of an asymmetric vibration (frequency 521 cm^{-1} and given the symbol e_{2g}) and 0, 1, 2, . . . etc. units of the symmetric vibration which causes the entire ring to expand and contract.

Triplet states are, however, formed to larger extents and play more important roles than would have been thought to be true twenty or thirty years ago. The classical work of Lewis and Kasha[129], which showed certain emissions to come from triplet states, opened the way for the rationalization of many phenomena which would otherwise prove to be quite incomprehensible. Perhaps, as Matsen et al.[130] have pointed out, an undue emphasis has been placed on electron spin and on the multiplicity of states. The symmetry of the entire wave function is really the important point, and the contributions to it of all states of the molecule must be considered. Viewed in this light the "triplet component", to use rather crude language, will depend on the vibrational quantum numbers in the excited state. If other isomers can exist, their contributions to the complete wave function must also be considered.

One factor must, however, be kept in mind. Triplet states invariably lie at lower energies than do the corresponding singlet states[131]. While it may be true that the multiplicity per se will not affect reactivity for many processes, the difference in energy may be decisive and some reactions can occur with singlet states which would not be possible with triplet states.

If contributions from all states, including those of different multiplicity, must be considered in developing the complete wave function, it is not possible in one sense to say that collisions or foreign fields "induce" crossovers from singlet to triplet states. If, however, a molecule is to be stabilized in a triplet state enough energy must somehow be removed so that the wave function can be mainly triplet. Thus, to say that collisions play no role in the crossover process would be fallacious.

The photochemist is always confronted with the necessity of obtaining sufficient intensity of radiation to give measurable effects. The great majority of the work has been done with mercury arcs of various types. Very high pressure arcs give continua with a large gap extending from about 2530 A to nearly 2700 A due to absorption of radiation by the mercury vapor itself. Lower pressure lamps give radiation centered around the mercury emission lines and resonance lamps give almost exclusively the 2537 A line (together with the 1849 A line if the lamp and reaction vessel materials do not absorb it).

Since the complete wave function will have variable contributions from other energy states and even from isomers and will depend on the vibrational energy level, it would be very desirable to use light of such monochromaticity that one and only one vibrational level could be formed at a time. The importance of doing this may vary greatly from one compound to another but we will illustrate the nature of the problem with the following simplified mechanism

$$M + h\nu \rightarrow {}^1M_v \tag{164}$$

$${}^1M_v + M \rightarrow {}^1M_0 \tag{165}$$

$${}^1M_v \rightarrow I \tag{166}$$

$${}^1M_v \rightarrow {}^3M_y \tag{167}$$

$${}^1M_0 \rightarrow M + h\nu \text{ etc.} \tag{168}$$

Without going into detail the rate of formation of the isomer I is competitive both with crossover to the triplet state and with vibrational relaxation to lower vibrational energy levels. In the liquid phase relaxation of vibrational energy should be extremely rapid. If the quantum yield of the isomer I is dependent on wavelength one must draw one or the other of the following conclusions: (a) the rate of (166) is very rapid indeed; (b) a change in wavelength has in reality caused a change in the electronic transition. Possibility (b) implies that rates of isomer formation would be different from different electronic states.

While much of the theoretical treatment of the details of photochemical reactions has been based on work in the gas phase, the majority of interesting new reactions today are being found in condensed phases. For a more complete understanding of these processes highly monochromatic light should be used. Fortunately this is becoming possible.

3. Experimental problems

The desirability of using very monochromatic light has been stressed. Two developments make it possible to do this to a greater extent than formerly. The first is, of course, the advent of new light sources. The second is the development of analytical techniques which permit both the identification and quantitative determination of much smaller amounts of products than formerly.

Two methods of obtaining monochromatic light are used. The first involves color filters and the second either prism or grating monochromators. Color filters may be used to advantage with a mercury arc since the lines are well separated. However, no color filter system is really adequate for modern photochemistry.

Prism monochromators made of quartz have the great disadvantage of not being very useful below about 2500 A. As one goes to shorter wavelengths the transparency decreases and with it the resolving power. Much quartz also fluoresces in its own right.

Fortunately grating monochromators of high quality are now available. As we have already seen, the problem of exciting to only one vibrational level in the upper state is crucial. To take a specific but hypothetical example let us assume that at 2500 A we must cover a range of 5 A or about 80 cm^{-1}. Plasma arcs and xenon lamps are available which will emit roughly 2 kilowatts of energy in the range of 2000 to 4000 A under optimum conditions. If the intensity in $erg.cm^{-2}$ is uniform throughout this spectral range we desire to use the fraction 0.0025 of the energy. But it will not be possible to use radiation emitted in all directions, and in reality we will be fortunate to use more than 10 to 20 per cent of the total emitted in the right range. Thus the realistic fraction will be about 0.0005 of the total radiation in the 2000–4000 A range. This amounts to about 1.5×10^{18} $quanta.sec^{-1}$. If the absorption is as low as 0.1 and the quantum yield 0.01, 1.5×10^{15} molecules will react per second or about 5×10^{18} (10^{-5} mole) per hour. Many modern analytical devices such as vapor phase chromatography, nuclear magnetic resonance, mass spectrometry, and even ordinary spectrometric methods can be used with quantities as small as this.

The above calculation may also be used in reverse. With more old fashioned lamps operating on a few hundred watts with a few per cent of the energy input in the right wavelength range the photochemist is forced either to use polychromatic light or excessively long exposures. Long exposure times introduce other errors due to the unsteadiness of the lamp, the reliability of intensity measurements, the effects of impurities, and the appreciable occurrence of thermal reactions.

The *rotating sector* method has proved to be of immense value for certain types of work. It has been used to study chain reactions, in which chain termination steps depend on a different power of the radical concentration from the chain propagating steps. It has also been used to obtain the absolute rates of certain elementary processes. Thus Gomer and Kistiakowsky[132] determined the absolute rate coefficient of the methyl plus methyl reaction

$$CH_3 \cdot + CH_3 \cdot \rightarrow C_2H_6 \tag{169}$$

by contrasting its behaviour with that of the process

$$CH_3 \cdot + RH \rightarrow CH_4 + R \cdot \tag{170}$$

The rotating sector method has also been used to investigate the second order decomposition of biacetyl at room temperature at 4358 A

$$2(CH_3CO)_2'' \rightarrow 2CH_3 \cdot + 2CO + (CH_3CO)_2 \tag{171}$$

where $(CH_3CO)_2''$ is an excited biacetyl molecule. It was shown that the mean life of the excited state in question is about 1.2×10^{-3} sec, which is that of the triplet state as determined by Kaskan and Duncan[60].

The principle of the rotating sector method has been described many times and we will not discuss it in detail[133]. Briefly it is as follows (see also Chapter 1, Volume 1).

If an intermediate such as a radical or an excited molecule is produced by absorption of radiation and disappears at a rate proportional to the first power of the intensity, it is possible to interrupt the light beam and the rate of disappearance integrated over time will then vary merely as the fraction of the time the light reaches the reaction vessel. If this is 0.25, the net rate with interruption will be 0.25 times the rate with full illumination. If, on the other hand, the intermediates disappear by a second order reaction (*i.e.* the rate is proportional to the square of their concentration) the situation is different. At very high rates of interruption the light will appear to be on all of the time but with one quarter of the intensity. At very slow rates of interruption the light will appear to be on one quarter of the time with full intensity. In a plot of rate *vs.* duration of flash the asymptotes for very slow and for very rapid interruption will differ by a factor of four. The inflection point will come at about the mean life of the intermediate being studied.

Flash photolysis has given photochemists an extremely powerful tool for the study of the details of photochemical reactions. Since the original description of the method by Norrish and Porter[134] in 1949 hundreds if not thousands of papers based on it have been published. By charging a large battery of condensers and then discharging them through a lamp it is possible in a time even as short as a microsecond to introduce so many quanta into a reaction vessel that virtually all molecules will be excited or dissociated. Not only is it possible to determine the nature of the products under these high intensity conditions but it is also possible by absorption spectroscopy or other suitable devices to determine the rates of disappearance of excited molecules, free atoms and free radicals.

The lifetimes and kinetic behavior of many triplet states of molecules, particularly of aromatic compounds, have been studied by Porter and his coworkers. The identification of the absorption spectra of methyl radicals, of HCO radicals and of NH_2 radicals, mainly in Herzberg's laboratory at Ottawa, has permitted flash photolysis to be used directly to measure the rates of radical reactions.

An interesting example of the use of the flash technique with solutions is found in studies by Grossweiner and Matheson[135] with iodides. By the use of a sapphire flash lamp, which emits wavelengths shorter than one made of quartz, the iodide ion can be shown to undergo the reaction

$$I^- + h\nu \rightarrow I\cdot + \text{electron} \tag{172}$$

The reactions of the iodine atoms can then be studied by using absorption spectroscopy.

More recently still the electron produced in this and in other ways has been shown to have a finite independent existence as a hydrated species[136]. This is very similar to solvated electrons in liquid ammonia known since the early work of Franklin and Kraus[137]. The solvated electron in liquid ammonia disappears very slowly in the absence of a catalyst, but the life of the hydrated electron is much shorter; if they undergo no other reactions, they are removed as follows

$$H_2O + \text{electron} \rightarrow OH^- + H\cdot \tag{173}$$

Pulse radiolysis[136], whereby intense beams of high energy electrons and other charged particles may be introduced into reaction systems, do for the radiation chemist what flash photolysis has done for the photochemist. This interesting development belongs outside of the scope of this chapter.

4. Different radiation types. Conclusions

Nearly all of the previous discussion refers to electromagnetic radiation commonly called visible and ultraviolet, with energies ranging from about two electron volts up to perhaps ten electron volts. Radiation of reasonably good intensity can be obtained over most of this energy range although special experimental difficulties are encountered toward the upper end of the range. Since it was decided to limit the scope of this chapter by not considering energies above about ten electron volts, a few comments might be made about other forms of radiation.

Electron beams of a few volts energy and sufficiently monochromatic to give meaningful results can be obtained in a variety of ways. Electrons emitted from hot filaments can be regulated by grids and collimated to give relatively intense beams. While such studies have rather passed out of fashion, resonance and ionization potentials have been determined for atoms; for di- and polyatomic gases few data are now available. Ionization without dissociation may occur but often ionization fragments are produced.

Ionization chambers combined with mass spectrometers permit determination of the types of ions produced at definite voltages[138]. Molecular ions produced by electron bombardment

$$M + electron \rightarrow M^+ + 2\ electrons \tag{174}$$

may at higher voltages produce atomic or radical ions

$$M + electron \rightarrow A^+ + B + 2\ electrons \tag{175}$$

where A^+ and B are fragments of the parent molecule. The distinction between M^+ and A^+ can be made with the mass spectrometer and the minimum electron energy required to give rise to each process can be determined.

In electron bombardment studies the distinction must be made between "vertical" ionization potentials and what might be called "minimum" ionization potentials. Thus if the electron energy is such as to give a maximum current of A^+ the electron may be removed from M so rapidly that A^+ and B have no time to separate, and they may then do so in such a way as to give appreciable kinetic energies to A^+ and B. The "vertical" ionization potential will differ from the minimum potential required to produce A^+ and B by the amount of the total kinetic energy imparted to the fragments.

Bond dissociation energies may be determined by studies of the type just mentioned[139]. For a series of aliphatic hydrocarbons the voltage required to produce H^+ ions was determined. By measurement of the kinetic energies of H^+ and use of the laws of conservation of momentum and of energy it was possible to determine the minimum energy required to separate H atoms from the parent molecules.

At potentials below the ionization potential excitation and dissociation of gas molecules can be brought about. The same states formed by optical absorption can be produced by electron bombardment but the same restrictive selection rules will not apply so that more will be observed using electron bombardment. Thus, 6^1S_0 atoms of mercury may not form 6^3P_0 atoms by absorption of electromagnetic radiation but a potential corresponding to this transition is observed in the bombardment of mercury atoms by electrons[140]. Probably much interesting work can be done with polyatomic molecules raised to resonance or ionization states by electron bombardment, but at the present time the field is in its infancy and we can only hope that much work will be done in the future.

Studies using accelerated charged particles heavier than electrons, *viz.* H^+ and

He^+ (or He^{++}) are very difficult to pursue unless the energy is considerably higher than the 10–12 volts we have set as an upper energy limit. The reason for this is evident. Since protons have about 1836 times the mass of the electron the linear velocity after the same accelerating voltage will be about 1/43 as much. For beams of comparable intensity the charge density will be 43 times as high and the consequent repulsion will scatter the protons. It is thus difficult to obtain high intensities of monochromatic protons with energies of a few volts. This situation is unfortunate for otherwise it would be possible to measure directly the activation energies for processes such as

$$H^+ + CH_4 \rightarrow H_2^+ + CH_3 \tag{176}$$

To date studies of this type have been unsuccessful.

A brief mention should be made of gamma radiation and of effects of high energy particles although they will be more fully discussed in Chapter 2. The primary effect of such radiation must be mainly ionization.

Gamma rays are usually so slightly absorbed in gases that not many studies of their chemical effects on gaseous reactants have been made. Absorption of gamma radiation is of two types: (a) photoelectric absorption in which nearly all of the energy of the gamma quantum appears as kinetic energy of the ejected electron. The only loss would be that of the ionization potential of the gas molecule in question and usually this will be low compared to the energy in the quantum. (b) Compton effect absorption whereby part of the momentum of the gamma ray is imparted to electrons and hence the kinetic energy of the electron depends on the scattering angle. A varying fraction of the gamma ray energy appears as kinetic energy of the electron, but in fact the fraction is generally small.

Ions formed by high energy radiation recombine eventually with electrons and when they do they may form excited electronic states of molecules and fragments of molecules. These states lose energy in various ways and cascade down to states which are formed by the more customary methods of photochemistry. Often the chemical effects of high energy radiation are not too different from those found in photochemistry.

There may be a contrast between the effects of various types of ionization. With a heavy particle such as an alpha particle or a proton the specific ionization is very high, that is, many ions are formed per unit path length. Thus ion–ion, ion–radical, and radical–radical reactions may be very important because concentrations of these intermediates are high. With particles of lower mass, such as photoelectrons, the specific ionization is much lower and the chance of second order effects much less. Thus the effect of specific ionization bears some resemblance to that obtained with rotating sectors and pulse radiolysis.

In conclusion we wish to emphasize that while much of the basic theory of the primary effects of low energy radiation is well understood, there is room for substantial improvement in the experimental data by use of more carefully controlled conditions. The rapid growth of research in photochemistry during the past ten years has been phenomenal. This is due partly to the very real interest in the purely scientific aspects of the field and partly to the occurrence of many synthetic reactions not obtained by more conventional techniques. The future holds great promise.

REFERENCES

1 J. CALVERT AND J. N. PITTS, JR., *Photochemistry*, Wiley, New York, 1965, p. 55.
2 A. C. G. MITCHELL AND M. ZEMANSKY, *Resonance Radiation and Excited Atoms*, Cambridge University Press, New York, 1934.
3 R. J. CVETANOVIC, *Progr. Reaction Kinetics*, 2 (1964) 41.
4 Ref. 1, p. 101.
5 K. YANG, *J. Am. Chem. Soc.*, 86 (1964) 3941.
6 K. YANG, *J. Am. Chem. Soc.*, 87 (1965) 5294.
7 D. E. HOARE AND J. S. PEARSON, *Advan. Photochem.*, 3 (1964) 83.
8 R. GOMER AND J. B. KISTIAKOWSKY, *J. Chem. Phys.*, 19 (1951) 85.
9 R. A. MARCUS, *J. Chem. Phys.*, 20 (1952) 364.
10 S. W. BENSON, *Advan. Photochem.*, 2 (1964) 1.
11 T. L. HILL, *J. Chem. Phys.*, 17 (1949) 1125.
12 R. M. NOYES, *J. Am. Chem. Soc.*, 73 (1951) 3039.
13 K. T. COMPTON AND L. A. TURNER, *Phil. Mag.*, 48 (1924) 360.
14 M. L. POOL, *Phys. Rev.*, 33 (1929) 22; 35 (1930) 1419.
15 W. A. NOYES, JR. AND P. A. LEIGHTON, *Photochemistry of Gases*, Reinhold, New York, 1941; Dover Press, New York, 1966, p. 228.
16 C. C. MCDONALD AND H. E. GUNNING, *J. Chem. Phys.*, 23 (1955) 532.
17 H. E. GUNNING AND O. P. STRAUSZ, *Advan. Photochem.*, 1 (1963) 209.
18 R. G. DICKINSON AND M. S. SHERRILL, *Proc. Natl. Acad. Sci. U.S.*, 12 (1926) 175.
19 D. H. VOLMAN, *Advan. Photochem.*, 1 (1963) 43.
20 D. H. VOLMAN, *J. Am. Chem. Soc.*, 76 (1954) 6034.
21 J. FRANCK, *Trans. Faraday Soc.*, 21 (1926) 536.
22 V. HENRI AND M. C. TEVES, *Compt. Rend.*, 179 (1924) 1156; *Nature*, 114 (1924) 894.
23 (a) G. HERZBERG, *Spectra of Diatomic Molecules*, Van Nostrand, New York, 1950.
   (b) G. W. KING, *Spectroscopy and Molecular Structure*, Holt, Rinehart and Winston, New York, 1964.
   (c) G. M. BARROW, *Molecular Spectroscopy*, McGraw-Hill, New York, 1962.
24 A. FARKAS, *Ortho Hydrogen, Para Hydrogen and Heavy Hydrogen*, Cambridge University Press, New York, 1935.
25 J. I. STEINFELD AND W. KLEMPERER, *J. Chem. Phys.*, 42 (1965) 3475.
26 E. RABINOWITCH AND W. C. WOOD, *J. Chem. Phys.*, 4 (1936) 358.
27 L. A. TURNER, *Phys. Rev.*, 41 (1932) 627.
28 *Cf.* J. P. HOWE AND W. A. NOYES, JR., *J. Am. Chem. Soc.*, 57 (1935) 1262.
29 F. J. LIPSCOMB, R. G. W. NORRISH AND G. PORTER, *Nature*, 174 (1954) 785.
30 *Cf.* G. HERZBERG, Ref. 23(a), pp. 541, 553.
31 H. P. BROIDA AND T. CARRINGTON, *J. Chem. Phys.*, 38 (1963) 136.
32 Ref. 1, p. 199.
33 *Cf.* G. HERZBERG, Ref. 23(a).
34 A. G. GAYDON, *Dissociation Energies*, Dover Publications, New York, 1950, p. 61.
35 Ref. 1, p. 226; Ref. 37, p. 66.
36 R. J. DONOVAN AND D. HUSAIN, *Trans. Faraday Soc.*, 62 (1966) 1050.
37 *Cf.* H. OKABE AND J. R. MCNESBY, *Advan. Photochem.*, 3 (1964) 57.

38 Ref. 15, p. 265.
39 Ref. 15, p. 299.
40 O. P. STRAUSZ AND H. E. GUNNING, J. Am. Chem. Soc., 84 (1962) 4080;
   A. R. KNIGHT, O. P. STRAUSZ, S. M. MALM AND H. E. GUNNING, ibid., 86 (1964) 4243;
   H. A. WIEBE, A. R. KNIGHT, O. P. STRAUSZ AND H. E. GUNNING, ibid., 87 (1965) 1443;
   K. S. SIDHU, E. M. LOWN, O. P. STRAUSZ AND H. E. GUNNING, ibid., 88 (1966) 254.
41 H. E. GUNNING AND O. P. STRAUSZ, Advan. Photochem., 4 (1966) 143.
42 J. N. MURRELL, The Theory of the Electronic Spectra of Organic Molecules, Wiley, New York, 1963; Ref. 1, p. 126; Ref. 23(b) and (c).
43 M. KASHA, Discussions Faraday Soc., 9 (1950) 14.
44 J. N. PITTS, JR., F. WILKINSON AND G. S. HAMMOND, Advan. Photochem., 1 (1963) 1.
45 J. H. RUSH AND H. SPONER, J. Chem. Phys., 20 (1952) 1847.
46 H. E. ZIMMERMAN, Advan. Photochem., 1 (1963) 183.
47 W. GROTH, Z. Physik. Chem. (Leipzig), B37 (1937) 307.
48 LOUIS KASSEL, Kinetics of Homogeneous Gas Reactions, The Chemical Catalog Company, New York, 1932.
49 S. W. BENSON, The Foundations of Chemical Kinetics, McGraw-Hill, New York, 1960, p. 218.
50 W. A. NOYES, JR., G. B. PORTER AND J. E. JOLLEY, Chem. Rev., 56 (1956) 49.
51 Ref. 1, p. 393.
52 J. HEICKLEN AND W. A. NOYES, JR., J. Am. Chem. Soc., 81 (1959) 3858.
53 Ref. 1, p. 782.
54 A. O. ALLEN, J. Am. Chem. Soc., 63 (1941) 708.
55 G. H. DAMON AND F. DANIELS, J. Am. Chem. Soc., 55 (1933) 2363.
56 G. H. DAMON, Ind. Eng. Chem., Anal. Ed., 7 (1935) 133.
57 P. FUGASSI, J. Am. Chem. Soc., 59 (1937) 2092.
58 M. S. MATHESON AND J. W. ZABOR, J. Chem. Phys., 7 (1939) 536.
59 W. A. NOYES, JR., A. B. F. DUNCAN AND W. M. MANNING, J. Chem. Phys., 2 (1934) 717.
60 W. E. KASKAN AND A. B. F. DUNCAN, J. Chem. Phys., 16 (1948) 223.
61 G. W. LUCKEY AND W. A. NOYES, JR., J. Chem. Phys., 19 (1951) 227.
62 G. M. ALMY AND S. ANDERSON, J. Chem. Phys., 8 (1950) 805.
63 R. D. RAUCLIFFE, Rev. Sci. Instr., 13 (1942) 413.
64 C. W. LARSON AND H. E. O'NEAL, J. Phys. Chem., 70 (1966) 2475.
65 R. G. W. NORRISH AND M. E. S. APPLEYARD, J. Chem. Soc., (1934) 874.
66 C. H. BAMFORD AND R. G. W. NORRISH, J. Chem. Soc., (1935) 1504.
67 C. H. BAMFORD AND R. G. W. NORRISH, J. Chem. Soc., (1938) 1521.
68 C. H. BAMFORD AND R. G. W. NORRISH, J. Chem. Soc., (1938) 1531.
69 C. H. BAMFORD AND R. G. W. NORRISH, J. Chem. Soc., (1938) 1544.
70 R. G. W. NORRISH AND F. W. KIRKBRIDE, J. Chem. Soc., (1932) 1518.
71 Ref. 1, p. 371.
72 R. P. PORTER AND W. A. NOYES, JR., J. Am. Chem. Soc., 81 (1959) 2307.
73 A. TEVENIN AND H. WEUJMIN, J. Chem. Phys., 3 (1935) 436.
74 Ref. 15, p. 284.
75 Ref. 1, p. 371.
76 G. HERZBERG AND D. A. RAMSAY, Proc. Roy. Soc. (London), A233 (1955) 34.
77 G. HERZBERG, Proc. Chem. Soc., (1959) 116.
78 C. S. PARMENTER AND W. A. NOYES, JR., J. Am. Chem. Soc., 85 (1963) 416.
79 W. DAVIS, JR. AND W. A. NOYES, JR., J. Am. Chem. Soc., 69 (1947) 2153.
80 P. SIGAL, J. Phys. Chem., 67 (1963) 2660.
81 R. SRINIVASAN, J. Am. Chem. Soc., 83 (1960) 2590.
82 J. E. GUILLET AND R. G. W. NORRISH, Nature, 173 (1954) 625.
83 J. E. GUILLET AND R. G. W. NORRISH, Proc. Roy. Soc. (London), A233 (1955) 153, 172.
84 R. SRINIVASAN, J. Am. Chem. Soc., 81 (1959) 5061.
85 G. R. MCMILLAN, J. G. CALVERT AND J. N. PITTS, JR., J. Am. Chem. Soc., 86 (1964) 3602.
86 P. AUSLOOS AND R. E. REBBERT, J. Am. Chem. Soc., 83 (1961) 4897.
87 V. BRUNET AND W. A. NOYES, JR., Bull. Soc. Chim. France, (1958) 121.

88 J. L. MICHAEL AND W. A. NOYES, JR., *J. Am. Chem. Soc.*, 85 (1963) 1027.
89 P. AUSLOOS AND R. E. REBBERT, *J. Am. Chem. Soc.*, 86 (1964) 4512.
90 P. J. WAGNER AND G. S. HAMMOND, *J. Am. Chem. Soc.*, 87 (1965) 4009.
91 P. BORRELL AND R. G. W. NORRISH, *Proc. Roy. Soc. (London)*, A262 (1961) 19.
92 R. SRINIVASAN, *J. Am. Chem. Soc.*, 82 (1960) 775.
93 G. CIAMICIAN AND P. SILBER, *Ber.*, 43 (1910) 1340.
94 Ref. 15, p. 452, contains a complete bibliography.
95 See W. A. NOYES, JR., *Proc. Acad. Sci. Lisbon (Portugal)*, 3 (1964), for a discussion of these problems.
96 G. B. PORTER, *J. Chem. Phys.*, 32 (1961) 1587.
97 H. L. BACKSTROM AND K. SANDROS, *Acta Chem. Scand.*, 14 (1960) 48.
98 G. M. ALMY AND P. R. GILLETTE, *J. Chem. Phys.*, 11 (1943) 188.
99 W. A. NOYES, JR., W. A. MULAC AND M. S. MATHESON, *J. Chem. Phys.*, 36 (1962) 880.
100 H. ISHIKAWA AND W. A. NOYES, JR., *J. Chem. Phys.*, 37 (1962) 583; *J. Am. Chem. Soc.*, 84 (1962) 1502.
101 W. A. NOYES, JR., W. A. MULAC AND D. A. HARTER, *J. Chem. Phys.*, 44 (1966) 2100.
102 E. H. GILMORE, G. E. GIBSON AND D. S. MCCLURE, *J. Chem. Phys.*, 20 (1952) 829. The value for Q_p given in Table II must be multiplied by 1.43; see corrections *J. Chem. Phys.*, 23 (1955) 399.
103 T. S. GODFREY AND G. PORTER, *Trans. Faraday Soc.*, 62 (1966) 7.
104 K. E. WILZBACH AND L. KAPLAN, *J. Am. Chem. Soc.*, 88 (1966) 2066.
105 S. LEACH AND E. MIGIRDICYAN, *J. Chim. Phys.*, 54 (1957) 643; *J. Chim. Phys.*, 58 (1961) 762; *Bull. Soc. Chim. Belges*, 71 (1962) 845.
106 K. E. WILZBACH AND L. KAPLAN, *J. Am. Chem. Soc.*, 86 (1964) 2307.
107 A. W. BURGSTAHLER AND P. L. CHIEN, *J. Am. Chem. Soc.*, 86 (1964) 2940, 5281.
108 E. M. ARNETT AND J. M. BOLLINGER, *Tetrahedron Letters*, (1964) 3803.
109 L. KAPLAN, K. E. WILZBACH, W. BROWN AND S. H. YANG, *J. Am. Chem. Soc.*, 87 (1965) 675.
110 E. E. VAN TAMMELEN AND S. P. PAPPAS, *J. Am. Chem. Soc.*, 84 (1962) 3789.
111 G. W. ROBINSON, *J. Mol. Spectry.*, 6 (1961) 58.
112 J. W. DONOVAN AND A. B. F. DUNCAN, *J. Chem. Phys.*, 35 (1961) 1389.
113 M. R. WRIGHT, R. P. FROSCH AND G. W. ROBINSON, *J. Chem. Phys.*, 33 (1960) 934.
114 R. B. CUNDALL, *Progr. Reaction Kinetics*, 2 (1964) 165.
115 Convincing evidence for this is found in the work of Evans on the irradiation of a CHD=CHD (*trans*) and O_2 mixture. Direct absorption to the triplet (2900–3400 A) produced appreciable isomerization to *cis*-CHO=CHD. D. F. EVANS, *J. Chem. Soc.*, (1960) 1735.
116 A. R. ALSON AND F. L. HUDSON, *J. Am. Chem. Soc.*, 55 (1933) 1410.
117 P. A. LEIGHTON AND F. A. LUCY, *J. Chem. Phys.*, 2 (1934) 756.
118 Ref. 1, p. 479.
119 R. GOMER AND W. A. NOYES, JR., *J. Am. Chem. Soc.*, 72 (1950) 101.
120 T. K. LIU AND A. B. F. DUNCAN, *J. Chem. Phys.*, 17 (1949) 241.
121 R. SRINIVASAN, *Advan. Photochem.*, 1 (1963) 83.
122 R. SRINIVASAN, *J. Am. Chem. Soc.*, 83 (1961) 4344.
123 D. PHILLIPS, *J. Phys. Chem.*, 70 (1966) 123.
124 R. SRINIVASAN, *Advan. Photochem.*, 4 (1966) 113.
125 E. MIGIRDICYAN AND S. LEACH, *Bull. Soc. Chim. Belges*, 71 (1962) 845.
126 K. E. WILZBACH AND L. KAPLAN, *J. Am. Chem. Soc.*, 87 (1965) 4004.
127 D. L. CHAPMAN, *Advan. Photochem.*, 1 (1963) 323.
128 D. F. EVANS, *J. Chem. Soc.*, (1957) 3885.
129 G. N. LEWIS AND M. KASHA, *J. Am. Chem. Soc.*, 66 (1944) 2100.
130 (a) F. A. MATSEN, *Advan. Quantum Chem.*, 6 (1964).
     (b) F. A. MATSEN, *J. Phys. Chem.*, 68 (1964) 3282.
     (c) F. A. MATSEN, A. A. CANTU AND R. D. POSHUSTA, *J. Phys. Chem.*, 70 (1966) 1558.
131 For an interesting discussion on the applicability of Hund's Rule see Ref. 130 (a) and F. A. MATSEN AND R. D. POSHUSTA, *Algebras, Ideals and Quantum Mechanics with Applications from the Symmetric Group*, Technical Report, The University of Texas, Austin, Texas 78712.

132 R. GOMER AND G. B. KISTIAKOWSKY, *J. Chem. Phys.*, 19 (1951) 85.
133 Ref. 15, pp. 202–209.
134 R. G. W. NORRISH AND G. PORTER, *Nature*, 164 (1949) 658.
135 L. I. GROSSWEINER AND M. S. MATHESON, *J. Chem. Phys.*, 23 (1955) 2443.
136 For a review see L. M. DORFMAN AND M. S. MATHESON, *Progr. Reaction Kinetics*, 3 (1965) 237.
137 E. C. FRANKLIN AND C. A. KRAUS, *Am. Chem. J.*, 23 (1900) 277.
138 F. H. FIELD AND J. L. FRANKLIN, *Electron Impact Phenomena and the Properties of Gaseous Ions*, Academic Press, New York, 1957.
139 Ref. 138, p. 128.
140 H. S. W. MASSEY AND E. H. S. BURHOP, *Electronic and Ionic Impact Phenomena*, Oxford University Press, London, 1952, p. 161.

Chapter 2

Effect of High Energy Radiation

GORDON HUGHES

1. Interaction of high energy radiation with matter

Radiation chemistry is the chemistry induced by high energy ionising radiation. Such radiation includes α- and β-particles, γ- and X-rays, protons and neutrons. The earliest systematic observations were made at the beginning of this century and were concerned mainly with the effect of α-particles from radium and radon on substrates. The much wider availability of radioactive materials consequent upon the second world war and the need for a greater understanding of the effects of radiation upon materials for the prosecution of atomic energy programmes have both given considerable impetus to the subject. The advent of the cyclotron and other particle accelerators has enabled the study of the effects of highly energetic particles. In Table 1 are given typical sources of radiation and their principal radiations, with appropriate energies (in million electron volts).

1.1 MECHANISM OF ENERGY LOSS

The chemical effects of all high energy ionising radiations may be attributed to electrons produced within the medium.

TABLE 1

SOURCES OF HIGH ENERGY IONISING RADIATIONS

Source	Radiation	Energy (MeV)
^{226}Ra	α	4.777
^{222}Rn	α	5.49
^{3}H	β	0.018
^{90}Sr	β	0.544
^{137}Cs	β	0.52
	γ	0.6616
^{60}Co	γ	1.332
	γ	1.173
X-ray tube	X	~0.2
Linear accelerator	e	~10

In the case of X- and γ-rays these electrons are ejected from atoms by the quanta of radiation. In the photoelectric effect, the quantum of energy is completely absorbed by the atom and the energy of the electron ejected subsequently is thus that of the quantum less the electronic binding energy. Since however, the binding energy is generally much smaller than the quantum of energy, most of the energy of the incident quantum is transmitted to the ejected electron. In the Compton effect, only a fraction of the energy of the quantum is given up to the electron and the quantum continues through the medium when it may eject further electrons. The photoelectric effect is significant for low energy X-rays, < 0.1 MeV, but for high energy γ-rays, > 1 MeV, most of the energy is absorbed by Compton effect. For very high energy γ-rays, > 20 MeV, the quantum of energy may be transformed into an electron and a positron. The absorption of energy by the Compton effect is proportional to electron density, *i.e.* the number of electrons per unit volume, whereas photoelectric absorption and pair production increase with increasing atomic number of the substrate.

A fast neutron loses its energy by interaction with an atomic nucleus and ejection of a proton. A slow neutron is absorbed by the nucleus giving rise to a radioactive atom which may decompose by the emission of a β-particle or γ-ray. Charged particles in general lose their energy by inelastic collisions with the electrons within an atom leading to the ejection of an electron.

Secondary electrons, *i.e.* those that have been ejected from atoms by incident radiation, will cause further ionisations or excitations until their energy is reduced to $\simeq kT$, when they are said to be thermalised. They may then be captured by positive ions or neutral molecules. Since all ionising radiations then basically give rise to these secondary electrons, it is to be expected that their chemical effects will be essentially similar.

1.2 SPECIES PRODUCED IN AN IRRADIATED SYSTEM

It is convenient at this stage to summarise the species that may be produced by the interaction of high energy ionising radiation with matter, though the evidence for their existence will be considered in detail in Section 2.

As has already been discussed, the primary radiolytic act may be represented by the reaction

$$\text{M} \xrightarrow{} \text{M}^+ + e$$

(In radiation chemistry the symbol $\xrightarrow{}$ signifies "under the influence of ionising radiation". $G(\text{X})$ represents the yield of a species X and is the number of molecules of X produced per 100 eV energy absorbed.) The electron thus produced, after it is thermalised, may be captured either by a parent positive ion or by a neutral molecule

$$e + M^+ \rightarrow M^*$$

$$e + M \;\; \rightarrow M^-$$

Positive and negative ions and excited molecules are thus amongst the early products. These species may themselves undergo further reactions, *e.g.*

Charge transfer

$$M^+ + A \rightarrow M + A^+$$

$$M^- + B \rightarrow M + B^-$$

Energy transfer

$$M^* + A \rightarrow M + A^*$$

Bond breaking reactions

(*a*) yielding radical products

$$M^* \rightarrow R_1 + R_2$$

$$M^+ \rightarrow R + C^+$$

$$M^- \rightarrow R + D^-$$

(*b*) yielding molecular products

$$M^* \rightarrow E + F$$

It will be seen that consequent upon the initial radiolytic act, a wide variety of species may be produced. It must be emphasised that as a consequence of the mechanism of energy loss, the distribution of these species will not be initially homogeneous. Thus it has been estimated[1] that for ^{60}Co γ-rays traversing liquid water the distance between successive ionisations along the path of the incident radiation will be $\simeq 10^4$ A. The electrons emitted at these sites will then bring about other ionisations such that reactive intermediates will be produced in "spurs" initially of diameter approximately 20 A. Within the spur the concentration of intermediates will be quite high, and may be $\simeq 1\ M$. Recombination reactions within the spur are important and give rise to the so-called molecular products. However, the intermediates may also diffuse out of the spur into the bulk of the solution (where their concentration will fall to $\simeq 10^{-8}\ M$) when they may react

TABLE 2

EFFECT OF LET ON PRODUCT YIELDS IN WATER RADIOLYSIS

Radiation	LET (keV per micron)	$G(H)$	$G(OH)$	$G(H_2)$	$G(H_2O_2)$
20 MeV betatron X-rays	0.28				
1.25 MeV γ-rays	0.42	3.7	2.9	0.40	0.80
200 keV X-rays	2.8				
0.018 MeV β-particles	3.2	2.9	2.1	0.6	1.0
5.3 MeV α-particles	158	0.6	0.5	1.57	1.45

with other substrates. For more densely ionising radiations the distance between spurs is less and with α-particles may become so small that the spurs overlap. Under these conditions recombination reactions will predominate.

It is convenient to characterise different radiations by their Linear Energy Transfer, abbreviated to LET. This is the rate of loss of energy per unit distance. LET values for different types of radiation and liquid water are shown in Table 2 which also shows the effect of LET on product yields[2–4] in the radiolysis of water at pH $\simeq 0.5$. $G(H_2)$ and $G(H_2O_2)$ are the yields of molecular hydrogen and hydrogen peroxide respectively, these products being considered to arise by recombination reactions within the spur. $G(H)$ and $G(OH)$ are the yields of the corresponding radical species which have diffused out of the spur.

It will be seen that there is an increase in the molecular yields parallelled by a decrease in the radical yields with increasing LET.

However, although the distribution of reactive entities may not be homogeneous within the bulk of the solution, it is possible to use homogeneous type kinetics in some cases. Thus a radical, X, after diffusing out of the spur, may react with either of solutes A and B present in solution. Provided the concentrations of A and B are sufficiently greater than that of X so that the change in their concentrations is negligible, then the probability that X will react with A is given by $k_A[A]/(k_A[A]+k_B[B])$ where k_A and k_B are the rate coefficients for the reactions of X with A and B respectively. This is independent of the concentration of X provided that radical recombination reactions of X may be neglected, which will probably be true outside the spur. That X may not be homogeneously distributed throughout the bulk of the solution does not affect the probability.

It follows that

$$G(-A) = \frac{k_A[A]}{k_A[A]+k_B[B]} \cdot G(X)$$

Expressions of this type are usually used in the form

$$\frac{1}{G(-A)} = \frac{1}{G(X)} \cdot \left(1 + \frac{k_B[B]}{k_A[A]}\right)$$

and permit the ready calculation of the rate coefficient ratio, k_B/k_A, from experimental data.

1.3 TIME SCALE OF EVENTS

In Table 3 is shown the time scale in which some of the processes which have been discussed take place. These times are necessarily approximate and depend on the substrate. In some cases the time scale for different processes will overlap.

Recently much valuable information has been obtained using the technique of pulse radiolysis[5]. This is the equivalent in radiation chemistry of flash photolysis in photochemistry. A short pulse of ionising radiation, usually electrons, may be obtained using either a linear accelerator or a Van de Graaff machine, the electron energies being approximately 10 MeV and 2 MeV respectively. The total energy per pulse is in the region 0.2–10 joules and the lifetime of the pulse $\simeq 10^{-6}$ sec.

TABLE 3

TIME SCALE OF EVENTS IN RADIATION CHEMISTRY

Time (sec)	Event
10^{-17}	Passage of ionising radiation with subsequent ionisations.
10^{-15}	Electronically excited species appear.
10^{-14}	Vibrationally excited species appear. Ion–molecule reactions. Dissociation of electronically excited species to give radicals.
10^{-13}	Recapture of electron by parent ion.
10^{-12}	Onset of radical diffusion.
10^{-11}	Electrons solvated in polar media.
10^{-10}	Diffusion controlled reactions may be complete.
10^{-8}	Radiative decay of singlet excited species.
10^{-7}	Molecular products complete.
10^{-4}	Radical reactions in bulk of substrate.
10^{-3}	Radiative decay of triplet states.
1	Reactions essentially complete except for post-irradiation phenomena in some systems.

A high concentration of reactive intermediates is produced and the reactions of those intermediates with lifetime $> 10^{-6}$ sec may be followed by conventional kinetic spectrophotometry or flash spectroscopy. Work is proceeding on the production of pulses of even shorter duration and some preliminary observations have been made[6] using pulses of $\simeq 10^{-9}$ sec. This should eventually yield some very valuable information concerning some of the initial events in the passage of ionising radiation through matter.

1.4 COMPARISON WITH PHOTOCHEMISTRY

The initiation of chemical reactions by low energy radiation, *e.g.* ultraviolet light, and by high energy ionising radiation, is compared in Table 4.

2. Experimental evidence for species present

That ionic entities were important intermediates in radiation chemistry was first suggested by Lind[7]. By measuring both the ionisation and the ozone produced

TABLE 4

COMPARISON OF THE EFFECT OF UV AND HIGH ENERGY IONISING RADIATION

		UV	*Ionising radiation*
1.	Type of radiation	May be monochromatic.	Due to energy loss, it is impossible to obtain mono-energetic radiation within a sample.
2.	Absorption of radiation	Depends on optical absorption coefficients—all radiation may be absorbed by solute. Absorption of quantum is always complete.	Depends on electron fraction of substrate. Most energy will be deposited in solvent. Incomplete absorption of quantum.
3.	Species present	Only free radicals and excited states. With low wavelength UV in some cases ions may be produced but quantum yield of ion pair is 1.	Free radicals, excited molecules and ionic species. Yields variable but characteristic of substrate.
4.	Selection rules	Excited states produced must obey selection rules.	Excited states produced not restricted by optical selection rules.
5.	Distribution of reactive entities	Homogeneous.	Inhomogeneous.
6.	Reaction vessels	Must be transparent to UV, *e.g.* silica.	Metal vessels are "transparent" to γ-radiation.

in gaseous oxygen when irradiated by α-particles, he obtained a yield of $\simeq 0.5$ molecule of ozone per ion pair. The ratio of the number of molecules decomposed (M) to the number of ion pairs formed (N) was known as the ionic yield. Values of M/N considerably greater than unity, as observed in some systems[8], were explained by assuming a clustering of neutral molecules around a central ion. The energy liberated on charge neutralisation of this ion might then lead to decomposition of all the molecules within a cluster.

However, it was subsequently shown[9] that these observations could be more satisfactorily accounted for by the occurrence of free radical chain reactions. The free radicals might arise from ionic or excited species. The radiation chemistry of liquid water was claimed to be explicable[10] in terms of the reactions of the hydrogen atom and the hydroxyl free radical. Ionic mechanisms consequently fell into disrepute, but it is interesting that in recent years ionic processes have been recognised as of increasing importance. The wheel has indeed turned full circle!

In most radiation chemical systems it is likely that reactions of ionic, free radical and excited species are important. One of the outstanding problems of radiation chemistry is to separate the relative importance of the reactions of each of these species. In this section the evidence for the existence of each of these species will be considered separately.

2.1 IONS AND ELECTRONS

Although ionic species are undoubtedly produced as a result of the initial act of absorption of ionising radiation, there is considerable variation in the lifetime of such species. An electron ejected from a parent molecule will travel through the medium against the coulombic attraction of the parent ion until it is thermalised. The distance travelled by the electron will depend on its energy and its rate of energy loss. There are considerable theoretical difficulties in the treatment of this rate of energy loss.

Early theories of the radiolysis of liquid water usefully illustrate two extreme possibilities. Samuel and Magee[11] estimated that a 10 eV electron would travel approximately 20 A in 10^{-13} sec before being thermalised, after which, as it would still be effectively within the electrostatic field of the positive ion, charge neutralisation would take place to give an excited molecule.

$$H_2O^+ + e \rightarrow H_2O^*$$

Thus although ionic intermediates may be formed, their lifetime is too short for them to be chemically significant. Lea[12] had previously suggested that the electron could escape from the field of the parent ion. Calculations by Platzman[13] indicated that the electron might well travel a distance of 50 A before thermalisation.

At this distance the electron would be essentially free from the electrostatic attraction of the positive ion and would be solvated by surrounding water molecules forming the hydrated electron, e_{aq}^-, in which form it might react with added solutes.

Recent experimental work has demonstrated the existence of the hydrated electron as will be discussed subsequently. It has, however, been proposed[14] that this could still arise from capture of the thermalised electron by a positive ion according to the reaction scheme

$$H_2O^+ + H_2O \rightarrow H_3O^+ + OH$$

$$H_3O^+ + e \rightarrow H_3O^*$$

$$H_3O \xrightarrow{\hspace{1cm}} \begin{array}{l} H + H_2O \\ (H_3O^+)_{aq} + e_{aq}^- \end{array}$$

At present it is not possible to distinguish between this suggestion and the earlier one of Platzman.

The existence of ionic species sufficiently long-lived to enter into chemical reaction will thus depend on both the density and the polarity of the medium since these will determine the rate of energy loss and the extent of solvation of the electron. In gases it might be anticipated that extensive charge separation will occur before the electron is thermalised and hence reactions of ionic intermediates would be significant. In non-polar liquids charge separation will be extremely small because of the low dielectric constant and poor solvating power of the medium. It has been estimated[15] that in condensed phases most ejected electrons would lose their energy sufficiently rapidly as not to escape the coulombic field of the parent positive ion. However, a small yield, $G \simeq 0.1$, of separated ion pairs might be produced. The stability of the electron in solids will depend on the number of electron traps, which will be determined by the structure and purity of the solid.

It is convenient then to discuss the existence of ionic species in gaseous, liquid and solid phases separately.

2.1.1 Gaseous systems

Direct measurement of the number of ions in an irradiated system was first made by Essex[16]. The current flowing in gaseous NH_3 when exposed to α-particles from radium was measured as a function of voltage. The results are shown in Fig. 1. Beyond 5 kV the current was independent of voltage indicating that all the ions produced in the system were collected at the electrodes. The current flowing under these conditions is known as the saturation current and from its magnitude, the number of ions present in the system can be calculated. In the absence of radium, no current was obtained clearly indicating the production of charged particles in the ammonia gas by the radium α-particles. From measurements of the number of

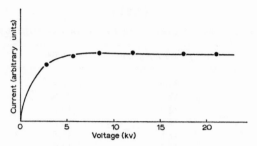

Fig. 1. Dependence of current on voltage in irradiated gaseous ammonia (from ref. 16).

molecules of NH_3 decomposed, M/N was found to be 1.09 at 22 °C and 1 atmosphere pressure.

Moreover by measuring the effect of applied electric field on the ionic yield, it is possible to estimate the importance of ionic processes[17]. Results obtained for ammonia are shown in Table 5.

TABLE 5

DECOMPOSITION OF NH_3 BY α-PARTICLES

| | Temperature | |
	30 °C	100 °C
Total decomposition yield	1.37	2.42
Ionic decomposition yield	0.40	0.72
Non-ionic decomposition yield	0.97	1.70

It is interesting that the ratio of ionic to non-ionic decomposition is not significantly dependent on temperature. The effect of applied electric field on the γ-radiolysis of hydrocarbon gases has also been investigated[18].

That reactions other than those involving ions should be important is not surprising. In Table 6 the values of the energy (W) required to create an ion pair within a gas, by electrons, are compared with the corresponding ionisation potential, *i.e.* the energy required simply to remove the electron thus producing an ion pair.

In all cases the ionisation potential is only $\simeq 2/5$ of the energy required to produce an ion pair by electrons in the gas, clearly indicating that a considerable fraction of the energy of the electrons must be dissipated in the formation of other species, possibly excited molecules.

The fragmentation of molecules by the action of slow electrons is the basic reaction proceeding in the mass spectrometer. Some attempts have been made to correlate the results of mass spectrometric observations with those of radiation

TABLE 6

ENERGIES REQUIRED FOR ION PAIR PRODUCTION

Gas	W (eV per ion pair)[19]	I (eV)[20]	I/W
H_2	36.3	15.4	0.42
He	42.3	24.6	0.42
N_2	34.9	15.6	0.45
O_2	30.8	12.2	0.40
CO_2	32.8	13.8	0.42
CH_4	27.3	13.0	0.48
C_2H_2	25.7	11.4	0.44
C_2H_4	26.1	10.6	0.41
C_2H_6	24.5	11.7	0.48

TABLE 7

2 MeV ELECTRON IRRADIATION OF METHANE

Species	$\dfrac{M}{(N)}_{\text{calculated}}$	$\dfrac{M}{(N)}_{\text{observed}}$
CH_4	−2.4	−2.5
H_2	1.6	1.9
C_2H_6	0.6	0.7
C_3H_8	0.08	0.05
C_4H_{10}	0.01	0.01
C_2H_4	0.01	0.01

induced decomposition. The ions formed in the mass spectrum of methane and their relative abundances are CH_4^+, 47%; CH_3^+, 40%; CH_2^+, 8%; CH^+, 4%; C^+, 1%. The reactions

$$CH_4^+ + CH_4 \rightarrow CH_5^+ + CH_3$$

$$CH_3^+ + CH_4 \rightarrow C_2H_5^+ + H_2$$

were found[21] to occur in the mass spectrometer. On the basis of these reactions and with certain simple assumptions regarding the fates of other ions and radicals, it was possible to obtain good agreement between observed and calculated yields of products, as shown in Table 7.

It was not necessary to include reactions of excited molecules. A similar correlation between mass spectral and radiolytic results was obtained for n-hexane[22]. If these correlations are significant it would indicate that, in these systems at least, excited species that may be produced decay without chemical decomposition.

Although some correlation is to be expected, particularly for the radiolysis of gases at low pressure where the conditions more closely resemble those obtaining in tne mass spectrometer, the correlation is surprisingly good. However, in condensed phases ion lifetimes will be considerably shorter than those in the mass spectrometer. The ions produced are more likely to disappear by bimolecular ion–molecule reactions rather than unimolecular fragmentation and it is doubtful whether any significant correlation could be obtained between the two conditions.

2.1.2 Liquid systems

Irradiated liquids do possess electrical conductivity, clearly indicating the existence of ionic species as in gaseous systems. Some results for cyclohexane[23] are shown in Fig. 2. Dielectric breakdown of the liquid occurs, however, before a saturation value of the current is reached.

In measuring the number of ions in an irradiated liquid it is important to ensure that the voltage applied is such that it will not accelerate the electrons

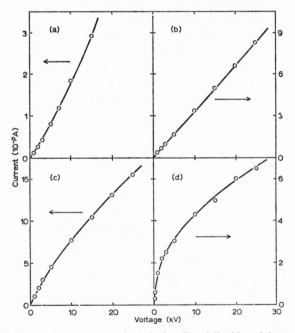

Fig. 2. Dependence of current on voltage in irradiated liquid cyclohexane (from ref. 23).

	P	Dose rate $(eV.ml^{-1}.sec^{-1})$	$t\ (°C)$
(a)	6	1.20×10^{15}	21
(b)	7	1.03×10^{14}	26
(c)	8	1.66×10^{13}	26
(d)	9	$8.9\ \times 10^{11}$	24

produced sufficiently to bring about further ionisations before they reach the electrode. Allen and Hummel[24] have determined the number of separated ion pairs produced in n-hexane irradiated by 1.5 MeV X-rays. The initial slope of the current–voltage curve is linear and is a measure of the ionic concentration provided that the current is so low that the number of ions going to the electrodes is much less than the number formed by the radiation and which disappear by recombination. G (separated ion pairs) was found to be 0.09, in striking agreement with the theoretical predictions of Magee. Similar results were obtained by Freeman[25,26] who showed that the yield did not change markedly for a series of paraffin hydrocarbons even although the viscosity changed by a factor of 540. It was concluded that most of the ion pairs formed initially undergo geminate recombination and do so within 10^{-10} sec.

Chemical evidence for the existence of electrons in irradiated cyclohexane was obtained from pulse radiolysis studies of solutions containing aromatic solutes[27]. Because of the lifetime of the pulse these experiments only allowed the determination of ions still surviving after 10^{-6} sec. With benzophenone and anthracene as scavengers transient absorption peaks at 700 nm and 730 nm respectively, were obtained. These were consistent with the known spectra of the benzophenone and anthracene radical ions and are most simply accounted for by assuming direct electron capture by these solutes. Positively charged ion radicals may also be produced since these are likely to have similar spectra. Ion yields can be calculated since the absorption coefficients are known, but these yields necessarily represent the sum of the positive and negative ion yields. Some results are shown in Fig. 3.

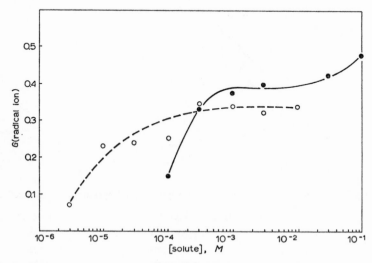

Fig. 3. Radical ion yields in irradiated liquid cyclohexane: (●–●), benzophenone; (○–○), anthracene (from ref. 27).

The limiting ionic yield at the higher solute concentrations is consistent with that obtained from conductivity studies.

Qualitative evidence that ionic species were significant intermediates was obtained from a study of the radiation induced polymerisation of isobutene[28, 29]. Since this monomer was known to be readily polymerised by ionic initiators, polymerisation by 2 MeV electrons at -80 °C seemed to indicate the existence of ionic intermediates. However, the polymerisation was inhibited by oxygen and benzoquinone which are known to be inhibitors for free radical polymerisations. It was subsequently suggested[30] that polymerisation was caused by the positive ion $(CH_3)_3C^+$ produced by the reactions

$$i-C_4H_8 \xrightarrow{\hspace{0.3cm}\wedge\wedge\wedge\hspace{0.3cm}} C_4H_8^+ + e$$

$$C_4H_8^+ + i-C_4H_8 \rightarrow (CH_3)_3C^+ + CH_2 = C - \overset{\cdot}{C}H_2$$
$$\overset{\mid}{C}H_3$$

The suppression of polymerisation by O_2 and benzoquinone was due to the fact that these species captured a significant fraction of the secondary electrons produced and not to their ability to scavenge free radicals. The negative ions thus produced underwent charge neutralisation with $(CH_3)_3C^+$ thus reducing the yield of polymerisation initiators. In confirmation of the fact that ionic rather than free radical processes were important was the observation that the production of free radicals within the solution by the photolysis of diacetyl failed to initiate polymerisation. Spectroscopic evidence for the existence of $(CH_3)_3C^+$ in the pulse radiolysis of isobutene was obtained subsequently[31].

Cyclopentadiene undergoes ionic polymerisation by Friedel–Crafts catalysts[32]. Its polymerisation by γ-rays is markedly suppressed by ammonia or amines but much less by diphenylpicrylhydrazyl or oxygen[33]. This again points to the ionic rather than the free radical nature of the radiation-induced polymerisation. The quenching effect of ammonia was postulated as due to reactions of the type

$$XH^+ + NH_3 \rightarrow X + NH_4^+$$

and

$$R^+ + NH_3 \rightarrow RNH_3^+$$

Proton transfer reactions were also postulated as important from a study of the effect of ND_3 on the hydrogen yield in the radiolysis of cyclohexane[33]. Yields of HD could only be accounted for by assuming the reactions

$$C_6H_{12} \xrightarrow{\hspace{0.3cm}\wedge\wedge\wedge\hspace{0.3cm}} C_6H_{12}^+ + e$$

$$C_6H_{12}^+ + ND_3 \rightarrow C_6H_{11}\cdot + ND_3H^+$$

followed by

$$e + ND_3H^+ \quad \overset{\longrightarrow}{\longrightarrow} \quad \begin{array}{l} ND_3 + H \\ ND_2H + D \end{array}$$

$$H + C_6H_{12} \rightarrow H_2 + C_6H_{11}$$

$$D + C_6H_{12} \rightarrow HD + C_6H_{11}$$

G(free ions) was 0.08, in agreement with the conductivity work. It was estimated[34] that although the majority of the ions produced by the radiation undergo geminate recombination, the lifetimes of about half of these ions lay in the range $10^{-9}-10^{-7}$ sec. This time scale allows ionic reactions with a very reactive scavenger at high concentrations ($> 10^{-2} M$) to proceed.

Much higher estimates of the yield of free electrons have been obtained using N_2O as scavenger[35]. It had already been shown from studies in water radiolysis[36] that N_2O was an efficient scavenger of electrons and that the reaction could be represented formally by

$$N_2O + e \rightarrow N_2 + O^-$$

Values of $G(N_2) \simeq 4$ corresponding to $G(e) \simeq 4$ were obtained for irradiated cyclohexane.

It is probable that all scavengers may capture electrons, which in the absence of scavenger would not escape the parent positive ions ($G = 2-3$), in addition to those electrons which have escaped the positive ion ($G \simeq 0.1$). Whereas the products of electron capture by most solutes still undergo charge neutralisation with neighbouring positive ions, it is assumed that N_2O^- decomposes before charge neutralisation can take place.

In non-polar liquids, then, the evidence seems conclusive that most of the electrons are recaptured by parent ions, at least in the absence of reactive scavengers, and that ionic processes therefore do not contribute significantly to the radiation-induced decomposition. However, in polar liquids where the electron could be solvated, the situation may be quite different.

Stein[37] proposed that the X-ray induced reduction of methylene blue in aqueous solution was due to its reaction with hydrated electrons. Calculations by Platzman[13] indicated that the hydrated electron could exist sufficiently long to react with other solutes. Such a species should have a blue colour and a hydration energy of $\simeq 2$ eV. Subsequent work[38-40] clearly indicated the existence of two reducing species, postulated as e_{aq}^- and H. Convincing evidence of the existence of the hydrated electron was obtained by the demonstration[41,42] that the major reducing species in neutral water had a charge of -1. The dependence of the rate coefficient on ionic strength for a reaction involving two ions is given by

$$\log k/k_0 = 1.02\, Z_A Z_B \frac{\mu^{\frac{1}{2}}}{1+a\mu^{\frac{1}{2}}}$$

where k, k_0 are the rate coefficients at ionic strength μ and infinite dilution respectively, Z_A, Z_B are the algebraic numbers of charges on the ions and a is a constant with a value $\simeq 1$ for water at room temperature. The reducing species was known to react with H_2O_2, O_2, H_3O^+ and NO_2^- in neutral solution.

$$e_{aq}^- + H_2O_2 \xrightarrow{k_1} OH + OH^-$$

$$e_{aq}^- + O_2 \xrightarrow{k_2} O_2^-$$

$$e_{aq}^- + H_3O^+ \xrightarrow{k_3} H_2O + H$$

$$e_{aq}^- + NO_2^- \xrightarrow{k_4} NO_2^=$$

It was possible to examine the effect of ionic strength on the rate coefficient ratios k_4/k_1, k_3/k_1 and k_2/k_1. In Fig. 4 the results are plotted in the form $\log K/K_0$ versus ionic strength where K, K_0, are the appropriate rate coefficient ratios at

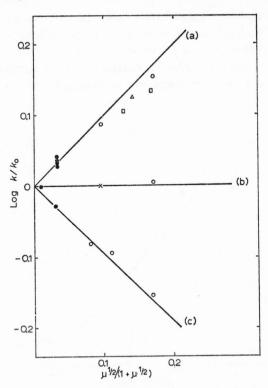

Fig. 4. Effect of ionic strength on reaction rates: (a) $K = k_4/k_1$; (b) $K = k_2/k_1$; (c) $K = k_3/k_1$ (from ref. 41).

ionic strengths μ and 0 respectively. The solid lines are theoretical, assuming unit negative charge on the reducing species.

Subsequently, the absorption spectrum of the transient hydrated electron was observed in pulse radiolysed deaerated water[43, 44]. Its spectrum is shown in Fig. 5. This absorption may also be observed with continuous radiolysis at high dose rates. Thus it has been observed[46] with hydrogen-saturated solutions of 10^{-3} M NaOH when irradiated in a 15,000 curie ^{60}Co source.

That this absorption is indeed due to the hydrated electron is confirmed by the following observations:

1. It is similar to that observed for the well-known solvated electron in ammonia alkali metal solutions[47].

2. It is suppressed by known electron scavengers, *e.g.* H_3O^+, N_2O.

3. The effect of ionic strength on the rate of reaction of this transient with $Fe(CN)_6^{3-}$ is consistent with a species of unit negative charge[48]

$$e_{aq}^- + Fe(CN)_6^{3-} \rightarrow Fe(CN)_6^{4-}$$

In the few years that have elapsed since these observations, a great deal of information has accumulated on the properties and rate coefficients of the hydrated electron. Some of its characteristic properties are shown in Table 8. The subscript f refers to formation from a hypothetical standard state for electrons having zero entropy, enthalpy and free energy and in a vacuum at the bulk electrostatic po-

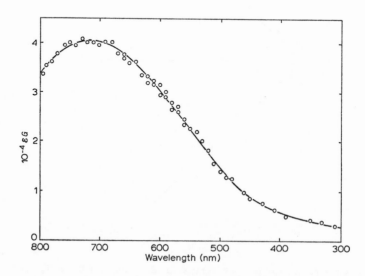

Fig. 5. Absorption spectrum of the hydrated electron (from ref. 45).

TABLE 8

PROPERTIES OF THE HYDRATED ELECTRON[49, 50]

λ_{max}	720 nm (1.72 eV)
ε_{720}	15,800 $M^{-1}.cm^{-1}$
pK	9.7
ΔG_f^0	-37.5 kcal.mole^{-1}
ΔH_f^0	-36.6 kcal.mole^{-1}
ΔS_f^0	3.1 cal.mole^{-1}.deg^{-1}
$\Delta G_{hyd.}^0$	-37.5 kcal.mole^{-1}
$\Delta H_{hyd.}^0$	-38.1 kcal.mole^{-1}
$\Delta S_{hyd.}^0$	-1.9 cal.mole^{-1}.deg^{-1}

tential of the interior of the water. The subscript hyd refers to hydration from the gas phase at a standard state of 1 atm fugacity. The small entropy of hydration is consistent with the electronic charge being dispersed over several water molecules.

Extensive compilations have been made of the absolute rate coefficients for the reactions of the hydrated electron with a wide variety of substrates[51]. Many of these are extremely rapid reactions with rate coefficients $\simeq 10^{10}\ M^{-1}.sec^{-1}$, the rates being in some cases diffusion controlled. The results of such studies are important not only for radiation chemistry but for much wider areas of chemistry where the rate coefficients may lead to an understanding of the electronic structure of the scavenging molecule[52].

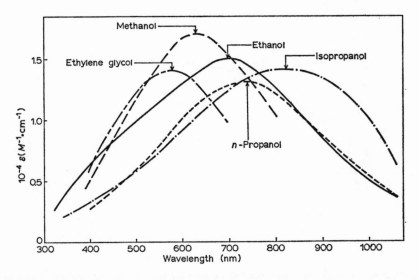

Fig. 6. Absorption spectra of the solvated electron in aliphatic alcohols (from ref. 53).

The solvated electron has also been shown to be produced in the pulse radiolysis of polar organic liquids, *viz.* amines, ethers and alcohols[53]. Some spectra in aliphatic alcohols are shown in Fig. 6. Decrease in the static dielectric constant shifts λ_{max} towards the red.

The existence of the solvated electron is well authenticated, then, in the radiolysis of polar liquids. The fate of the corresponding positive ions is less clear. Lea[54] suggested that in liquid water the ion H_2O^+ would decompose within the relaxation time of water $(10^{-11}$ sec) on becoming hydrated

$$H_2O^+ + H_2O \rightarrow H_3O^+ + OH$$

Although it has been suggested[55] that a similar species, the positive polaron, might have a sufficiently long lifetime in liquid water to undergo chemical reaction, no conclusive evidence for such reactions has yet been put forward. Positive ions formed in organic liquids probably decompose rapidly to free radicals with the elimination of a proton.

2.1.3 Solid systems

The production of trapped electrons in inorganic crystals by the action of ionising radiation at room temperature is well known[56]. Electrons are ejected from the ions of the crystal by the ionising radiation. Although most of these will subsequently be recaptured, some may be trapped and held in negative ion vacancies

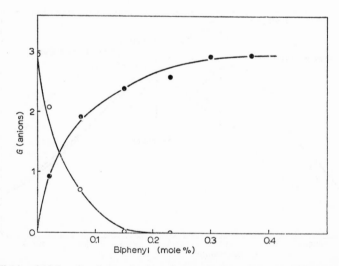

Fig. 7. Yields of biphenyl anion and solvated electron for solutions of biphenyl in ethanol at 77 °K: (●-●), G(biphenyl anion); (O-O), normalised G(solvated electron) (from ref. 57).

in the lattice structure forming F centres. In the alkali halides these give rise to absorption bands in the visible and ultraviolet regions. ESR measurements are consistent with a strong interaction between the electron and its six surrounding alkali metal ions.

Of more direct chemical interest are the results obtained from the irradiation of frozen liquids. Thus it is possible that at suitably low temperatures, some of the transients that react rapidly in radiolyses at room temperature may be sufficiently long lived to be characterised by the usual ESR and spectrophotometric techniques.

The γ-irradiation of organic glasses containing small concentrations of naphthalene and biphenyl has been investigated[57] at 77 °K. Irradiation of these solutes in tetrahydro-2-methylfuran, ethanol and hydrocarbon glasses gave rise to spectra similar to those obtained by reaction of the solutions with sodium, when the corresponding organic anion is known to be produced. Irradiation of glasses in the absence of solutes gave rise to intense visible absorption with ethanol and tetrahydro-2-methylfuran but not with hydrocarbons. These absorptions could be readily bleached by visible light or by warming the sample. At small concentrations of solute in tetrahydro-2-methylfuran both solvent and anion spectra were obtained. On bleaching there was a corresponding increase in the anion spectra. Some results are shown in Fig. 7. These observations are consistent with the production of electrons by the radiation. These may react with suitable scavengers or, in a polar medium, be solvated or trapped.

Trapped electrons have also been identified in the radiolysis of glassy methanol[58]. The yield of trapped electrons in methanol glasses decreases from $\simeq 3$ for γ-rays to $\simeq 0.1$ for α-particles produced by a $^{10}B(n, \alpha)$ reaction[59]. This has been attributed to local heating in the tracks of more densely ionising radiation since trapped electrons are known to react on warming. With halogen containing scavengers, dissociative electron attachment was observed[60]

$$RX + e \rightarrow R + X^-$$

There is much less direct evidence for the existence of positive ions in the products of the radiolysis of glasses. When N,N-dimethyl-p-phenylenediamine (NNDA) was used as scavenger, the spectrum of the corresponding positive ion was observed[60] with hydrocarbon glasses but not with those of polar solvents. Although positive charge exchange may occur in hydrocarbon solvents

$$RH \xrightarrow{} RH^+ + e$$

$$RH^+ + NNDA \rightarrow RH + NNDA^+$$

proton transfer reactions probably occur in polar solvents

$$CH_3CH_2OH \xrightarrow{} CH_3CH_2OH^+ + e$$

$$CH_3CH_2OH^+ + CH_3CH_2OH \rightarrow CH_3\dot{C}HOH + CH_3CH_2OH_2^+$$

References pp. 103–106

Irradiation of perylene in tetrahydro-2-methylfuran glass gave rise to an absorption at 580 nm which was attributed to its positive ion[61].

Aqueous glasses can only be obtained from solutions containing high concentrations of acid or alkali. Trapped electrons are not detected in acid glasses although the ESR spectrum of the hydrogen atom is observed[62]. Presumably the reaction

$$e + H^+ \rightarrow H$$

occurs rapidly even in the glassy state. That electrons are formed initially, however, has been demonstrated using N_2O as scavenger[63]. With alkali glasses, ESR measurements indicate the existence of both trapped electrons and the $O^-\cdot$ ion radical[64]. This latter species is produced by the reaction of a positive hole with the hydroxide ion

$$(H_2O)^+ + OH^- \rightarrow OH + H_2O$$

followed by removal of a proton

$$OH + OH^- \rightarrow O^-\cdot + H_2O$$

It is suggested that in irradiated pure ice the electrons and the positive hole initially are coupled to each other in an exciton-like bound state.

2.2 FREE RADICALS

It has already been pointed out (p. 73) that the postulate of free radical chain reactions provided a reasonable explanation for early results of the radiolyses of gases. It was later suggested[10] that the radiation chemistry of aqueous solutions could best be explained by the production of H atoms and OH radicals. Subsequently, the results of a large variety of radiolyses were explained in terms of radical reactions occurring therein. Although such experiments did not provide conclusive evidence for the existence of free radicals in the systems, the results obtained, e.g. product analysis, rate coefficients, were not inconsistent with the occurrence of free radical processes.

More recently the use of ESR techniques has allowed the unequivocal identification of free radicals in radiolyses. It will therefore be convenient, in this section, first to discuss this evidence and then to summarise some of the other radical characteristics, e.g. structure, spectra, reactivities, which have been deduced from systems where the supposition of free radical processes provides the simplest explanation of results.

Radicals produced in the solid phase are likely to be trapped, except in the case of very small radicals, and can be examined at leisure by the usual ESR and spectrophotometric techniques. Only at very high doses, when the radical concentration becomes such that radicals are produced in juxtaposition, are radical–radical reactions likely to be significant. In the liquid phase most radicals are likely to have short lifetimes and to disappear either by radical–radical recombination or reaction with solutes. However, the use of pulse radiolysis techniques has enabled these reactions to be studied. In the gaseous phase the rate of deposition of energy for a given radiation is considerably less than that in the liquid phase. Overall radical concentrations will, therefore, be lower. Moreover, radicals once formed are more easily able to diffuse away since they are not held within a solvent cage. It might be anticipated therefore, that radical–radical reactions should be less important in the gaseous phase. This should lead to an increase in the so-called free radical yield with a corresponding decrease in the so-called molecular yield. In agreement with this, it has been shown[65] that, in the ^{60}Co γ-radiolysis of water, whereas the total radical yield, $G_w(H) + G_w(OH)$, for liquid water is $\simeq$ 7, for water vapour it is $\simeq$ 15.

In general, however, the effect of phase is much less marked than for ionic species and results for different phases will not be considered separately in this section. Since, in fact, more experiments have been carried out on the radiation chemistry of liquids than of gases or solids, most of the results discussed in this section refer to the liquid state.

2.2.1 ESR data

Unambiguous evidence that hydrogen atoms were produced in the radiolysis of water was obtained[66] from ESR measurements on irradiated frozen solutions of H_2SO_4, $HClO_4$ and H_3PO_4 at 77 °K. The assignment of the pair of lines observed to the H atom was confirmed by substitution of H_2O by D_2O when three new lines were obtained (corresponding to a D atom with a nuclear spin of 1). However, with 16 % H_2SO_4, $G(H)$ was only 0.04. It was shown that the hydrogen atoms disappeared on warming by a second order process without developing electrical conductivity. H atoms were not observed[67] in pure ice at 77 °K but could be detected at 4 °K. The hydroxyl radical has similarly been identified[68] in irradiated ice at 77 °K but disappears in the temperature range 103–133 °K. Irradiation by γ-rays is now widely used as a standard method for the production of radicals in solid matrices prior to their examination by ESR techniques[69, 70].

That free radicals are produced in the radiolysis of liquid systems was first shown conclusively by the ESR experiments of Fessenden and Schuler[71]. The ESR spectra were observed during the continuous irradiation of the liquid samples by an electron beam from a Van de Graaff accelerator. The spectrum obtained

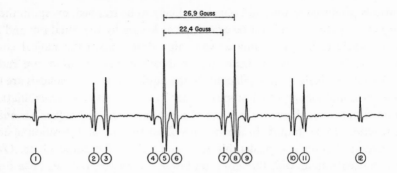

Fig. 8. ESR spectrum of irradiated liquid ethane at 93 °K (from ref. 71).

from liquid ethane is shown in Fig. 8. The 12 line spectrum is consistent with that expected from the C_2H_5 radical. A wide variety of liquid hydrocarbons was examined in this way and a wealth of information on the radicals produced was obtained. This information is valuable not only for its contribution to radiation chemistry but for the knowledge of radical structure gained thereby.

The results obtained indicate that for n-alkanes and for C_5–C_7 cycloalkanes most radicals arise by the loss of a hydrogen atom from the parent molecule. Although some C–C bond rupture occurs in the lower hydrocarbons, its importance decreases with increasing molecular weight. More extensive C–C bond rupture was observed in branched chain hydrocarbons.

Although in some cases radicals were observed as the primary products of radiolysis, they might also be produced in secondary reactions. Thus the production of the t-butyl radical in the radiolysis of neopentane, neohexane and 2,2,4-trimethylpentane was attributed to the addition of hydrogen atoms to isobutylene

$$\begin{array}{c} CH_3 \\ | \\ CH_3-C-CH_3 \\ | \\ CH_3 \end{array} \longrightarrow \begin{array}{c} CH_3-C=CH_2+CH_4 \\ | \\ CH_3 \end{array}$$

$$\longrightarrow \begin{array}{c} CH_3 \\ | \ \cdot \\ CH_3-C-CH_2+H \\ | \\ CH_3 \end{array}$$

$$H+CH_3-C=CH_2 \longrightarrow CH_3-\overset{\cdot}{C}-CH_3$$
$$\qquad\quad | \qquad\qquad\qquad |$$
$$\qquad\quad CH_3 \qquad\qquad\quad CH_3$$

With liquid ethane it was found that although the ethyl radical is the major radical species, lines attributable to methyl and vinyl radicals were observed.

The vinyl radical concentration was only 2–3 % of the ethyl radical concentration but was unaffected by the addition of deuteroethylene as scavenger. It seems likely that it arises as a primary product of the radiolysis, *viz.*

$$C_2H_6 \xrightarrow{} C_2H_3 + H_2 + H$$

In oxygenated solution, the ESR signals characteristic of alkyl radicals disappear, giving rise, in the case of higher hydrocarbons only, to a single broad line attributed to the alkylperoxy radical

$$R + O_2 \rightarrow RO_2$$

The stationary state concentration of radicals was shown to be proportional to the square root of the dose rate indicating that, as might be expected, the radicals, which are produced at a uniform rate, disappear by a second order recombination reaction

$$RH \xrightarrow{} R_1$$

$$R_1 + R_1 \rightarrow \text{stable products}$$

Absolute radical concentrations could be determined by reference to the signal obtained from the stable free radical galvinoxyl. Hence it was possible to determine the absolute rate coefficient for the recombination of ethyl radicals in liquid ethane as 3×10^8 l.mole^{-1}.sec^{-1} at 98 °K. The activation energy for this reaction was 780 cal.mole^{-1}, which is essentially that for the diffusion controlled process.

2.2.2 Spectra

The first application of the technique of pulse radiolysis led to the detection of the benzyl radical[72]. Its spectrum is shown in Fig. 9. The identification of this

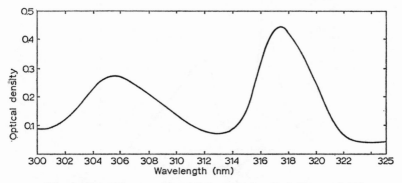

Fig. 9. Absorption spectrum of the benzyl radical (from ref. 72).

spectrum with $C_6H_5\dot{C}H_2$ was based on the absorption maxima at 318 and 306 nm as observed by Porter and Windsor in the flash photolysis of benzyl chloride solutions[73]. The spectrum was obtained on the radiolysis of solutions of benzyl chloride, alcohol or formate in 33 % ethanol–67 % glycerol as solvent. When cyclohexane or ethanol were used as solvent no spectrum was observed since the apparatus precluded measurement before 25 μsec after beam cut-off, and in these less viscous solvents recombination of benzyl radicals was too rapid to permit their detection. It was shown that in ethanol–glycerol the radicals decayed by two competitive second order processes

$$C_6H_5\dot{C}H_2 + C_6H_5\dot{C}H_2 \rightarrow (C_6H_5CH_2)_2$$

$$C_6H_5\dot{C}H_2 + CH_3\dot{C}HOH \rightarrow C_6H_5CH_2CH(CH_3)OH$$

Although the extinction coefficient of the benzyl radical was not known, product analysis gave an approximate total yield of benzyl radicals, from whence it could be deduced that $\varepsilon_{318} \simeq 1,100$ l.mole^{-1}.cm^{-1} and the rate coefficients for the second order processes were 4×10^7 and 2×10^8 l.mole^{-1}.sec^{-1} respectively. These rate coefficients are almost equal to those for the diffusion-controlled bimolecular reaction.

Irradiation of cyclohexane–O_2 solutions gave rise[74] to the transient spectrum shown in Fig. 10. Identification of this spectrum with the cyclohexyl peroxy radical was based on the following observations:

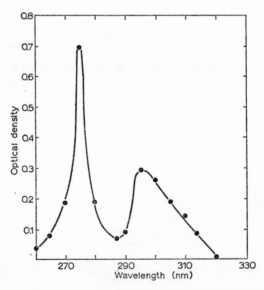

Fig. 10. Absorption spectrum of the cyclohexyl peroxy radical (from ref. 74).

1. Irradiation of deoxygenated cyclohexane at 83 °K gave the ESR spectrum of cyclohexyl.

2. No spectroscopically detectable transient which absorbed above 240 nm was observed in the irradiation of liquid deoxygenated cyclohexane.

3. Cyclohexanol, cyclohexanone and cyclohexylhydroperoxide were identified as major oxidation products.

Their formation was attributed to the reactions

$$C_6H_{11}O_2\cdot + C_6H_{11}O_2\cdot \rightarrow C_6H_{11}OH + C_6H_{10}O + O_2$$

$$C_6H_{11}O_2\cdot + HO_2\cdot \rightarrow C_6H_{11}OOH + O_2$$

The values obtained for the rate coefficients of these reactions were $(1.6 \pm 0.6) \times 10^6$ and $(3 \pm 1) \times 10^6$ l.mole^{-1}.sec^{-1} respectively. Similar values of the rate coefficients for the recombination of peroxy radicals were obtained from studies of the photo-oxidation of cyclohexene and other olefins by the rotating sector technique[75].

Production of peroxy radicals was presumably *via* the reactions

$$C_6H_{12} \rightsquigarrow C_6H_{11}\cdot + H\cdot$$

$$C_6H_{11}\cdot + O_2 \rightarrow C_6H_{11}O_2\cdot$$

$$H\cdot + O_2 \rightarrow HO_2\cdot$$

The observed rate coefficients for recombination are approximately 10^{-4} times the diffusion-controlled rate coefficients. There was, however, no measurable change in the rates of reaction over the temperature range 25–71 °C, indicating that the low rate coefficients arise because of steric hindrance to the recombination rather than because of an energy barrier. Rate coefficients have been evaluated for a variety of radical recombinations[76], although in fact the assignment of radicals is by no means unambiguous.

Spectra of the $CH_3\dot{C}HOH$ radical have been obtained from the radiolysis of pure and aqueous ethanol[77]. These are shown in Fig. 11. The assignment is based on a combination of ESR studies in the solid state, the effects of isotopic substitution, and product analysis. In aqueous solution the radical arises *via* the reactions

$$H_2O \rightsquigarrow H\cdot, \cdot OH, e_{aq}^-, H^+, H_2O_2, H_2$$

$$H\cdot + CH_3CH_2OH \rightarrow H_2 + CH_3\dot{C}HOH$$

$$\cdot OH + CH_3CH_2OH \rightarrow H_2O + CH_3\dot{C}HOH$$

and disappears by the reaction

$$2CH_3\dot{C}HOH \rightarrow \underset{\underset{\displaystyle CH_3CHOH}{|}}{CH_3CHOH}$$

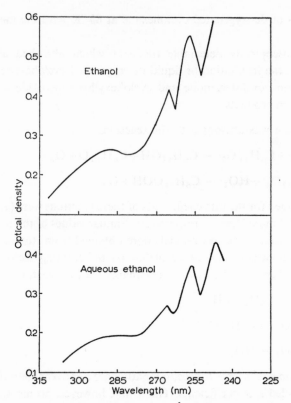

Fig. 11. Absorption spectra of the CH$_3\dot{\text{C}}$HOH radical (from ref. 77).

2.2.3 Chemical inferences of radical production

The decolorisation of solutions of the stable free radical diphenylpicrylhydrazyl (DPPH) in a wide variety of organic solvents upon irradiation has been known for some time and attributed to capture of radicals produced by the action of the radiation on the solvent by DPPH

$$RH \xrightarrow{\text{---\hspace{-2pt}\wedge\hspace{-2pt}---}} R\cdot + H\cdot$$

$$\dot{R} + DPPH \rightarrow DPPHR$$

Since the extinction coefficient for the radical is known, it was possible[78] from the extent of decolorisation to calculate G(total radicals). However, more recent work[79] has shown that excited molecules as well as free radicals are capable of destroying DPPH so that decolorisation is not uniquely diagnostic of the presence of free radicals. Characterisation of the radicals produced is difficult although some attempt has been made to separate the products of radical addition by column chromatography[78].

Extensive investigations have been made of the radiolysis of organic liquids containing iodine. Decolorisation of these solutions on irradiation is most simply accounted for by capture of the radicals produced by iodine

$$R + I_2 \rightarrow RI + I\cdot$$

and estimates of the radical yield have been made based on the consumption of iodine[80]. At low concentrations of iodine, the yield is independent of iodine concentration. However, ionic processes, *e.g.*

$$RH \xrightarrow{\makebox[1.5em]{\rule[0.5ex]{1.2em}{0.4pt}}} RH^+ + e$$

$$e + I_2 \rightarrow I^- + I\cdot$$

$$RH^+ + I^- \rightarrow RH^* + I\cdot$$

cannot be entirely ruled out. The radiolysis of aromatic compounds containing dissolved iodine appears to be much more complex[81].

By using radioactive iodine-131 as scavenger and identifying the alkyl iodides formed, by isotopic dilution analysis, much information can be obtained as to the nature of the radicals produced[82-84]. Some typical product distributions are shown in Table 9.

These distributions indicate that the radicals are produced predominantly by C–H rather than C–C bond rupture in agreement with the conclusions from ESR experiments. Similar conclusions have been reached using O_2 as scavenger with subsequent identification of the peroxides produced[86].

TABLE 9

PRODUCT DISTRIBUTION IN HYDROCARBON–I_2 RADIOLYSIS

Alkyl iodide	G(RI)	
	n-*butane*	n-*hexane*[†85]
Methyl	0.4	0.10
Ethyl	0.9	0.40
n-Propyl	0.1	0.70
i-Propyl	0.0	
n-Butyl	1.0	0.30
sec-Butyl	2.1	
n-Hexyl		0.70
sec-Hexyl		2.60

† analysis by gas–liquid chromatography.

The incorporation of ^{14}C in unsaturated hydrocarbons on irradiation in the presence of $^{14}CH_3I$ is most easily explained by recombination of radicals produced by the radiation[87]

$$RH \xrightarrow{} R\cdot + H\cdot$$

$$^{14}CH_3I \xrightarrow{} {^{14}CH_3}\cdot + I\cdot$$

$$R\cdot + {^{14}CH_3}\cdot \rightarrow {^{14}CH_3R}$$

In this way detailed information has been obtained as to the nature of the radicals produced in unsaturated hydrocarbons. In propylene the predominant radicals and their yields are allyl (45 %), isopropyl (33 %) and n-propyl (12 %) showing that, as with saturated hydrocarbons, most radicals arise as a result of C–H rather than C–C bond reorganisations. It was suggested that unsaturated radicals arose *via* ion–molecule reactions

$$C_4H_8 \xrightarrow{} C_4H_8^+ + e$$

$$C_4H_8^+ + C_4H_8 \rightarrow C_4H_9^+ + C_4H_7\cdot$$

and saturated radicals as a result of subsequent charge neutralisation

$$C_4H_9^+ + e \rightarrow C_4H_9\cdot$$

In Table 10 are shown values of radical yields for selected compounds, obtained using different scavengers[88,89]. In general, reasonable agreement is found amongst values with different scavengers.

TABLE 10

RADICAL YIELDS FOR ORGANIC COMPOUNDS

Compound	G(radical) using scavenger		
	$FeCl_3$	I_2	DPPH
Diethyl ether	12.5		15
α-Methyltetra-hydrofuran	4.1		
Dioxan	3.5		12
Diphenylmethane	0.9		
Phenetole	0.85		
Benzene	0.86	0.66	0.75
Diphenyl ether	0.6		
n-Pentane		7.9	
n-Hexane		7.6	
n-Octane		7.4	
Cyclohexane		6.2	8.7

In the radiolysis of saturated hydrocarbons, the observed yield of hydrogen evolved may be separated into two components, one of which is sensitive to the effect of added scavengers and one which is not [90]. This is most simply accounted for by the reactions

$$RH \xrightarrow{G_1} H_2 + products$$

$$RH \xrightarrow{G_2} H\cdot + R\cdot$$

In the presence of added scavenger, hydrogen atoms will be competed for

$$H\cdot + RH \xrightarrow{k_3} H_2 + R\cdot$$

$$H\cdot + S \xrightarrow{k_4} HS$$

If the product of the addition of hydrogen atoms to the scavenger does not ultimately give hydrogen then the dependence of hydrogen yield on scavenger concentration is given by

$$\frac{1}{G(H_2)_0 - G(H_2)} = \frac{1}{G_2}\left(1 + \frac{k_3\,[\text{hydrocarbon}]}{k_4\,[\text{scavenger}]}\right)$$

where $G(H_2)_0$ $(= G_1 + G_2)$ and $G(H_2)$ are the yields in the absence and presence of scavenger respectively. Results for the effect of methyl methacrylate in n-hexane are shown in Fig. 12 and are seen to be in good agreement with the above equation.

Values of k_4 are known from gas phase studies for a variety of unsaturated compounds[91] and with these non-polar solvents it may be assumed that the values will be the same in the liquid phase. Hence it has been possible to determine values of the rate coefficients for a wide series of hydrogen atom abstraction reactions in solution. On the basis of these, it has been possible to determine values of k_4 for other scavengers[92]. Some of these rate coefficients are shown in Table 11. It has however, been suggested that alternative physical mechanisms may be responsible for the decrease in hydrogen yield observed in the presence of scavengers[93].

ESR measurements on irradiated ice have demonstrated the existence of the hydrogen atom and the hydroxyl radical (see p. 87) and it seems reasonable to assume that these species will also exist in irradiated water.

That the hydrogen atom is an important intermediate in the radiolysis of acidic solutions is based on the following evidence. Experiments indicate the existence of two reducing species in water radiolysis[94]. These exist in acid–base equilibrium[95,96].

One of them has been identified as the hydrated electron[43] and consequently it seems likely that the other must be a hydrogen atom

$$H\cdot + OH^- \rightleftharpoons e_{aq}^- + H_2O$$

Qualitatively the reactions of this species are similar to those of hydrogen atoms produced by an electrodeless high frequency discharge in hydrogen gas and passed into aqueous solution[97,98]. Moreover at pH 2 the reducing agent is uncharged[42].

Ionic strength experiments have shown that in neutral solution the oxidising species is uncharged[99] but at pH 13 has unit negative charge[100]. It seems likely therefore that the oxidising agent is the OH radical and that the following equilibrium exists

$$OH^- + \cdot OH \rightleftharpoons H_2O + O^-$$

pK_{OH} has been measured[101] as 11.9 ± 0.2.

Rate coefficients for the reactions of H atoms and OH radicals with a large number of solutes have been determined[102,103]. A selection of these is presented

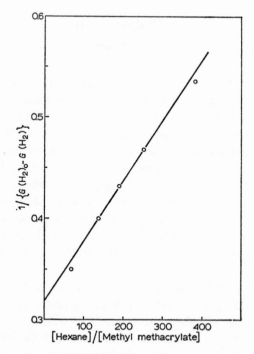

Fig. 12. Kinetic analysis of hydrogen yields for solutions of methyl methacrylate in *n*-hexane (from ref. 90).

TABLE 11

RATE COEFFICIENTS FOR HYDROGEN ATOM REACTIONS

$T = 23 \pm 1 \ °C$

Scavenger	Hydrogen abstraction $10^{-6} \ k_3$, $l.mole^{-1}.sec^{-1}$	Hydrogen addition $10^{-6} \ k_4$, $l.mole^{-1}.sec^{-1}$
n-Hexane	4.9	
Cyclohexane	6.6	
Ethyl acetate		83
Acetic acid	150	610
Benzene		180
Toluene		290
Aniline		400
Cyclohexene	110	490
Methyl iodide	720	6000†
Cyclohexyl iodide	900	7600†

† refers to the reaction $H + RI \rightarrow R + HI$

TABLE 12

RATE COEFFICIENTS FOR REACTIONS OF H AND OH IN LIQUID WATER

$T = 25 \ °C$

Substrate	H $10^{-7} \ k$, $l.mole^{-1}.sec^{-1}$	OH $10^{-7} \ k$, $l.mole^{-1}.sec^{-1}$
Acetic acid	0.01	1.4
Methanol	0.16	44
Formaldehyde	0.3	200
Ethyl acetate	0.58	24
t-Butanol	0.58	25
Ethanol	1.5	98
Isopropanol	3.6	390
Acetone	350	5.8

in Table 12. For those reactions involving simple abstraction of hydrogen the results for H and OH reactivities correlate fairly well.

Summarising, there is unambiguous evidence from ESR measurements that free radicals are produced in many systems on irradiation and a wealth of information from scavenger experiments is consistent with this. The radicals may arise from dissociation of excited molecules

$$RH^* \rightarrow R_1\cdot + R_2\cdot$$

or of ionic species

$$RH^+ \rightarrow R\cdot + H^+$$
$$H_2O^- \rightarrow H\cdot + OH^-$$

References pp. 103–106

It is likely that in any system, radicals are produced by more than one process, but much more work needs to be done before it will be possible to separate out the contributions of the individual reactions.

2.3 EXCITED MOLECULES

Excited molecules in radiolyses may arise by ion neutralisation

$$A \rightsquigarrow A^+ + e$$

$$A^+ + e \rightarrow A^*$$

or by direct excitation

$$A \rightsquigarrow A^*$$

Charge neutralisation leads to the production of both singlet and triplet excited states[104]. Whereas excitation by highly energetic electrons produces only singlet excited molecules, both singlet and triplet states may be produced by slow moving electrons. Since the probability of electron capture by the parent ion is less likely in the gaseous phase, it follows that direct excitation is likely to be the predominant reaction leading to the production of excited states in the gaseous phase, whereas both charge neutralisation and direct excitation may be important in condensed phases.

The excited states produced may or may not lead to chemical change. Their reactions may be summarised as follows

No chemical reaction

$$A_S \rightarrow A_G + h\nu \quad \text{fluorescence}$$

$$A_S \rightarrow A_G$$

$$A_S \rightarrow A_T$$

$$A_T \rightarrow A_G + h\nu \quad \text{phosphorescence}$$

Chemical reaction

$$A^* \rightarrow M_1 + M_2$$

$$A^* + B \rightarrow A_G + B^* \quad (B^* \rightarrow \text{ products})$$

(In this section A^* represents any excited state of A. A_G, A_S and A_T refer to the ground state, singlet excited, and triplet excited state of A, respectively.) The products of chemical reaction may be stable molecules or free radicals. Energy transfer processes will be important in systems of more than one component and will depend very much on the state of aggregation.

It will be convenient to review firstly the evidence for the production of specific excited states in radiation chemistry and then to examine the details of some of their reactions.

In addition to the benzyl radical produced in the pulse radiolysis of solutions of benzyl chloride and formate (see p. 89) McCarthy and MacLachlan[72] observed a second longer-lived species which was not detected when the solvents alone were irradiated. This species decayed by a first order process with a half life of 36 msec in cyclohexane. The suggestion that this transient was the triplet state of stilbene was supported by the observation that a similar transient of half life 40 msec was observed in the flash photolysis of *trans*-stilbene in cyclohexane solution. The triplet state was not produced in the flash photolysis of solutions of benzyl chloride or formate indicating that in the radiolysis, its production must arise from states not excited by uv light of wavelength > 200 nm. Somewhat surprisingly, the decay of the triplet was not greatly affected by the viscosity or polarity of the solvent.

Since this first identification of a triplet state in pulse radiolysis, numerous other triplet states have been identified. The triplet state of anthracene has been observed in pulse radiolysis studies of anthracene in benzene[105], acetone[106] and dioxan[107] in the liquid phase. Identification was based on the known spectrum of the triplet state of anthracene from flash photolysis studies[108]. The spectra of some triplet states observed in acetone solution are shown in Fig. 13. The rate of formation of the triplet state of anthracene indicated its production from an acetone precursor which was estimated to have a half life greater than 5 μsec for first order decay in pure acetone. This long life for the precursor suggests that this too is a triplet rather than singlet excited species. Thus

$$(CH_3COCH_3)_T + C_{14}H_{10} \rightarrow CH_3COCH_3 + (C_{14}H_{10})_T$$

The second-order rate coefficient for the above reaction was found to be $(6.2 \pm 0.6) \times 10^9$ l.mole^{-1}.sec^{-1} at 23 °C. Scavenger studies indicated that $G(CH_3COCH_3)_T$ = 1.1. The anthracene negative ion was also observed.

By contrast, in the pulse radiolysis of naphthalene and xenon mixtures in the gaseous phase, the triplet state of naphthalene was produced almost instantaneously (< 0.5 μsec)[109]. This was attributed to energy transfer by direct impact of sub-excitation electrons.

The effect of solvent on the production of triplet states has been examined[110]. The yield of triplet states was relatively high in those non-polar solvents where they were not able to abstract hydrogen atoms. In solvents of intermediate polarity there is a decrease in the yield of triplet states accompanied by increased yields of

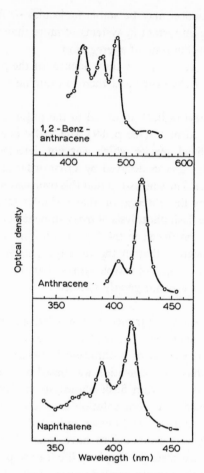

Fig. 13. Absorption spectra of triplet states of aromatic molecules in acetone solution (from ref. 106).

radicals and anions. In aqueous solutions no significant production of triplet states was observed.

Kinetic evidence for the importance of triplet states has also been obtained. 4-Pentenal is an important product both in the photolysis[111] and radiolysis[112] of liquid cyclopentanone. The yield is not affected by conventional radical scavengers such as DPPH or FeCl$_3$ but is decreased by oxygen or 1,3-pentadiene. Phosphorescence is observed in the photolysis of cyclopentanone[113] indicating the formation of a triplet state. The effect of 1,3-pentadiene on the yield of 4-pentenal is quantitatively consistent with the scheme

$$C + h\nu \rightarrow C_S$$

$$C_S \rightarrow C_T$$

$$C_T \xrightarrow{k_3} \text{4-pentenal}$$

$$C_T + D \xrightarrow{k_4} C + D_T$$

$$D_T \rightarrow \text{isomerisation of D}$$

where C = cyclopentanone and D = 1,3-pentadiene. Moreover, identical values of $k_4/k_3 \simeq 21$ were obtained from the results of both the photolysis and radiolysis.

From studies of the radiation-induced geometrical isomerisation of *cis*- and *trans*-2-butene in hydrocarbon solvents it has been concluded[114,115] that whereas in paraffin solutions isomerisation is brought about by ionic intermediates, in benzene solutions it is due to triplet–triplet energy interchange from an excited state of benzene. In benzene solution the isomerisation yields were not greatly affected by conventional radical scavengers but could be decreased by addition of unsaturated compounds and by paramagnetic species, *e.g.* oxygen, galvinoxyl. The total yield of the benzene triplet ($^3B_{1u}$) was estimated as 4.23. From studies of the effect of temperature on the rate coefficient for energy transfer it appears that the excitation is not localised in single molecules, but is delocalised in domains.

Evidence for the production of specific singlet excited states is much more sparse because of their much shorter lifetimes. However, some progress has been made in the study of energy transfer processes in the nanosecond region[116], and should yield valuable information. Thermoluminescence studies on γ-irradiated solid solutions of naphthalene in crystalline biphenyl show the fluorescence characteristic of the singlet–singlet transition in biphenyl on warming up, in addition to the triplet–singlet phosphorescence of naphthalene[117].

Much valuable information regarding the relative contributions of excited molecule and ionic reactions can be obtained from a study of the radiolysis in an applied electric field as outlined by Essex[17]. Ausloos *et al.*[118] have used this technique and also studied the effect of density, isotopic labelling and the addition of free radical scavengers in order to separate out the different processes occurring. Products arising from ion–molecule reactions are not likely to be significantly affected by an applied electric field whereas an enhanced yield of products might be expected from excited molecule reactions in so far as the collision of accelerated electrons with the substrate is likely to increase the yield of excited molecules. The applied voltage however, is not sufficiently great to induce secondary ionisation. Thus in the γ-radiolysis of C_2H_6–C_2D_6–NO mixtures in the gaseous phase the yields of C_2D_3H and CD_3H remain essentially constant while the yields of C_2D_4 and CD_4 increase nearly fourfold over the range 0–1300 V. (NO is added to scavenge all free radical products.) Photolysis experiments had already indicated[119,120] that the major modes of decomposition of excited ethane molecules are

$$C_2H_6^* \rightarrow H_2 + CH_3CH:$$

$$C_2H_6^* \rightarrow C_2H_4 + H\cdot + H\cdot$$

$$C_2H_6^* \rightarrow CH_4 + CH_2:$$

although the relative importance of each of these modes depends on the energy of the UV quantum. Isotopic analysis of the products of radiolysis indicated that on electron impact the energy imparted to an ethane molecule was > 10 eV.

In Table 13 are shown values for the number of excited molecules produced per ion pair in γ-radiolysis in the gaseous phase[121].

TABLE 13

RELATIVE YIELDS OF EXCITED MOLECULE AND IONIC PROCESSES IN γ-RADIOLYSIS

Molecule	Excited molecule/Ion pair
C_2H_4	1.0
C_2H_6	0.5
C_3H_8	0.3
$n\text{-}C_5H_{12}$	0.3

It is clear that in the gaseous phase the contribution from excited molecule reactions does not exceed that from ion–molecule reactions. This is probably to be attributed to the fact that the electron may more readily escape its parent ion in the gaseous phase and consequently the production of excited states by charge neutralisation is diminished. It would be interesting to examine the effect of LET on the relative yields of these processes.

Studies on a range of saturated hydrocarbons indicate that the main decomposition reactions of electronically excited alkane molecules formed by electron impact are[118, 122–124]

$$C_nH_{2n+2}^* \begin{cases} \nearrow C_nH_{2n} + H_2 \\ \nearrow C_nH\cdot_{2n+1} + H\cdot \\ \searrow C_xH_{2x} + C_yH_{2y+2} \\ \searrow C_xH\cdot_{2x+1} + C_yH\cdot_{2y+1} \end{cases}$$

Excited molecules produced by the radiation may also transfer their energy to other substrates. The yield of hydrogen in the radiolysis of cyclohexane–benzene solutions is considerably lower than would be expected from the electron fraction of each component and the known hydrogen yields for the pure compounds[125].

This was attributed to the transfer of energy from cyclohexane to benzene, which was then able to dissipate its energy without significant chemical reaction. This now classic piece of work gave rise to the idea of "sponge-type" protection. Since this concept was first advanced, much attention has been paid to the cyclohexane–benzene system and it is now clear that processes other than energy transfer may be operative[126].

Ferric chloride has been used as a radical scavenger in a variety of organic systems (see p. 94). At low concentrations in liquid benzene $G(-FeCl_3) \simeq 0.8$, which is consistent with the radical yield obtained using other scavengers. However, at high concentrations[127] or in crystalline benzene[88] there is a marked increase in $G(-FeCl_3)$. This was attributed to the interaction of excited benzene molecules. In the solid state it was concluded that energy transfer proceeded via exciton states, involving the $^1B_{2u}$ state of benzene, in which the energy was delocalised over a domain of $\simeq$ 50–100 A. Similar observations have been made with dimethylaniline, phenetole and cumene[128].

3. Conclusion

There is definite evidence that ions, electrons, free radicals and excited molecules are produced in systems exposed to ionising radiation. Qualitatively, it has been possible precisely to identify some of these intermediates in selected systems. Quantitatively it has been possible to separate out the yields arising from each of these products in only a limited number of systems and much more work needs to be done. More information is also desirable on the inter-relationship of these species in a particular system. A vast amount of kinetic data characterising some of these intermediates has also been accumulated.

It is to be anticipated that, as in the past, the answers to some of these problems will be of interest not only for their particular implications in radiation chemistry but also for their wider contribution to chemistry in general.

REFERENCES

1 C. J. HOCHANADEL, in M. BURTON, J. S. KIRBY SMITH AND J. L. MAGEE (Eds.), *Comparative Effects of Radiation*, Wiley, New York, 1960, p. 158.
2 C. J. HOCHANADEL AND S. C. LIND, *Ann. Rev. Phys. Chem.*, 7 (1956) 83.
3 E. COLLINSON, F. S. DAINTON AND J. KROH, *Nature*, 187 (1960) 475.
4 M. LEFORT AND X. TARAGO, *J. Phys. Chem.*, 63 (1959) 833.
5 M. EBERT, J. P. KEENE, A. J. SWALLOW AND J. H. BAXENDALE (Eds.), *Pulse Radiolysis*, Academic Press, London, 1965.
6 J. K. THOMAS, *3rd International Congress of Radiation Research, Abstracts*, 1966, p. 5.
7 S. C. LIND, *Am. Chem. J.*, 47 (1911) 397.
8 S. C. LIND, *J. Phys. Chem.*, 16 (1912) 564.
9 H. EYRING, J. O. HIRSCHFELDER AND H. S. TAYLOR, *J. Chem. Phys.*, 4 (1936) 479, 570.
10 J. WEISS, *Nature*, 153 (1944) 748.

11 A. H. SAMUEL AND J. L. MAGEE, *J. Chem. Phys.*, 21 (1953) 1080.
12 D. E. LEA, *Actions of Radiation on Living Cells*, Macmillan, New York, 1947, p. 47.
13 R. L. PLATZMAN, *Basic Mechanisms in Radiobiology*, National Research Council Publication No. 305, Washington D.C., 1953, p. 22.
14 J. L. MAGEE, *Radiation Res. Suppl.*, 4 (1964) 20.
15 J. L. MAGEE, *Ann. Rev. Phys. Chem.*, 12 (1961) 389.
16 H. ESSEX AND D. FITZGERALD, *J. Am. Chem. Soc.*, 56 (1934) 65.
17 H. ESSEX, *J. Phys. Chem.*, 58 (1954) 42.
18 T. W. WOODWARD AND R. A. BACK, *Can. J. Chem.*, 41 (1963) 1463.
19 R. L. PLATZMAN, *Intern. J. Appl. Radiation Isotopes*, 10 (1961) 116.
20 R. W. KISER, *Tables of Ionisation Potentials*, U.S. Atomic Energy Commission TID-6142, Washington D.C., 1960.
21 G. G. MEISELS, W. H. HAMILL AND R. R. WILLIAMS, JR., *J. Phys. Chem.*, 61 (1957) 1456.
22 J. H. FUTRELL, *J. Am. Chem. Soc.*, 81 (1959) 5921.
23 G. R. FREEMAN, *Discussions Faraday Soc.*, 36 (1963) 250.
24 A. O. ALLEN AND A. HUMMEL, *Discussions Faraday Soc.*, 36 (1963) 95.
25 G. R. FREEMAN, *J. Chem. Phys.*, 39 (1963) 988.
26 G. R. FREEMAN AND J. M. FAYADH, *J. Chem. Phys.*, 43 (1965) 86.
27 J. P. KEENE, E. J. LAND AND A. J. SWALLOW, *J. Am. Chem. Soc.*, 87 (1965) 5284.
28 W. H. T. DAVISON, S. H. PINNER AND R. WORRALL, *Chem. and Ind.* (*London*), (1957) 1274.
29 W. H. T. DAVISON, S. H. PINNER AND R. WORRALL, *Proc. Roy. Soc.* (*London*), A 252 (1959)187.
30 E. COLLINSON, F. S. DAINTON AND H. A. GILLIS, *J. Phys. Chem.*, 63 (1959) 909.
31 E. J. BURRELL, *J. Phys. Chem.*, 68 (1964) 3885.
32 P. J. WILSON, JR. AND J. H. WELLS, *Chem. Rev.*, 34 (1944) 1.
33 W. R. BUSLER, D. H. MARTIN AND F. WILLIAMS, *Discussions Faraday Soc.*, 36 (1963) 102.
34 F. WILLIAMS, *J. Am. Chem. Soc.*, 86 (1964) 3954.
35 G. SCHOLES AND M. SIMIC, *Nature*, 202 (1964) 895.
36 F. S. DAINTON AND D. B. PETERSON, *Proc. Roy. Soc.* (*London*), A267 (1962) 443.
37 G. STEIN, *Discussions Faraday Soc.*, 12 (1952) 227.
38 J. H. BAXENDALE AND G. HUGHES, *Z. Physik. Chem.* (*Frankfurt*), 14 (1958) 306.
39 N. F. BARR AND A. O. ALLEN, *J. Phys. Chem.*, 63 (1959) 928.
40 E. HAYON AND J. WEISS, *Proceedings of the 2nd United Nations International Conference on the Peaceful Uses of Atomic Energy, Geneva, 1958*, United Nations, Geneva, 1958, volume 29, p. 80.
41 G. CZAPSKI AND H. A. SCHWARZ, *J. Phys. Chem.*, 66 (1962) 471.
42 E. COLLINSON, F. S. DAINTON, D. R. SMITH AND S. TAZUKÉ, *Proc. Chem. Soc.*, (1962) 140.
43 J. W. BOAG AND E. J. HART, *Nature*, 197 (1963) 45.
44 J. P. KEENE, *Nature*, 197 (1963) 47.
45 J. P. KEENE, *Discussions Faraday Soc.*, 36 (1963) 307.
46 S. GORDON AND E. J. HART, *J. Am. Chem. Soc.*, 86 (1964) 5343.
47 E. J. HART AND J. W. BOAG, *J. Am. Chem. Soc.*, 84 (1962) 4090.
48 S. GORDON, E. J. HART, M. S. MATHESON, J. RABANI AND J. K. THOMAS, *J. Am. Chem. Soc.*, 85 (1963) 1375.
49 M. S. MATHESON, in R. F. GOULD (Ed.), *Solvated Electron*, Advances in Chemistry Series No. 50, American Chemical Society, Washington D.C., 1965, p. 45.
50 J. JORTNER AND R. M. NOYES, *J. Phys. Chem.*, 70 (1966) 770.
51 L. M. DORFMAN AND M. S. MATHESON, *Progr. Reaction Kinetics*, 3 (1965) 237.
52 M. ANBAR, ref. 49, p. 55.
53 L. M. DORFMAN, ref. 49, p. 36.
54 D. E. LEA, *Brit. J. Radiol., Suppl. No. 1*, (1947) 59.
55 J. WEISS, *Nature*, 186 (1960) 751.
56 D. J. E. INGRAM, *Free Radicals*, Butterworths, London, 1958, p. 170.
57 M. R. RONAYNE, J. P. GUARINO AND W. H. HAMILL, *J. Am. Chem. Soc.*, 84 (1962) 4230.
58 F. S. DAINTON, G. A. SALMON AND J. TEPLY, *Proc. Roy. Soc.* (*London*), A286 (1965) 27.
59 J. WENDENBURG AND A. HENGLEIN, *Z. Naturforsch.*, 196 (1964) 995.

60 W. H. HAMILL, J. P. GUARINO, M. R. RONAYNE AND J. A. WARD, *Discussions Faraday Soc.*, 36 (1963) 169.
61 N. CHRISTODOULEAS AND W. H. HAMILL, *J. Am. Chem. Soc.*, 86 (1964) 5413.
62 R. LIVINGSTON, H. ZELDES AND E. H. TAYLOR, *Phys. Rev.*, 94 (1954) 725.
63 F. S. DAINTON AND F. T. JONES, *Trans. Faraday Soc.*, 61 (1965) 1681.
64 P. N. MOORTHY AND J. J. WEISS, ref. 49, p. 180.
65 J. H. BAXENDALE AND G. P. GILBERT, *Discussions Faraday Soc.*, 36 (1963) 186.
66 R. LIVINGSTON, H. ZELDES AND E. H. TAYLOR, *Discussions Faraday Soc.*, 19 (1955) 166.
67 L. H. PIETTE, R. C. REMPEL, H. E. WEAVER AND J. M. FLOURNOY, *J. Chem. Phys.*, 30 (1959) 1623.
68 J. KROH, B. E. GREEN AND J. W. T. SPINKS, *Can. J. Chem.*, 40 (1962) 413.
69 B. SMALLER AND M. S. MATHESON, *J. Chem. Phys.*, 28 (1958) 1169.
70 P. B. AYSCOUGH AND C. THOMSON, *Trans. Faraday Soc.*, 58 (1962) 1477.
71 R. W. FESSENDEN AND R. H. SCHULER, *J. Chem. Phys.*, 39 (1963) 2147.
72 R. L. McCARTHY AND A. MACLACHLAN, *Trans. Faraday Soc.*, 56 (1960) 1187.
73 G. PORTER AND M. W. WINDSOR, *Nature*, 180 (1957) 187.
74 R. L. McCARTHY AND A. MACLACHLAN, *J. Chem. Phys.*, 35 (1961) 1625.
75 L. BATEMAN AND G. GEE, *Proc. Roy. Soc. (London)*, A195 (1948) 391.
76 R. L. McCARTHY AND A. MACLACHLAN, *Trans. Faraday Soc.*, 57 (1961) 1107.
77 I. A. TAUB AND L. M. DORFMAN, *J. Am. Chem. Soc.*, 84 (1962) 4053.
78 A. PREVOST-BÉRNAS, A. CHAPIRO, C. COUSIN, Y. LANDLER AND M. MAGAT, *Discussions Faraday Soc.*, 12 (1952) 98.
79 S. CIBOROWSKI, N. COLEBOURNE, E. COLLINSON AND F. S. DAINTON, *Trans. Faraday Soc.*, 57 (1961) 1123.
80 P. F. FORSYTH, E. N. WEBER AND R. H. SCHULER, *J. Chem. Phys.*, 22 (1954) 66.
81 A. T. FELLOWS AND R. H. SCHULER, *J. Phys. Chem.*, 65 (1961) 1451.
82 R. R. WILLIAMS, JR. AND W. H. HAMILL, *J. Am. Chem. Soc.*, 72 (1950) 1857.
83 L. H. GEVANTMAN AND R. R. WILLIAMS, JR., *J. Phys. Chem.*, 56 (1952) 569.
84 C. E. McCAULEY AND R. H. SCHULER, *J. Am. Chem. Soc.*, 79 (1957) 4008.
85 H. A. DEWHURST, *J. Phys. Chem.*, 62 (1958) 15.
86 G. DOBSON AND G. HUGHES, *J. Phys. Chem.*, 69 (1965) 1814.
87 R. A. HOLROYD AND G. W. KLEIN, *J. Phys. Chem.*, 69 (1965) 194.
88 E. COLLINSON, J. J. CONLAY AND F. S. DAINTON, *Discussions Faraday Soc.*, 36 (1963) 153.
89 A. J. SWALLOW, *Radiation Chemistry of Organic Compounds*, Pergamon, Oxford, 1960, Table 1.5.
90 T. J. HARDWICK, *J. Phys. Chem.*, 65 (1961) 101.
91 P. E. M. ALLEN, H. W. MELVILLE AND J. C. ROBB, *Proc. Roy. Soc. (London)*, A218 (1953) 311.
92 T. J. HARDWICK, *J. Phys. Chem.*, 66 (1963) 117, 291, 2246.
93 P. J. DYNE, J. DENHARTOG AND D. R. SMITH, *Discussions Faraday Soc.*, 36 (1963) 135.
94 N. F. BARR AND A. O. ALLEN, *J. Phys. Chem.*, 63 (1959) 928.
95 J. H. BAXENDALE AND G. HUGHES, *Z. Physik. Chem. (Frankfurt)*, 14 (1958) 323.
96 M. S. MATHESON AND J. RABANI, *J. Phys. Chem.*, 69 (1965) 1324.
97 G. CZAPSKI, J. JORTNER AND G. STEIN, *J. Phys. Chem.*, 65 (1961) 964.
98 J. JORTNER AND J. RABANI, *J. Phys. Chem.*, 66 (1962) 2078.
99 A. HUMMEL AND A. O. ALLEN, *Radiation Res.*, 17 (1962) 302.
100 G. HUGHES AND H. A. MAKADA, *Trans. Faraday Soc.*, 64 (1968) 3276.
101 J. RABANI AND M. S. MATHESON, *J. Am. Chem. Soc.*, 86 (1964) 3175.
102 M. ANBAR AND P. NETA, *Intern. J. Appl. Radiation Isotopes*, 16 (1965) 227.
103 G. E. ADAMS, J. W. BOAG, J. CURRANT AND B. D. MICHAEL, ref. 5, p. 131.
104 J. L. MAGEE, ref. 1, p. 130.
105 J. M. NOSWORTHY AND J. P. KEENE, *Proc. Chem. Soc.*, (1964) 114.
106 S. ARAI AND L. M. DORFMAN, *J. Phys. Chem.*, 69 (1965) 2239.
107 J. H. BAXENDALE, E. M. FIELDEN AND J. P. KEENE, *Science*, 148 (1965) 637.
108 G. PORTER AND M. W. WINDSOR, *Proc. Roy. Soc. (London)*, A245 (1958) 238.
109 M. C. SAUER, JR. AND L. M. DORFMAN, *J. Am. Chem. Soc.*, 86 (1964) 4218.

110 J. P. KEENE, T. J. KEMP AND G. A. SALMON, *Proc. Roy. Soc. (London)*, A 287 (1965) 494.
111 P. DUNION AND C. N. TRUMBORE, *J. Am. Chem. Soc.*, 87 (1965) 4211.
112 D. L. DUGLE AND G. R. FREEMAN, *Trans. Faraday Soc.*, 61 (1965) 1174.
113 S. R. LAPAGLIA AND B. C. ROQUITTE, *J. Phys. Chem.*, 66 (1962) 1739.
114 R. B. CUNDALL AND P. A. GRIFFITHS, *Discussions Faraday Soc.*, 36 (1963) 111.
115 R. B. CUNDALL AND P. A. GRIFFITHS, *Trans. Faraday Soc.*, 61 (1965) 1968.
116 M. A. DILLON AND M. BURTON, ref. 5, p. 259.
117 J. BULLOT AND F. KIEFFER, *Compt. Rend.*, 260 (1965) 4721.
118 H. H. CARMICHAEL, R. GORDON JR. AND P. AUSLOOS, *J. Chem. Phys.*, 42 (1965) 343.
119 R. F. HAMPSON JR., J. R. MCNESBY, H. AKIMOTO AND I. TANAKA, *J. Chem. Phys.*, 40 (1964) 1099.
120 H. OKABE AND J. R. MCNESBY, *J. Chem. Phys.*, 34 (1961) 668.
121 P. AUSLOOS, *Ann. Rev. Phys. Chem.*, 17 (1966) 205.
122 P. AUSLOOS AND S. G. LIAS, *J. Chem. Phys.*, 41 (1964) 3962.
123 P. AUSLOOS, S. G. LIAS AND I. B. SANDOVAL, *Discussions Faraday Soc.*, 36 (1963) 66.
124 S. G. LIAS AND P. AUSLOOS, *J. Chem. Phys.*, 43 (1965) 2748.
125 J. P. MANION AND M. BURTON, *J. Phys. Chem.*, 56 (1952) 560.
126 W. V. SHERMAN, *J. Chem. Soc.*, (1965) 5402.
127 E. A. CHERNIAK, E. COLLINSON AND F. S. DAINTON, *Trans. Faraday Soc.*, 60 (1964) 1408.
128 F. S. DAINTON, I. KOSA-SOMOGYI AND G. A. SALMON, *Trans. Faraday Soc.*, 61 (1965) 871.

Chapter 3

The Chemical Production of Excited States†

TUCKER CARRINGTON‡

AND

DAVID GARVIN

1. Introduction

In this chapter we treat gas phase chemical reactions that form identifiable energy-rich products. We deal almost exclusively with reactions near room temperature in which the reactants are at least approximately in an equilibrium distribution over translational and internal energy states. An excess of reactants, or an inert gas, provides a heat bath for maintaining this equilibrium. The products are produced either in electronically excited states, or in the ground electronic state with vibrational energy considerably in excess of that corresponding to equilibrium at the temperature of the reactants. Most of the reactions treated here have been studied by spectroscopic techniques. Vibrational excitation may be observed either in emission or absorption, whereas electronic excitation is usually observed in emission. Although spectroscopic methods are the most specific, other methods for detecting excited reaction products have been used. They may be detected by their high reactivity in kinetic experiments[1], or by a kinematic analysis of scattering in beam experiments[2].

We have not attempted to catalog all known reactions producing excited states. Instead, we have attempted to emphasize systems which are reasonably well understood, or which present particularly interesting qualitative features.

1.1 DISTRIBUTION OF REACTION PRODUCTS OVER INTERNAL ENERGY STATES

Spectroscopic methods naturally focus attention on the distribution of products over internal energy states. It seems clear that virtually every elementary reaction produces products in a non-equilibrium distribution. Even phase space models[3], which assume a "strong coupling" complex in a certain region of phase space, give

---

† Contribution of the U. S. Department of Commerce, National Bureau of Standards, not subject to copyright.
‡ Now at Department of Chemistry, York University, Toronto, Canada.

References pp. 174–181

non-equilibrium energy distribution in products. Equilibrium distributions are only to be expected when the region of strong coupling is arbitrarily large.

Of course, any molecule not in its lowest electronic, vibrational and rotational state is, strictly speaking, in an excited state. Hence the title includes all chemical reactions. Since we do not intend to discuss all chemical reactions, we have restricted the discussion to gas phase reactions with the reactants distributed in approximately an equilibrium way over their internal energy states.

In most observable systems the product energy distribution is at least partially relaxed by radiation or inelastic collisions following the reactive collision. The production of excited molecules will be observed, or not, according to the efficiency of relaxation processes in the experimental system. In this chapter we are primarily interested in the initial distribution of products over internal energy states, undistorted by partial relaxation. This distribution is best obtained by skillful design of the experiment so as to minimize relaxation effects, as in beam experiments and some chemiluminescent systems[4]. If this is impractical, one may attempt to "subtract" the effects of relaxation by inverting the appropriate rate equations for the relaxation of the original distribution. Highly accurate estimates of the relaxation rate coefficients used in this procedure are not required[4, 5].

1.2 MODELS OF REACTION[†]

The reactions to which most attention will be directed in this chapter are simple two, three or four atom systems. Thus they are in or on the border of the domain to which scattering theory has been applied and in which direct kinematic experiments (molecular beam scattering) have been undertaken. We do not propose to review the developments in either of these fields in any detail; that has been done by others[2, 6]. Instead, a description will be given of some of the theoretical models that have been applied to exothermic reactions that produce excited products.

Two classes of models can be identified. These are kinematic descriptions of the collision that leads to reaction, and potential energy surface representations of that interaction. Of course, these types overlap; the former finds its full expression in the calculation of trajectories on potential surfaces[7].

1.2.1 Kinematic models

Two types of reactive collisions are considered in the kinematic models: the collision complex and the direct interaction. The complex model visualizes the

---

† See also Chapter 3, Volume 2.

reactants coming together and remaining together for a period of time at least as long as a rotational period. This complex survives long enough so that only the total energy and angular momentum are significant. The properties of the product (scattering angles and energy distribution) are characteristic of this collision *complex*. The terms "strong-coupling", "snarled trajectories" and "phase space-model" are associated with this behavior. Crossed molecular beam experiments on the reactions

$$Cs + RbCl \rightarrow CsCl + Rb$$

$$K + RbCl \rightarrow KCl + Rb$$

fall in this class[8].

The direct model is characterized by simple, smooth trajectories with little "waste motion". For the reaction $A + BC \rightarrow AB + C$, B moves from C to A and the reactants separate within a vibrational period. The outcome of the collision depends on the details of the initial conditions, not just on the total energy and angular momentum.

The scattering in a direct interaction has two limiting types, stripping and rebound. In stripping, the product AB leaves the reaction region in a trajectory that is an approximate continuation of that of A. This type is characterized by large cross sections and conversion of reaction energy into internal energy of the products. Rebound (backward scattering of AB) is characterized by small cross sections and appreciable release of reaction energy into kinetic energy.

The extreme stripping case is "spectator stripping", in which A snatches the atom (group) B from molecule BC with no alteration in the velocity or internal energy of the fragment C. The molecular beam studies of the reactions of Cs, K and Rb with halogens and of Ar^+ and N_2^+ with D_2 have been treated in this way (refs. [9,10,10a]).

When extended from three atom to larger systems the complex model predicts that any internal excitation will be distributed broadly between the products, AB and CD in the process $A + BCD \rightarrow AB + CD$, while the stripping model emphasizes excitation of AB. In its extreme limit, however, the spectator stripping model would predict some excitation in CD since the fragment –CD leaves the reaction unchanged, in an electronic (bonding) environment that may be significantly different from that in the molecule CD. This would result, for example, if the CD bond length in BCD were much different from that in the molecule CD.

Other pictures have been suggested to explain the appearance of large amounts of excitation in the product AB in the reaction $A + BC \rightarrow AB + C$. For example, if an atom A approaches a molecule BC with sufficient velocity to compress the incipient A–B bond prior to scission of the B–C bond (which must be the weaker), preferential excitation of AB may occur as the products separate[11]. Another idea is that AB excitation may occur readily if the A–B bond is extended (under tension) as a repulsion between B and C "kicks" B toward its new partner[12]. This last is a simple picture of a type of "mixed energy release" discussed later. It has also been suggested that the relative masses of the reactants and products (in a linear collision) set limits on the excitation of the newly formed AB molecule (ref. [13]).

1.2.2 Potential energy surfaces

The most satisfactory treatment of the reactions of interest in this chapter is in terms of classical trajectories on potential energy surfaces. They provide a detailed consideration of the reactive interaction (for which the kinematic models are limiting cases[7]), and provide ample scope for the theoretician to apply his intuition in explaining reactive molecular collisions. Reactions are naturally divided into those which take place on a single surface, usually leading to vibrational excitation, and those which involve two or more surfaces, often leading to electronic excitation.

The reader should be warned that the calculation of a realistic surface is far from easy, since many crucial approximations have to be made. The interpretation of the results of the study (computed trajectories, efficiency of reaction, distribution of energy) requires a dedicated surface watcher. We shall emphasize qualitative conclusions, in the firm belief that these are the important end products. This use of surfaces is entirely analogous to the use of mechanisms in macroscopic kinetics: a model is chosen, its results are determined and these are compared with those derived by the experimenter from his raw data.

The concept of a potential energy curve or surface is reasonably useful when the coupling between electronic and vibrational motion is small. The potential energy function gives the electronic energy of the system for arbitrary fixed positions of the nuclei.

All of the *surfaces* for reactions have more than three dimensions. For a triatomic system there are three independent coordinates $(3N-6)$ and the potential energy function $V(r_1, r_2, r_3)$ is a surface in a four dimensional space. The potential function usually shown for a triatomic system ABC is a three dimensional projection of this four dimensional space, the ABC angle being held fixed. Motion restricted to such a projected surface allows no rotation of BC relative to A at large distances and no bending vibration of ABC at short distances.

In terms of spectroscopic observables, the potential energy function is that function $V(r)$ which, when inserted into the quantum mechanical formulation of the vibration problem, gives the observed vibrational levels. In beam experiments, it is the potential which gives the observed scattering. In chemical excitation processes, it is the surface which predicts the observed total cross section and the observed distribution of products over internal energy states. Potential energy functions may be calculated from first principles[14], or they may be constructed empirically to fit given experimental data[15]. In any case there are two parts to the problem. One part is to derive the surface, the other is to use the surface to predict the results of experiment.

Fortunately, fine details of the surface are unlikely to be important in interpreting current experiments. A few qualitative features, such as the range of forces, the relative slopes of the reactant and product regions of the surface, and the energetics, will have a large effect on product energy distributions[7].

1.2.3 Reactions on a single potential surface

The easiest way to try to understand the dynamics of these reactions is to calculate classical trajectories on empirical or approximate quantum mechanical surfaces[12,15-17]. The most recent of these classical trajectory calculations include motion in all degrees of freedom except center of mass translation. One can then adjust the qualitative characteristics of the potential surface and observe the corresponding changes in predicted rotational and vibrational distributions of products, and in the differential and total reactive cross sections. One hopes to arrive in this way at simple generalities relating certain types of surfaces to certain types of scattering and product energy distributions. For example in treating reactions of the type $A+BC \rightarrow AB^v+C$, Polanyi et al.[7,12,18] have discussed the importance of the fraction of the reaction energy which is released during various stages of the reaction.

They have introduced the terms attractive-, mixed-, and repulsive-energy release. These have been defined sharply enough to permit quantitative statements to be made about the nature of a surface. For the reaction $A+BC \rightarrow AB+C$, attractive energy release is the liberation of reaction energy while A is approaching BC, the latter retaining its original bond distance

$$\rightarrow A \ldots B-C$$

Repulsive energy release is the liberation of reaction energy during the "second half" of the reaction when C is separating from AB which is in approximately its final state

$$A-B \ldots C \rightarrow$$

Mixed energy release is the concerted process in which energy is liberated as A approaches B with simultaneous departure of C

$$\rightarrow A \ldots B \ldots C \rightarrow$$

In terms of a potential energy surface, attractive release occurs in the valley of the reactants, as judged from the energy contours, repulsive release in that of the products, and mixed in the transition region. These concepts are illustrated in the reaction coordinate diagram, Fig. 1.

A rough correlation has been found[12,19] between the fraction of attractive plus mixed energy release and the fraction of energy going into vibration of the product AB. Bunker and Blais[15] have found somewhat similar results. Surfaces in which virtually all the energy release is on the reactant side of the transition state tend to produce motion with relatively long interaction time, suggesting the strong interaction visualized in phase space theories[3]. This type of interaction tends to produce a broad distribution of vibrational and rotational energy in the products[4]. Further trajectory calculations have been made for $K+CH_3I$[17,17a] and for $K+C_2H_5I$[20].

For reactive scattering of $H+H_2$, quantum mechanical and classical treatments for the same surface have been compared[21,22], and the agreement is good. This is encouraging. It leads one to have faith in classical calculations since this system is the least classical of all. A quantum mechanical treatment of reactive scattering of $K+HBr$ on an empirical potential surface[23] has shown that agreement with experiment can be obtained by suitable adjustment of parameters describing the surface. Child[24] has done a quantum mechanical treatment of $K+HI$ based on the model of planar motion with the I atom treated as infinitely massive. He treated two empirical energy surfaces and found that removal of an activation barrier considerably increased the rotational excitation of the product KI. Marcus[25] has given formal analytical treatments, both classical and quantum mechanical, of reactive scattering in linear collision in which there is little change of vibrational energy.

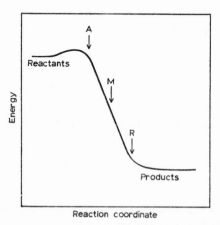

Fig. 1. Reaction coordinate diagram showing general locations of regions for repulsive (R), mixed (M) and attractive (A) energy release. For a mixed energy release surface the transition state will be near M, but for an attractive (repulsive) surface the transition state will be near R(A).

1.2.4 Reactions involving more than one potential surface

In reactions involving electronically excited states, the interaction of two potential energy surfaces is likely to be involved. This interaction is reasonably well understood and easy to visualize in the case of potential curves for diatomic systems, but for systems containing more than two atoms, the situation is considerably less tractable. One tries to develop a description, starting from the observed kinetics of a reaction and using whatever spectroscopic data may be available.

In discussing molecular systems which must be described in terms of more than one potential surface, it is desirable to have a clear definition of the variously used term *crossing*. It is also important to distinguish between (*a*) interaction of potential surfaces, and (*b*) transitions from one adiabatic surface to another induced by coupling between nuclear and electronic motions (failure of the Born–Oppenheimer approximation).

A potential surface gives the electronic energy of a system with the positions of the clamped nuclei as parameters. In this solution of the electronic problem, motions of the nuclei are assumed to be arbitrarily slow. A set of potential surfaces can be calculated from a given approximate electronic Hamiltonian. Two such surfaces are said to cross when, for a given point or set of points in configuration space, they have the same energy. At the crossing point (or region) the two surfaces do not interact, in the given approximation, for if they did they would repel each other and the crossing would be avoided. If two surfaces correspond to electronic states with the same symmetry species, they will interact if calculated with a sufficiently complete electronic Hamiltonian, and the crossing will be avoided in this higher approximation (non-crossing rule[26]). The situation is illustrated for diatomic potential curves in Fig. 2, where A_0, B_0 are the approximate curves which cross, and A, B are the higher order curves in which crossing is avoided. Frequently, spectroscopic data give information about the approximate curves only, since these are adequately valid everywhere but near the region of interaction. The interaction of potential curves just discussed is caused by the detailed structure of the electronic Hamiltonian, and has nothing to do with nuclear motion.

Suppose now that we have calculated a set of potential curves, A and B in Fig. 2, with sufficient accuracy in the electronic Hamiltonian so that all crossings which, by symmetry, will be avoided are in fact avoided. We may discuss the behavior of a diatomic molecule in terms of these curves. If the motion of the nuclei is arbitrarily slow, and there is no external perturbation, the molecule will "stay" on one of these adiabatic curves indefinitely. However, if the splitting between A and B is small, effects of nuclear motion may not be negligible. To illustrate this effect, suppose we are discussing radiative recombination of ground state atoms in terms of the curves of Fig. 2, a typical inverse predissociation involving two excited electronic states. As the atoms, approaching on curve A, reach the region of interaction, the system may with high probability undergo a transition from

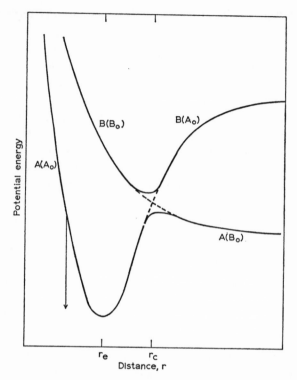

Fig. 2. Schematic potential curves showing avoided crossing in higher approximation. The full curves A and B are electronic eigenvalues (with r as parameter) for the complete Hamiltonian, neglecting coupling with nuclear motion. The zero order curves A_0 and B_0, including dashed parts through the crossing point, intersect because they are derived from an incomplete Hamiltonian. The arrow represents the possibility of spontaneous radiation from the A state to the ground state, not shown.

adiabatic curve A to B if the separation of the curves is small, and if the nuclear motion, which induces the transition, is fast. After reaching the turning point on the repulsive part of B, the atoms separate and the transition between B and A may again occur in the interaction region. The result in the case of small splitting and rapid nuclear motion is that there is little recombination into state A_0, where the recombination could be confirmed by emission of radiation, and the behavior of the system appears to be adequately described by the approximate curve B_0, which is everywhere repulsive. If, on this same set of potential curves, we consider molecules initially in the approximate state A_0, in a vibrational level just above the maximum of A, a similar argument shows that dissociation on curve A will occur (predissociation), but it will be slow if the splitting is small and the vibrational motion on A_0 is fast.

The distinction between interaction of different zero order electronic states for clamped nuclei, and interaction of different Born–Oppenheimer states due to the velocity of the nuclei, is clear theoretically, but it is by no means easy to distinguish these interactions in kinetic experiments since the second tends to mask the first.

The interaction of electronic with nuclear motions has been discussed as a time dependent problem[27], an explicit expression for the operator coupling the two being derived. This permits analysis of the Born–Oppenheimer approximation, and of the approximation in which the motion of the nuclei is treated classically in terms of an explicitly time dependent Hamiltonian. This latter approximation has been applied[28] to electron rearrangement collisions such as $H + Be^{++} \rightarrow H^+ + Be^+$. The result is a considerable improvement on the much used Landau–Zener treatment[29]. Another treatment applying time dependent perturbation theory to Born–Oppenheimer wave functions[30] has been used to give transition probabilities for bound–bound, bound–continuum, and continuum–continuum transitions.

The above discussion of curve crossing applies to potential surfaces as well as to curves. There is however one simple and important qualitative difference between the two cases. In the diatomic case, curves of the same symmetry cannot cross, if calculated in sufficiently high approximation. With surfaces, crossing of such states is possible because there are several coordinates which can be varied in order to find a point or region in configuration space at which the energies of two surfaces are equal and simultaneously their interaction is zero[31]. These intersections thus have two fewer dimensions than the dimensions of the surfaces themselves, since they are defined by two simultaneous conditions on the coordinates. Surfaces of the same symmetry species which do nevertheless cross without interaction on some locus will in fact interact and avoid each other in regions immediately surrounding that locus, (conical intersection[31]), so that crossing here does not mean the absence of interaction, as it does in the diatomic case.

1.3 CORRELATION RULES

Correlation rules relate the symmetry of reactants to the symmetry of products. More precisely, they give the symmetry of the fragments which can result when a molecule or transition state is distorted in the direction of reactants or products[32,33]. A familiar example is the correlation of the states of a diatomic molecule with those of its constituent atoms. Within the Born–Oppenheimer separation we can deal with strictly electronic correlation rules, valid when there is negligible coupling between electronic and vibrational wave functions. When such coupling is important, correlations forbidden on a strictly electronic basis may be allowed, so the validity of purely electronic correlation rules is hard to assess for polyatomic molecules with strongly excited vibration.

It has been emphasized[32] that an association or atom transfer reaction will in general go through configurations of lowest possible symmetry: one plane for three atom systems, no symmetry elements for systems of four or more atoms. When configurations of low symmetry are involved, the symmetry correlations will impose few restrictions on the accessible products.

The correlation with dissociation products will often depend on the path the molecule follows in dissociating. For example, in the unlikely event that a molecule ABC dissociates through linear configurations, the adiabatic products may be different from those for dissociation through bent configurations[31]. An important use of correlation rules is in predicting regions of intersection of potential surfaces required by symmetry. For example, a degenerate Π state of a linear molecule will split into two electronic states when the molecule is bent. Hence we know that these two states must intersect at the linear configuration[31,33].

1.4 ROTATIONAL EXCITATION

While excitation of vibrational and electronic states is limited only by conservation of energy, excitation of rotation is limited in addition by conservation of total angular momentum. One part of the total angular momentum of a collision complex is made up of the orbital angular momentum of the two reactants corresponding to their translational motion relative to their common center of mass. The rest of the total is contributed by the internal, rotational angular momentum of the reactants. When (or if) the complex breaks up, the total may be divided up quite differently among the orbital and rotational parts. How this happens will of course depend on the potential energy surface involved, particularly on its range, and angle dependence. Classical trajectory computations on various kinds of potential energy surfaces have been used to investigate rotational excitation of products in atom transfer reactions[12,15,21].

In two-body radiative association reactions, the angular momentum of interest is simply the total angular momentum of the reactants, rotational plus orbital. The rotational contributions will typically be the angular momentum associated with energy kT. Hence, for a diatomic molecule, a typical value will be $J(J+1) = kT/B$ where B is the rotational constant. The orbital contribution will be of the same order of magnitude since the linear velocity and the impact parameter (the distance of closest approach if there were no interaction) will be comparable to tangential velocity and bond length in a rotating diatomic molecule. The orbital contribution can, however, be considerably larger if heavy particles and long range forces are involved in the collision. Unless this is the case, the angular momentum of the association product will be comparable to what it would have in equilibrium at the temperature of the reactants.

In association reactions stabilized by a third body, the third body is often a neutral partner which simply removes excess energy or angular momentum. Hence associations of this type are also unlikely to produce excess rotational excitation. If the third body is more intimately involved, a discussion similar to that given below for atom transfer reactions is more appropriate.

Consider the angular momentum of the product AB of the atom transfer reaction $A + BC \rightarrow AB + C$. The total angular momentum of the intermediate ABC is made up in the way discussed above. When this complex breaks up, the rotational angular momentum of the product AB may considerably exceed the total for the complex if the departing C has also a large orbital angular momentum roughly equal in magnitude and opposite in direction to that of AB. The limit here is not conservation of total angular momentum, but rather it is the maximum orbital angular momentum of the outgoing atom C. This is determined by the product of its linear momentum and its outgoing impact parameter. This last is determined roughly by the range of forces between C and AB. More precisely, it is the relative masses and the detailed structure of the potential surface which determine the result. The splitting off of a fragment with high rotational angular momentum from a complex with small total angular momentum is illustrated by the photodissociation of H_2O^{34}. Here the initial complex H_2O^e has angular momentum of about $3\hbar$ while the dissociation product OH^e has rotational angular momentum exceeding $20\hbar$. Rotational excitation and the partitioning of angular momentum have been discussed in detail by Herschbach[2].

In beam experiments, where attention is focused on translational energy and scattering angle, internal energy is measured only indirectly and it is difficult to deduce how it is divided between rotation and vibration. Some progress in this direction has however been made using electric deflection analysis[2]. On the other hand, spectroscopic observations clearly distinguish between rotational and other forms of excitation. They suffer from the fact that collisions will relax non-equilibrium rotational distributions if the lifetime of the excited molecule is much longer than the time between collisions. At practical pressures this will usually be the case for vibrational, but may not be so for electronic transitions.

2. Vibrational excitation

2.1 ATOM- OR GROUP-TRANSFER REACTIONS

The general formulation of a group transfer reaction is

$$A + B-C \rightarrow A-B + C$$

where A, B and C may represent either atoms or multiatomic fragments[†]. The

---

[†] In equations in this section and in Table 1 the bonds that are formed or broken during reaction are shown explicitly.

characteristic feature of the process is that the bonding between B and C is destroyed and bonds are formed between A and B.

In this section, group transfer reactions in which the product molecule A–B is vibrationally excited but still in its ground electronic state are considered. (Transfer reactions that produce electronic excitation are discussed in Section 3.4.) The available experimental evidence is tabulated. Only typical examples are described. The principal points discussed are the limitations that experimental technique has imposed on observation and interpretation of this type of chemi-excitation and the extraction of generalizations concerning this class of reaction.

Although the novelty of observing chemically produced vibrational excitation provided an initial impetus, the main purpose of the studies to date has been to determine in detail the relative proportions of excited molecules in the various energy states, the fraction of the reaction energy that goes into internal excitation, which products are excited, and the fate of the excited molecules. Such data are used as aids in the construction of potential energy surfaces to be used, in turn, to describe the dynamics of the reactions. In short, the studies have been in the hands of kineticists. As interest in the subject has spread, more attention has been paid to applications: laser action and the reactions of the excited molecules.

A number of bimolecular group transfer reactions in which vibrational excitation has been observed are recorded in Table 1. These are almost exclusively atom-transfers. The reactions are grouped by the atom transferred, and the molecules are written to show the change in bonding between reagents and products. Stated more strictly, Table 1 summarizes studies of certain experimental systems in which direct evidence has been obtained, by spectroscopic methods, of the presence of vibrationally excited molecules. The cause of this chemi-excitation has been assigned to the reactions chiefly on the basis of kinetic analyses of the systems. These assignments are sound, but two caveats are needed. In no system studied is the reaction reported the sole one occurring, nor is the evidence on all possible types of excitation complete. Future work will refine and probably complicate the conclusions presented here.

Several generalizations may be made about observed excitation in this class of reactions. They are presented here as a guide to the discussion; not all are apparent from the data in Table 1.

1. The primary location of vibrational excitation is in the product molecule in which a new bond is formed, *i.e.*, A–B. Excitation of the other product, C, occurs but to a lesser extent.

2. Rotational excitation may accompany the vibrational excitation, but is less evident in the observations (because rotational energy relaxation is rapid).

3. Excitation of molecules occurs to the maximum extent permitted by the energetics of the reaction and the need for the products to separate. This maximum excitation is, however, neither the exclusive nor the most probable reaction path.

4. When the energetics of the reaction permits, electronic excitation can occur

in atom transfer reactions. It competes successfully with vibrational excitation.

5. Details of the observed excitation are strongly dependent upon experimental conditions and instrumental limitations, either as to which product may be observed, or as to the influence of subsequent chemical and energy transfer processes.

The reactions in Table 1 are transfer of H, O, N, S, halogen and alkali metal atoms. They are also reactions of atoms (H, O, N, S, halogen and alkali metal atoms) with small molecules. They are exothermic and have high specific rates (low activation energies and normal steric factors). These features are desirable for the study of chemi-excitation. High heat release permits substantial excitation. Small, low moment of inertia product molecules have spectra that *may* be resolved with moderate power spectrographic and spectrometric instruments. High speed provides the necessary number of reaction acts per unit time.

High speed results from a combination of high intrinsic rate and high reactant concentration. Here the limitation is atom concentration. Two methods have been widely used to attain relatively high concentrations: flash photolysis and atom production either in electric discharges or by specific preparative reactions.

2.1.1 Flash photolysis: molecular oxygen and hydroxyl (see also Vol. 1, p. 118)

The flash photolyses of ozone, nitrogen dioxide and chlorine dioxide were among the first systems studied. They have provided information on the production of vibrationally excited oxygen molecules. In each case the method has been to introduce a gaseous mixture of the decomposable molecule and a large excess of inert gas into a long tube, transparent to ultraviolet light, that is flanked by (or surrounded by) one or more "flash lamps". These light sources are noble gas filled discharge tubes. These are close relatives of those used in high speed photography. A large quantity of energy, several thousand joules, stored in a condenser bank, is discharged through the lamps in a period of 10 to 250 microseconds. The resulting flash floods the reaction cell with a high intensity, short duration pulse of photons, many of which are capable of dissociating the substrate molecule. The lamp spectrum is a continuum, peaked in the quartz ultraviolet region, but with substantial contribution from both shorter and longer wavelengths. Some atomic emission lines are superimposed on the continuum. At a known time after the photolytic flash (25 to 1000 microseconds) a second, much weaker lamp, placed in line with the reaction cell is flashed. Its light traverses the cell and is focused on the entrance slit of a spectrograph. This second lamp is the analysis source. Its spectrogram is examined for absorption by the reacting mixture. Repeated experiments of this type yield either a time history of the reaction, or data on the effect of varying the experimental parameters.

In each of the three cases, photolysis of O_3, NO_2 and ClO_2, the reaction pattern is similar. The case of ozone is considered here. It has been studied repeatedly

          (text continued on p. 126)

TABLE 1

VIBRATIONAL EXCITATION IN GROUP TRANSFER REACTIONS

Reaction	ΔH, kcal.mole⁻¹	Highest vibrational level possible[a]	Highest vibrational level observed	Distribution of excitation, remarks[b]	Experimental conditions[c]	References[d]
Hydrogen halide formation						
$H+Cl-Cl \rightarrow HCl^v+Cl_2$	−47	$v=6$	$v=6$	Maximum at $v=2, 3$, rotation non-equilibrium, population inversion, $k=3\times10^9$	Flow mixing, emission, and laser in H_2+Cl_2 photolysis. 4×10^{-4} torr	4, 18, 46, 65, 287, 288, 291, 293–297, 43a
$Cl+H-I \rightarrow HCl^v+I$	−34	$v=4$	$v=4$	Maximum at $v=3$, rotation non-equilibrium, laser action	Flow mixing, emission. 1.5×10^{-4} torr	4, 46
$Cl+H-Br \rightarrow HCl^v+Br$	−16	$v=2$	$v=2$		Flow mixing, emission. 1×10^{-4} torr	46
$H+Cl-NO \rightarrow HCl^v+NO$	−65	$v=9$	$v=9(10?)$	$T_v \sim 3$–4000 °K, rotation non-equilibrium, $T_r \sim 2000$	Flow mixing, emission. 0.01–1.8 torr	298, 299
$Br+H-I \rightarrow HBr^v+I$	−16	$v=2$	$v=2$		Flow mixing, emission. 3×10^{-4} torr	4
$H+Br-Br \rightarrow HBr^v+Br$	−43	$v=6$	$v=6$	Maximum at $v=3, 4$, $Br(^2P_{\frac{1}{2}})$ observed in secondary reaction, $k=1\times10^4$	Flow mixing, emission. 0.001 torr	4, 46, 49, 240, 241
$H+F-F \rightarrow HF^v+F$	−90	$v=9$	$v=9$	Doubtful identification of reaction	H_2–F_2 flames, emission	300

Reaction	v	Conditions	Method	Ref.
$O(^1D)+H\text{-}H \rightarrow OH^v+H$ -44	$v = 2$	$v = 2(3?)$ $T_v > 2000\ ^\circ K$	Flash photolysis of O_3, absorption	36, 38
$O(^1D)+H\text{-}OH \rightarrow OH^v+OH$ -29	$v = 2$	$v = 2$ 3 is at 29.2 kcal, $T_v > 2000\ ^\circ K$	Flash photolysis of O_3, absorption	36, 38, 56
$O(^1D)+H\text{-}Cl \rightarrow OH^v+Cl$ -45	$v = 4$	$v = 1$ $T_v > 2000\ ^\circ K$. $v = 5$ is at 46 kcal	Flash photolysis of O_3, absorption	36, 38
$O(^1D)+H\text{-}NH_2 \rightarrow OH^v+NH_2$ -45	$v = 4$	$v = 2$ $T_v > 2000\ ^\circ K$	Flash photolysis of O_3, absorption	36, 38
$O(^1D)+H\text{-}CH_3 \rightarrow OH^v+CH_3$ -46	$v = 5$	$v = 2$ $T_v > 2000\ ^\circ K$	Flash photolysis of O_3, absorption	36, 38
$H+O\text{-}HO \rightarrow OH^v+OH$ -38	$v = 4$	$v = 3$ $T_v \sim 2000\ ^\circ K$. HO_2^*, H_2O^v also observed	Flow mixing of H, O_2, emission. 0.3 torr	301–303
$H+O\text{-}O_2 \rightarrow OH^v+O_2$ -78 $\rightarrow OH+O_2^e$	$v = 9$	$v = 9(10)$ $T_v \sim 9250\ ^\circ K$, $p > 0.1$ torr, $O_2(^1\Sigma_g^+)$, $O_2(^1\Delta)$, scant evidence, OH^e observed, secondary reaction, $k = 1.5\times10^{10}$ Probable maximum at $v = 8, 9$, $p < 0.001$ torr	Flow mixing, emission. 3×10^{-4}–5 torr	304–313, 326
$H+O\text{-}NO \rightarrow OH+NO$ -29	$v = 3$	$v = 3$ No excitation of OH. H_2O^v from secondary reaction. $k = 3\times10^{10}$	Flow mixing, emission. 2–4 torr	314–316

(continued on p. 122)

TABLE 1 (*continued*)

Reaction	ΔH, kcal.mole⁻¹	Highest vibrational level possible[a]	Highest vibrational level observed	Distribution of excitation, remarks[b]	Experimental conditions[c]	References[d]
Molecular oxygen						
$O + O\text{-}ClO \rightarrow O_2^v + ClO$	-61	$v = 15$	$v = 8$	Flat maximum at $v = 5\text{–}7$	Flash photolysis of ClO_2, absorption	42
$O + O\text{-}NO \rightarrow O_2^v + NO^v$	-46	$v = 11$	$v = 13(O_2)$ $v = 2?(NO)$	Maximum at $v = 8$, $T_v(NO) \sim 2000$. Energy transfer may be important. $k = 2 \times 10^9$	Flash photolysis of NO_2, absorption; and flow mixing, emission	37, 38, 42, 58, 317, 318
$O(^1D) + O\text{-}O_2 \rightarrow O_2^v + O_2$	-138	$v \sim 35$	$v = 23(29?)$	Maximum at $v = 11\text{–}13$, rotation cold. $k > 2 \times 10^9$	Flash photolysis of O_3	36–38, 40, 40a, 40b
$O(^3P) + O\text{-}O_2 \rightarrow O_2^v + O_2$	-94	$v = 29$	$v = 17?$	Maximum $v = 13?$, rotation cold. $k = 3 \times 10^8$	Flash photolysis and thermal decomposition of O_3, absorption	41, 319
Alkali metal halide[e]						
$M + X_2 \rightarrow MX^v + X$				$M = Na, K, Rb, Cs$ $X_2 = Br_2, I_2, ICl, IBr$	Molecular beam	2
$K + X_3 \rightarrow KX^v + X$	< -40			$X_3 = Cl_2, Br_2, I_2, CNCl,$ $CNBr.$ D-line emission (37 kcal.) Continuum	Flow mixing	50
$X + Na_2 \rightarrow NaX^v + Na$ $\rightarrow NaX + Na(^2P)$	< -50			$X = Cl, Br, I.$ D-line emission	Flow mixing of Na, X_2	50

Reaction			Method	Ref.
→ KX + K(5²P)		K(5²P) emission (71 kcal.)		
M + XH → MX + H	≧ −5	M = Na, K X = Cl, Br, I. D-line emission (H+H+M?) (H+MH → ?)	Molecular beam and flow mixing	2, 50
Na + HgCl → NaCl^v + Hg	−65	D-line radiation, similar results with COCl₂, PCl₃	Flow mixing of Na, HgX₂, or molecular beam	2, 50
Na + MX$_n$ → NaX^v + MX$_{n-1}$	> −48	MX$_n$ = GeCl₄, TiCl₄, SCl₂, S₂Cl₂, POCl₃, CrO₂Cl₂. Continuum, secondary reactions involving MX$_{n-1}$	Flow mixing or molecular beam	2, 50
M + RX$_n$ → MX^v + RX$_{n-1}$		RX$_n$ = organic polyhalides. D-line emission from excitation reactions involving RX$_{n-1}$ rearrangement. Continuum (electronic excitation?)	Molecular beam and flow mixing	2, 50, 51, 327
NaO + Na₂ → Na₂O^v + Na		D-line emission	Flow mixing of Na, N₂O	50
Na + NO₂ → NaO^v + NO		D-line emission	Flow mixing and molecular beam	2, 50
M + RbCl → MCl^v + Rb		M = K, Cs	Molecular beam	8

(continued on p. 124)

TABLE 1 (*continued*)

Reaction	ΔH, kcal.mole^{-1}	Highest vibrational level possible[a]	Highest vibrational level observed	Distribution of excitation, remarks[b]	Experimental conditions[c]	References[d]
Miscellaneous						
$OH+H-O \rightarrow H_2O^v+O$	-17		1 kcal?	Possible secondary reaction	Flow mixing of H, NO$_2$, emission	309, 313, 315, 316
$Br+O_3 \rightarrow BrO^v+O_2$	-19	$v \sim 10$	$v=4$		Flash photolysis of Br$_2$, absorption	35, 319, 320
$Cl+O_3 \rightarrow ClO^v+O_2$	-40		$v=5$	F+O$_3$ also studied, no excitation	Flash photolysis of Cl$_2$, absorption	35, 319, 320
$Cl+ClNO \rightarrow Cl_2+NO$	-20			Excitation of NO sought, not found $k=1\times10^9$	Flash photolysis of NOCl	39, 57
$N+NO \rightarrow N_2^v+O$	-75	$v=12$	$v=1$	Also thermal measurement of N$_2$, excitation > 25 kcal.mole^{-1}. $k=2\times10^9$	Flow mixing at 1 torr, electric discharge through N$_2$	321–324
$N+NO_2 \rightarrow N_2O^v+O$	-42			ν_1, ν_3 of N$_2$O excited, maximum at 17 kcal.mole^{-1}. $k=1\times10^{10}$	Flow mixing, emission. 0.06 torr	44, 325

Reaction	$-\Delta H$			Remarks		Ref.
$N+O-NO \rightarrow NO''+NO$	-78	$v \simeq 16$	$v = 2$	Weak emission, minor reaction. $k = 1 \times 10^{10}$	Flow mixing, emission. 0.06 torr	325
$O+S-CS \rightarrow SO''+CS''$	-30	$v = 9(SO)$ $v = 8(CS)$	$v = 4(SO)$ $T_v(SO) \sim 2870\ ^\circ K.$ $v = 3(CS)$ $T_v(CS) \sim 1775\ ^\circ K.$ Secondary reactions important for SO_2 excitation. $k = 5 \times 10^8$	Flash photolysis of CS_2, NO_2 mixtures, absorption	19
More complex cases						
$CS+SO \rightarrow CO''+S_2$	-53	$v = 10(CO)$	$v = 14$	Stimulated emission reaction is one possibility; an exchange reaction?	Flash photolysis of CS_2, O_2 mixtures, emission	61
$CN+O_2 \rightarrow CO+NO''$	~ -110	$v = 1$	$v = 1(NO)$	Minor reaction in system, NCO and O major products; an exchange reaction?	Flash photolysis of C_2N_2, O_2 mixtures, absorption	43
$S+S_2Cl_2 \rightarrow S_2''+SCl_2$	-36	$v \sim 18$	$v = 12$	Possible source of excitation may be a rearrangement	Flash photolysis of S_2Cl_2, absorption	62

[a] This is the vibrational level lying next below the energy indicated by the ΔH of reaction. Any activation energy could increase the excitation.

[b] T_v, T_r indicate Boltzmann distributions with shape parameters corresponding to T. These are very sensitive to experimental conditions. Rate coefficients (k) in units of l. mole^{-1}.sec^{-1} at 300 °K.

[c] Where the reaction is a step in a sequence, the reactants are indicated here. The statements here do not correspond, line for line, to the remarks columns.

[d] The references for the table are included in the general list. Some do not report observations of excitation, but are important for interpretation of the experimental system.

[e] References for this section are mainly to review papers. The examples given here do not show all the alkali metal atom reaction studies. Luminescence (D-line or continuum) is produced by steps following those tabulated. Radiation from MX has not been identified in these studies. Molecular beam studies show internal excitation (vibration plus rotation).

(refs. [35-40, 40a, 40b]) and has provided important information on vibrational chemi-excitation. At the same time it illustrates the complexities and large number of reactions that must be considered in evaluating such experiments.

The reactions of interest are[†]

$$O(^3P) + O_3 \rightarrow O_2^v + O_2 \quad \Delta H = -93 \text{ kcal.mole}^{-1}$$

$$O(^1D) + O_3 \rightarrow O_2^v + O_2 \quad \Delta H = -138$$

$$O(^1D) + HR \rightarrow OH^v + R \quad \Delta H < 0$$

which can follow the production of oxygen atoms in the photolyses

$$O_3 + h\nu \rightarrow O(^3P) + O_2$$

$$O_3 + h\nu \rightarrow O(^1D) + O_2(^1\Delta)$$

Excited ground state oxygen molecules, $X^3\Sigma_g^-$, have been observed in levels up to $v = 23$ and possibly up to $v = 29$. The highest level, $v = 29$, could on energetic grounds be excited by the less exothermic reaction of $O(^3P)$. The spectra show that the maximum in the population distribution of excited molecules is in $v = 12$, 13 or 14 and that both higher and lower levels have progressively lower populations.

Photons of sufficient energy are present in the ultraviolet flash to produce $O(^1D)$. Indeed, these should be the principal product from ozone photolyzed by wavelengths shorter than 2500 A. Their presence has been deduced from competition experiments in which ozone is photolyzed in the presence of molecules containing hydrogen atoms (H_2, H_2O, NH_3, CH_4, HCl). In these cases vibrationally excited OH is formed[36,38]. This is energetically possible for reactions of $O(^1D)$ but not for $O(^3P)$, as for example

$$O(^3P) + H_2O \rightarrow OH + OH \quad \Delta H = +16 \text{ kcal.mole}^{-1}$$

$$O(^1D) + H_2O \rightarrow OH^v + OH \quad \Delta H = -29$$

However, ground state atoms, $O(^3P)$, cannot be ignored. That they can produce vibrationally excited O_2 is proved by shock tube studies of the rate of thermal decomposition of ozone[41]. This work, at reaction temperatures up to 900 °K, included absorption spectrography of the decomposing gas. The excited O_2^v were found with an intensity distribution (for the levels observed) similar to that obtained in photolytic experiments. Production of $O(^1D)$ in the thermal decom-

---

[†] The conventions used in equations are: vibrational excitation is indicated by superscript v, electronic excitation by superscript e, specific states or levels of excitation in parentheses after the molecule, and ΔH for the thermal reaction with no consideration of excitation. Molecules are often written with the group transferred from the second reactant nearest to the receiving molecule.

position is unlikely. In addition, the flash photolysis of NO_2 has been shown to produce O_2^v when the photons needed to produce $O(^1D)$ had been filtered out[42, 43]. Diagnostic experiments for the presence of $O(^1D)$ have not been reported for this system.

2.1.2 Atom–molecule reactions studied in flow systems: the hydrogen halide system

The principal reaction discussed above forms oxygen molecules in high vibrational levels of the ground state. This is chemi-excitation but is not chemilumines-cence: vibration–rotation transitions of homonuclear molecules are forbidden. For such cases electronic absorption spectroscopy is the required technique. For reactions in which a heteronuclear diatomic (or a polyatomic) molecule is excited these transitions are allowed. They are overtones of the molecular transitions that occur in the near infrared. These excited products emit spontaneously. The reactions are chemiluminescent, their emission spectra may be obtained and analyzed in order to deduce the detailed course of the reaction.

A crucial step in both emission and absorption studies is the conversion of intensities of bands or lines to concentrations of molecules. This is not easy. The transition probabilities for bands connecting excited states are not often known to any reasonable precision. Also, one further difficulty plagues the emission studies. In contrast to absorption studies the population of molecules in the lowest state, $v = 0$, is not measurable. Relative populations are usually reported. Self absorption measurements can be used to overcome this difficulty[43a].

The favorite technique for studying chemiluminescent reactions has been to mix the reagents in a low pressure flow system and then to use the resulting luminous zone as a spectral light source. This technique has been exploited for combination reactions, for group transfer reactions and, quite generally, for the excitation of molecular spectra.

Production of the reactive atom or molecular fragment has been achieved by several techniques. The "cleanest" method is decomposition of a molecule (H_2, O_3) on a heated surface. A second method is to use a very fast chemical reaction. A favorite one for oxygen atoms is

$$N + NO \rightarrow N_2 + O$$

which, near its stoichiometric point, provides a good source of ground state atoms. The chemiluminescence of this reaction makes it easy to locate this point. When N atoms are in excess all the NO is consumed, and a blue luminescence results from

$$O + N + M \rightarrow NO^e + M$$

$$NO^e \rightarrow NO + hv$$

When NO is in excess, a green–white emission due to

$$O + NO + M \rightarrow NO_2^e + M$$

$$NO_2^e \rightarrow NO_2 + h\nu$$

occurs. The reaction is non-luminous in between. Although a specific source, this reaction suffers from the disadvantage that the N_2 is vibrationally excited, providing a potential second reactant.

Both the thermal decomposition and the reaction methods provide usable but low yields. Somewhat higher yields of atoms are obtained by passing a gas mixture containing the decomposable molecule through an electric discharge. This is a more widely applicable method, but it is a dirty, shotgun approach. The discharge, which may be continuous or pulsed, d.c., low frequency a.c. or microwave, acts by a combination of electron–molecule collisions leading to dissociation and ionization and later ion neutralization steps. It is foolhardy to assume that the desired species is the only highly reactive one present in the resultant mixture. Nevertheless this is often the technique of choice. Attempts are made to eliminate ions by trapping them on charged grids or by providing a long path between the discharge and the reaction zone. Excited molecules in active nitrogen have been deactivated by passing the mixture through glass wool (to provide a high area surface for energy exchange)[44]. Conversely, when excited molecules have been wanted in the absence of atoms, the latter have been removed (from oxygen) by reaction with mercury deposits[45]. The efficacy of these techniques is unknown. The art of "cleaning up" the products of an electric discharge verges on witchcraft. A more satisfactory procedure, one that has occasionally been used, is to check the discharge results using a more specific method for producing the desired reactive species.

The most important systems studied by the flow method are $H + X_2$, $H + HX$ and $X + HX$ where X represents a halogen atom. These reactions are of prime interest because they are three-atom systems. Thus, there is hope that they may be treated successfully in terms of potential energy surfaces. They have been studied in considerable detail, both experimentally and theoretically, by Polanyi and his coworkers over the past decade. The studies have been aimed at determining the initial distribution of product molecules in the available vibrational and rotational levels, at finding the amount of reaction energy that goes into internal excitation and at establishing a satisfactory theoretical model.

The recent work is the most important. Vibrational and rotational energy exchange among the products and with the rest of the mixture have been minimized by running the reactions at lower pressures (10^{-2} to 10^{-4} torr) than is normal and by cooling the reaction vessel to liquid nitrogen temperatures. Gaseous collisions are thus reduced and wall collisions are used to remove excited molecules. Where required, quantitative corrections, based on experiment, have been made for the remaining energy transfer processes. As a result the observed excitation can be made to yield a good approximation to that produced by the reaction[4, 46]. The cases of interest are

$$H + Cl-Cl \rightarrow HCl^v + Cl \quad \Delta H = -45 \text{ kcal.mole}^{-1}$$

and

$$Cl + H-I \rightarrow HCl^v + I \quad \Delta H = -34 \text{ kcal.mole}^{-1}$$

These exhibit population inversion at low pressures ($H + Cl_2$ at 4×10^{-4}–2×10^{-2} torr; $Cl + HI$ at 1.5×10^{-4} torr). A population distribution plot, $N(v)$ versus v, shows a rise in N up to $v = 3$ and then a decrease. The highest accessible vibrational levels in each case have a small but observable population at these low pressures. For both reactions strongly non-Boltzmann rotational population distributions have been found. Two maxima were found for HCl ($v = 2$), one probably due to relaxed molecules, the other reflecting the initial distribution. Stimulated emission has been found for both reactions, that for $Cl + HI$ being strong enough to provide laser action. These are spectacular demonstrations of excitation by chemical reaction. Other halogens which have been reacted with H atoms and other halogen atom–hydrogen halide combinations show qualitatively similar results.

Estimates have been made for the fraction of the energy of reaction that goes into internal excitation. The very low pressure work has revised earlier estimates upward. Vibrational excitation accounts for roughly 50 % of the energy. For $Cl + HI$ most of the remaining energy goes into rotation, thus, by implication, leaving the products with translational energy comparable to that of the reactants.

2.1.3 Comparison of molecular oxygen and hydrogen halide excitation

The vibrational population distributions of excited molecular oxygen and of hydrogen chloride that were described in (2.1.1) and (2.1.2) are similar. They peak at vibrational levels well below the maximum accessible level and near half the excitation energy. Rotational excitation is observed for HCl but not for O_2. The vibrational distribution for HCl is probably very close to that initially produced by reaction. An argument based on the similarity of the HCl and O_2 distributions and on Smith's kinematic analysis[13] would suggest that the O_2 distribution also reflects the initial condition even though the experimental conditions are quite different. However, the evidence is poor. The factors that can modify the initial distribution of energy in the excited product are discussed below.

First, collisional relaxation of rotation is fast. One to ten collisions are sufficient for rotation–translation energy transfer. At 1 torr at least one collision per microsecond will occur for both O_2 and HCl. In contrast, rotational relaxation by radiation, when allowed, is very slow, of the order of 10^2 seconds. The absence of observed rotational excitation in the higher pressure photolysis experiments is understandable: sampling occurs after relaxation. Early studies on the hydrogen halides at pressures of 1 torr and above also showed rotational equilibration.

Vibration–translation energy transfer in collisions is less efficient than for rotation–translation. It is sharply different for the two cases. For oxygen at room temperature about 8×10^5 collisions (with oxygen) are necessary[47]. Hydrogen chloride may require only 500 to 1500 collisions[4,47]. For oxygen there is no allowed radiative relaxation mechanism. For HCl the radiative relaxation time should be in the range 10^{-4} to 10^{-2} seconds. Were these the only effects, a distribution reflecting that produced by reaction could be observed at higher pressures for O_2 than for HCl. However, vibration–vibration energy exchange can be very important, and can modify a population drastically. It is more efficient than vibration–translation energy exchange.

The importance of vibration–vibration exchange is now apparent in the results of the early experiments on HCl and on hydroxyl (from $H + O_3 \rightarrow OH^v + O_2$) at pressures above 0.1 torr. These showed vibrational distributions that were nearly Boltzmann, but with shape parameters corresponding to "temperatures" of several thousand degrees, far greater than the translational temperatures. These distributions were samples taken after rotational relaxation and *after* the efficient vibration–vibration energy exchange

$$HCl(v = i) + HCl(v = j) \rightarrow HCl(v = k) + HCl(v = l)$$

but during the slower vibration–translational energy transfer process[18,48].

A similar exchange has been shown to occur rapidly in the ozone photolysis system[40a,40b], *viz.*

$$O_2^v + O_3 \rightarrow O_2^{v'} + O_3$$

Because of this, the O_2^v population distributions should be considered as the result of (at least) two reactions of similar speeds

$$O(^1D) + O_3 \rightarrow O_2^v + O_2$$
$$O_2^v + O_3 \rightleftarrows O_2^{v'} + O_3$$

Since the observed distributions change only slowly with time, the experimental problem of separating the two effects is difficult, and at present unsolved.

2.1.4 Excitation of alkali metal salts

The reactions of sodium and potassium atoms with halogens, nitrogen oxides and alkyl halides are the classic examples of chemi-excitation. They stand out in discussions of the subject[49], have been reviewed periodically[50-52] and continue to be studied[2,53]. Research on these reactions began before 1930. Chemiluminescence has been examined since that time using the low pressure diffusion technique developed by M. Polanyi[54]. In recent years these reactions have been the staple diet for crossed molecular beam scattering experiments.

The reactions are fast. Activation energies greater than 5 kcal.mole^{-1} are unusual (and not well established). Observed chemiluminescence is often that of the alkali metal atom (D-line radiation), although for some cases (Na reactions with inorganic polyhalides), diffuse radiation from a molecular product has been observed.

The mechanism that explains the chemiluminescence is, for the traditional example

$$Na + Cl_2 \rightarrow NaCl^v + Cl$$

$$Cl + Na_2 \rightarrow NaCl^{v'} + Na$$

$$Cl + Na_2 \rightarrow NaCl + Na(^2P)$$

$$NaCl^{v'} + Na \rightarrow NaCl + Na(^2P)$$

$$Na(^2P) \rightarrow Na(^2S) + h\nu$$

For sodium (but not for potassium) the halide produced in the initial reaction has insufficient energy to excite the metal atom by energy transfer. The reaction of Cl with the dimer releases enough energy to provide this excitation either directly or *via* a subsequent energy transfer. The temperature dependence of the luminescence can be related to the equilibrium concentration of dimers. The results of quenching experiments give evidence of the importance of the energy transfer step.

The sodium D-line radiation dominates the system because the NaCl^v is long-lived, the vibrational–electronic energy transfer is efficient and the excited atom radiates in 10^{-8} seconds. The multistep process bleeds off the excitation energy. This behavior probably is common in systems containing atoms with low-lying energetically accessible electronic states[29,55].

Crossed molecular beams have been used to study nearly as wide a range of alkali metal atom reactions as has been examined by diffusion flames. An excellent review has been provided by Herschbach[2]. The multi-step mechanism displayed for chemiluminescence studies does not apply to the scattering experiments. Only the initial bimolecular reaction is important at the low pressures used.

This method provides the most direct, detailed information about reactive collisions of any yet devised. Two beams, each with one reactant, are arranged to intersect at a large angle in a small region within an evacuated chamber. The scattering of the reactants and products is observed, that is, their intensities at various angles relative to the initial directions of the beams. The reactant beams may have thermal velocity distributions or may include only a small range of velocities. Total scattered intensities (at a particular angle), or separate intensities for individual species may be observed. The velocity distribution for scattered molecules may be determined. What may be done in a particular case is strongly dependent upon available experimental techniques, in particular upon methods for detection (see Chapter 2, Volume 1).

Excitation of the internal degrees of freedom of a molecule is not observed directly, in contrast to spectroscopic studies. Instead, excitation is deduced from the analysis of the scattering pattern, by comparison with that which occurs in the absence of reaction and with consideration of the possible energy release during reaction. Although indirect, the excitation evidence can be made unambiguous by measurement of the product translational energies. Thresholds for reaction (analogous to activation energies) and energy dependent cross sections are obtained. The work has led to the development and refinement of the kinematic models discussed in Section 1.2.1, and is becoming important for the development of realistic potential energy surfaces.

2.1.5 Distribution of excitation between the reaction products

In the study of chemiluminescence in elementary transfer reactions the behavior of the new-bond molecule has been emphasized over the past decade. This is understandable. The initial successful experiments showed new-bond excitation. Also, the work was conditioned by the observation of a highly non-thermal emission from OH in the upper atmosphere[59] and the demonstration[60] that a likely cause was the reaction $H + O_3 \rightarrow OH^v + O_2$. Indeed, early work aimed at a test of the possibility that the principal excitation channel was *via* the highest accessible level. This hypothesis has not been proven. The available facts contradict it. Many levels of the new-bond molecule are populated by reaction. Ample evidence for this is provided by the reactions listed in Table 1.

Unfortunately, most of these experiments do not define the energy distribution between the products. In many cases it was possible to examine only one of the products. The problem has been to devise experiments in which it is possible to observe the spectra of both products simultaneously, preferably with the same technique. The present state of our knowledge is that there is ample evidence for excitation of the new-bond molecule up to levels corresponding to the energy released by reaction, and there are data, fewer in number but no less certain, showing some excitation in the old-bond product. These are considered below.

Ozone was flash photolyzed in the presence of ^{18}O enriched water and added argon $(O_3 : H_2O : Ar = 1 : 1 : 8)$. Vibrationally excited OH was observed[56].

This is produced by the reaction

$$^{16}O(^1D) + H^{18}OH \rightarrow H^{16}O^v + H^{18}O \quad \Delta H = -29 \text{ kcal.mole}^{-1}$$

The results show $^{16}OH(v = 2, 1, 0)$ and $^{18}OH(v = 1, 0)$ with the ratio $^{18}OH(v = 1)/^{16}OH(v = 1) \sim 0.01$. This ratio is larger than can be attributed to ^{18}O atoms formed in the photolysis due to the natural abundance (0.002) of ^{18}O, but still is very low compared to the ^{16}OH excitation. Rotational excitation was not observed.

When a mixture of chlorine and NOCl (in N_2) was flash photolyzed through a Pyrex filter (to suppress NOCl photolysis) appreciable NOCl decomposition occurred[57]. The nitric oxide was observed in absorption but with virtually no excitation. The reaction is

$$Cl + Cl-NO \rightarrow Cl_2 + NO \quad \Delta H \sim -20 \text{ kcal.mole}^{-1}$$

Excitation of the Cl_2 was not studied. The negative results for NO could be due to insufficient sensitivity of the method, but they do indicate that the fraction of reaction energy going into NO is slight. The reaction

$$H + ClNO \rightarrow HCl + NO$$

has also been examined for excitation of NO with similar results [298,299]. The occurrence of old-bond excitation in these two reactions should be considered doubtful.

Excitation of both products of the reaction

$$O(^3P) + S-CS \rightarrow O-S^v + CS^v \quad \Delta H \sim -23 \text{ kcal.mole}^{-1}$$

has been shown by an absorption spectrographic study of the photolysis of NO_2 in the presence of (much larger) quantities of carbon disulfide[19]. Population distributions have been determined for both products. These are approximately Boltzmann, with a higher vibrational "temperature" (flatter distribution) for SO. The highest levels observed for each product correspond to about the same energy (approximately half the energy of reaction). The fraction of the available energy going into all levels of SO was 18 %, and 8 % for CS. Rapid removal of SO by various reactions complicates the analysis of the system. Excitation of higher levels than those observed cannot be ruled out.

Old-bond excitation may have been observed in infrared region emission studies of the reaction of O atoms with NO_2[58]. Interference spectrometry techniques have been used to examine the infrared emission that occurs when O (presumably ^3P) is mixed with NO or NO_2, the reactions being

$$O + NO + M \rightarrow NO_2^e + M$$

$$O + NO_2 \rightarrow O_2^v + NO^v$$

In these two systems unresolved emission corresponding to the $\Delta v = 1$ sequence of NO was observed. Based on its variation with experimental conditions it was assigned to the second reaction, above, rather than to an energy transfer process. This work has not been correlated with the flash photolysis experiments, examination of the O_2 excitation not being possible with the technique used.

Clearly, our present knowledge of the distribution of energy among the products of transfer reactions is fragmentary. The work summarized in this section suggests strongly that much greater emphasis on the study of the old-bond molecule is desirable.

2.2 FOUR-CENTER EXCHANGE REACTIONS

The general form for this type of elementary step is

$$A–B + C–D \rightarrow A–C + B–D$$

There are no certain cases of this type for which excitation has been observed. The best evidence comes from the reaction

$$CN + O_2 \rightarrow CO + NO^v \quad \Delta H < -100 \text{ kcal.mole}^{-1}$$

which has been postulated to occur following the flash photolysis of $(CN)_2$ and CNBr in the presence of oxygen[43]. It is a minor reaction, the principal oxygen consuming step ($> 85 \%$) being

$$CN + O_2 \rightarrow NCO + O$$

There is a possibility that the CO transitions observed in laser activity in $CS_2–O_2$ mixtures[61] could be formed in an analogous reaction.

Excited diatomic sulfur is observed in the flash photolysis of S_2Cl_2 and has been attributed[62] to

$$S + S_2Cl_2 \rightarrow S_2^v + SCl_2$$

If the structure of S_2Cl_2 is S–SCl$_2$ (analogous to thionyl chloride) this is an atom transfer reaction. To date only the form Cl–S–S–Cl (hydrogen peroxide structure) has been found[63]. Both forms are known for S_2F_2[64]. Unless the flash photolysis work stimulates discovery of the unsymmetrical form of S_2Cl_2, this reaction must be classed either as a rearrangement or be considered suspect.

2.3 COMBINATION REACTIONS PRODUCING VIBRATIONAL EXCITATION

The general form is

$$A+B \rightleftarrows AB^v$$

$$AB^v+M \rightarrow AB+M'$$

which is the reverse of the Lindemann mechanism for unimolecular reactions. It is explained in the discussion of combination reactions that produce electronic excitation that successful combination of atoms in the absence of a third body (M) is improbable because of rapid redissociation of the diatomic collision complex. The same is true for more complex molecules, but the time-scale is longer. As the complexity of the molecule increases so does its lifetime, that is, it may survive many oscillations before decomposing. Thus the two step mechanism is the appropriate one.

The situation with regard to combination reactions can be summarized briefly. They should produce excited products. The formation of a chemical bond is invariably exothermic. It would come as a great surprise to the kineticist to find a combination reaction that did not show excitation.

Nevertheless, spectroscopic evidence for this state of affairs is very limited. One example is the H+HX system[65]. Chemiluminescence of the HX is found. For the diagnostic case with D atoms, the reaction

$$D+HCl \rightarrow HD+Cl$$

is followed by

$$D+Cl+M \rightarrow DCl(v \leqq 7)+M$$

$$2D+HCl \rightarrow D_2+HCl(v = 1)$$

These not only show the excitation of the product, but the transfer of energy to the third body. Additional evidence is provided by the combination[66-70]

$$H+NO+M \rightarrow HNO^v+M$$

and possibly by the reaction of NO with fluorine[71].

By far, most of the evidence for excitation upon combination comes from an analysis of the kinetics of the reaction. The rate depends upon the pressure of the "third body" — that is, the newly formed molecule decomposes unless it is deactivated. In theory this deactivation need not remove a large amount of energy on each collision — merely enough to put the new molecule below its dissociation energy. From that point on, the higher efficiency for deactivation versus reactivation assures stable product formation. Of course, where M is the product molecule, vibration–vibration energy transfer is likely to speed up the process of energy redistribution, as mentioned for the group-transfer reaction systems.

Combination and addition reactions have been used effectively for the study of excited species. In effect, chemi-excitation reactions have been used for synthesis of reagents of known excitation energy[1,72-81]. A major effort has been made to use such excited molecules as tools for the exploration of the details of unimolecular decomposition reactions (see Rabinovitch and Setser[82]).

2.3.1 Addition of hydrogen atoms to alkenes

There have been many studies of reactions of the type[72-81]

$$H + >C=C< \rightarrow >CH-\overset{\cdot}{C}<$$

Our example is the series of reactions of H and D atoms with *cis*- and *trans*-butene-2 and with butene-1. They provide a set of *sec*-butyl radicals, $CH_3CH_2\overset{\cdot}{C}HCH_3$, with minimum excitation ranging[83] from 39 to 43 kcal.mole^{-1}. Relative to their decomposition into methyl radicals and propylene, these radicals have average excess energies ranging from 9 to 13 kcal.mole^{-1} at room temperature and about 1 kcal.mole^{-1} less at -78 °C. "Excess energy" toward reaction means energy above the thermochemical requirement for the not-yet-studied thermal reaction. Rates of decomposition (to radical and propylene) for this graded set of butyl radicals increase with excess energy. A good correspondence between experiment and the Rice–Ramsperger–Kassel–Marcus theory (see Chapter 3, Volume 2) has been established. Energy transfer to inert molecules (stabilization) has been studied; the "quanta" transferred are large, ranging from 1 to 9 kcal.mole^{-1} per collision.

The emphasis in this work has been on the behavior of the excited alkyl radical. This may change. New techniques for precision measurement of the rates of H- and D-atom reactions have been reported[84,85]. Taken together with the knowledge gained about the subsequent reactions of the alkyl radical, a more detailed examination of the excitation process may be possible and fruitful.

2.3.2 Addition and insertion reactions of methylene

The carbenes are also effective excitation reagents in addition reactions. Methylene, $H_2C:$, is the parent compound in this series. It is also the most reactive. Its ground state is a triplet[86], the reactions of which are principally hydrogen abstractions and non-stereospecific additions to carbon–carbon double bonds.

Singlet methylene is an electronically excited molecule the reactions of which are of interest here. These are insertion into a carbon–hydrogen bond and stereospecific addition across a carbon–carbon double bond, *i.e.* production of a cyclopropane with retention of the original *cis* or *trans* substituent configuration. Methylene radicals, particularly those produced by photolysis, may have internal excitation. Very often both singlet and triplet radicals are produced in the same system. There are extensive reviews of the field[87–89]. The products formed by singlet methylene addition to alkenes have appreciably more excitation than do the alkyl radicals produced by H-atom addition. Again, a reagent which is excited to a fairly narrow band of states high above the ground level, is synthesized by addition[90–97]. It may isomerize, split or be stabilized by collision. The rates of these secondary reactions can be measured and subjected to theoretical analysis.

Similarly, the reactions of methylene with saturated compounds and hydrogen require postulation of excited products to explain the kinetics[98–100]. The production of excited methyl cyclopropane

$$CH_2 + CH_3CH{=}CH_2 \rightarrow \underset{\displaystyle\bigvee}{CH_3{-}CH{-}CH_2}{}^*$$
$$\underset{CH_2}{}$$

$$CH_2 + \underset{\displaystyle\bigvee}{CH_2{-}CH_2} \rightarrow \underset{\displaystyle\bigvee}{CH_3{-}CH{-}CH_2^*}$$
$$\underset{CH_2}{} \qquad \underset{CH_2}{}$$

is one example of the effective use of methylene[101, 102]. The excited molecules differ in excitation by 7.9 kcal.mole^{-1}, the difference in the enthalpy of formation of cyclopropane and propylene. Both isomerize to a mixture of the butenes. The lifetimes of the hot cyclopropanes depend upon their excitation energy. When formed using diazomethane as the methylene source, the molecules have lifetimes of 8.3 and 2.3×10^{-10} seconds respectively. Those produced in the thermal isomerizations have an average lifetime of 10^{-7} seconds. This energy dependence has been explained satisfactorily by the Rice–Ramsperger–Kassel–Marcus theory for this and for similar isomerizations of cyclic compounds[88, 92]. Similar comparative studies have been made for other cyclic compounds.

Although highly excited and energetically able to decompose, the molecules formed either by H-atom or by H_2C: addition survive many vibrations. The energy of reaction is distributed over the entire molecule prior to isomerization or bond rupture. This has been deduced from analysis of the proportions of the various products that are formed (when more than one path is available). The proportions change very little, if at all, with excitation energy. Thus, for complex excited molecules, their history, *i.e.* mode of formation, becomes irrelevant in the analysis of subsequent reactions.

2.3.3 *Combination of free radicals*

An excited product is to be expected from radical–radical combination. The ethane produced in the recombination of methyl radicals must contain the net of the energy released by the carbon–carbon bond formation and the change in configurations of the methyl group. Unless stabilized, this hot molecule will revert to the reactants, *viz.*

$$CH_3 + CH_3 \rightleftarrows C_2H_6^v$$

$$C_2H_6^v + M \rightarrow C_2H_6 + M$$

This is a case for which only stabilization is of interest. It has not been treated explicitly as an excitation reaction. Its rate has been measured directly[103-105] at low temperatures (300–450 °C) as have those for several other radical combinations. Without doubt, the combination of methyl radicals is the most important reaction of this type for chemical kinetics. It serves as a reference standard for the measurement of rates of many hydrogen abstraction reactions, but very little is known about the temperature dependence of its rate coefficient.

Radical combinations become important for chemi-excitation only when the product can undergo reactions other than stabilization and reformation of the reactants. The alternative reactions require that weaker bonds be present in the new molecule than in the one formed in the combination. The alkyl halides are useful in this respect. Elimination of HX is a competitive secondary reaction[106-110]. Our final example of the use of chemi-excitation is drawn from this field[111]. It shows the application of three techniques: group transfer reactions, radical combination, and methylene insertion

$$C_2H_5 + F_2 \rightarrow C_2H_5F^v + F$$

$$CH_3 + CH_2F \rightarrow C_2H_5F^v$$

$$CH_2 + CH_3F \rightarrow C_2H_5F^v$$

producing, in the order shown, a series of fluorides with about 70, 90 and 115 kcal.mole^{-1} excess energy. A selection of mono- and poly-fluorinated alkanes in the C_2–C_4 series has been studied (and some mixed F, Cl halides). These molecules eliminate HX unless stabilized by collision, *viz.*

$$C_2H_5F^v \rightarrow C_2H_4 + HF$$

$$C_2H_5F^v + M \rightarrow C_2H_5F + M$$

Rates have been determined and analyzed in terms of the type of C–X bond broken, and have been treated by the Rice–Ramsperger–Kassel theory with reasonable success[111].

2.4 EXCITATION IN DECOMPOSITION REACTIONS

Decomposition reactions, $A \to B + C$, provide two extreme cases. The unimolecular decomposition that involves rupture of a single bond, $A-B \to A + B$, usually has an activation energy almost equal to the bond dissociation energy. Excitation is absent, although either A or B may be unstable relative to other products and may isomerize in elementary steps with little or no activation energy. Decompositions which involve considerable bond rearrangement (bond shortening is the simplest example) may produce excited molecules.

Electronic excitation is found in the thermal decomposition of diazomethane

$$CH_2NN \to CH_2^e + N_2$$

The methylene is produced in the singlet state, the one characterized by insertion reactions[112]. It is this state, not the ground triplet state, that correlates with the reactant.

Singlet carbenes are also favored as intermediates in the decomposition of diazirines[113,114]. Apart from the decomposition of diazirine itself, $H_2C\begin{smallmatrix}N\\\|\\N\end{smallmatrix}$, the clearest evidence comes from work on methylethyldiazirine. Methylcyclopropane is a minor product, the formation of which is probably by an intramolecular C–H insertion (cyclization) in the carbene.

Vibrational excitation of triplet methylene appears to be possible. Photolysis of diazomethane in the presence of alkenes shows product distributions as a function of pressure that could indicate a change in the methylene singlet/triplet proportions in the mixture[97]. Internal conversion from singlet to triplet has been suggested[115]. These triplet methylene molecules should be excited.

We do not find evidence for (or fail to understand arguments about) the production of vibrationally excited singlet methylene or higher carbenes in thermal decompositions.

3. Electronic excitation

3.1 THE RADIATIVE RECOMBINATION OF ATOMS

It is clear both experimentally and theoretically that the radiative recombination of two atoms, without a third body, is an improbable process. Experimentally, such radiation is seldom observed in the laboratory. Theoretically, the process is improbable because the duration of a collision is a small fraction of the radiative lifetime of the molecule thus formed. The duration of the collision at a velocity of 5×10^4 cm.sec^{-1} will be of the order of 10^{-8} cm$/5 \times 10^4$ cm.sec$^{-1} = 2 \times 10^{-13}$ sec, although non-central collisions will have a somewhat greater duration. The radiative lifetime for an allowed transition in the visible or near ultraviolet is likely to be in the range 10^{-7}–10^{-8} sec, so the probability of radiative recombination will typically be of the order of 10^{-5} per collision. The probability may be much less if the radiative transition moment is a strongly decreasing function of internuclear distance. This is likely to be the case when the recombining atoms are in their ground states, or in metastable states. Calculation of the intensity distribution of the light emitted in 2-body recombination has been discussed by Akriche et al.[116] and more rigorously by Mies and Smith[117]. When the probability of two-body radiative recombination is low, three-body recombination will dominate at pressures higher than some transition value which is correspondingly low. These remarks apply to recombination on a single potential curve. If there is another curve to which the system may make a transition during the collision the probability of radiative recombination may be somewhat increased. Barth[118] has reviewed a number of three-body radiative recombination reactions. Palmer and Carabetta[119] have discussed radiative recombination in terms of transition state theory, which they consider to be particularly appropriate in this case. Palmer[120] has more recently given an equilibrium theory in which the radiative transition rate as a function of internuclear distance is derived from the absorption coefficient.

In the study of any radiative recombination process, one tries to answer a number of fairly well defined questions, mostly related to potential curves. From what electronic states is emission observed? With what atomic states do these molecular states correlate? Does the recombination take place on a single potential curve, or is a transition between two curves involved? Is a potential curve with a significant maximum involved? Is a third body necessary, either to stabilize the atom pair on a single curve, or to induce a transition to another curve? In the case of a transition between two electronic states, is there an approximate equilibrium? What is the vibrational and rotational distribution of newly formed molecules? What is the recombination rate coefficient as a function of temperature or cross section as a function of energy? In principle these questions can be answered either theoretically or experimentally. In fact, they have been answered experimentally in most cases, but the answers are seldom as certain or as numerous as one would wish. This becomes clear in the following discussion of particular cases.

3.1.1 Helium atom recombination

A simple and well studied process which illustrates the interaction of two atoms on a single potential curve is the emission of the 600 A bands of helium in d.c. discharges and pulsed afterglows; the processes being

$$He(2\,^1S) + He \rightarrow He_2(A^1\Sigma_u^+) \rightarrow 2He + h\nu \ \text{(600 A bands)} \tag{1}$$

The ground state of He_2 is of course repulsive, but the A state has a well depth of about 2.5 eV and a small maximum (0.03 eV) at about 3 times the equilibrium separation. According to the work of Mies and Smith[117], and Smith[121], the radiation is emitted from continuous states just above this maximum. The radiation is a structured continuum showing six maxima of decreasing intensity in the range 600–620 A, with still weaker maxima at longer wavelengths. Calculation of Franck–Condon factors, using a continuous upper state wave function having about 17 nodes above the potential well, gives good agreement with the observations[121,122] on the position and intensity of maxima in the continuum. The recombination is relatively favorable for radiative interaction because the short wavelength tends to give a high transition probability. Even so, the cross section is only 2×10^{-20} cm^2 at room temperature[123]. The transition moment is approximately independent of internuclear distance in the region of interest in spite of the fact that $He(2\,^1S)$ is metastable[123]. It is perhaps worth emphasizing that the helium recombination continuum is highly structured, though upper and lower states are both continuous.

Although molecular emission from helium afterglows is mainly due to ion–electron recombination[124,125], atom recombination processes in addition to that leading to the 600 A bands may be important in the positive column of a d.c. discharge. Among the processes which have been proposed to explain the strong molecular emission from helium discharges are[126,127]

$$He(2\,^3S) + He \xrightarrow{M} He_2(2s\,^3\Sigma_u^+)\text{(metastable)}$$

the triplet analog of Eqn. (1) and

$$He(2\,^3P) + He \xrightarrow{M} He_2(2p\pi\,^3\Pi_g) \tag{3}$$

The first of these has to go over a hump of more than 0.2 eV, while the second has no hump.

3.1.2 Halogen recombination

Recent work on these systems includes Cl_2, Br_2, and several of the interhalogen molecules. The results can be summarized in terms of the potential curves in Figs. 3 and 4. The excited electronic states of the halogens are described in Hund's case c coupling, so that spin is not a good quantum number, and reasonably strong

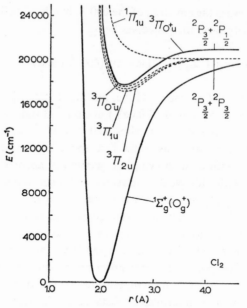

Fig. 3. Potential energy curves for Cl₂ (Bader and Ogryzlo[130]). The broken curves are extrapolated or estimated.

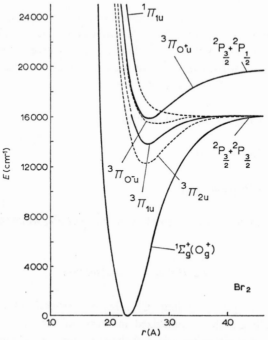

Fig. 4. Potential energy curves for Br₂ (Gibbs and Ogryzlo[138]). The broken curves are extrapolated or estimated.

radiative transitions between singlet and triplet states may be expected. Spin–orbit coupling is also large in the atoms, so that the multiplet splitting is relatively large, especially for the heavier halogens. In Cl, the excited $^2P_{\frac{1}{2}}$ state lies 881 cm^{-1} above the ground state and is populated to the extent of about 0.7 % at 300 °K. The splitting in Br is 3685 cm^{-1} and at room temperature the equilibrium fraction in the excited state is about 10^{-8}.

The states from which banded emission has been observed in the halogens are $^3\Pi_{0^+u}$, correlating with a ground state atom plus an excited $^2P_{\frac{1}{2}}$ atom (the lighter atom being excited in the interhalogens), and $^3\Pi_{1u}$, correlating with two ground state atoms. Emission from this latter state is therefore the result of recombination on a single potential curve. Emission from $^3\Pi_{0^+u}$ on the other hand must require the interaction of two curves, unless a relatively large number of halogen atoms are in the excited state. Continuous emission may arise from the $^1\Pi_{1u}$ state, but this will occur in the visible only for rather high kinetic energy of the atoms.

The most recent and thorough investigation of the radiative recombination of chlorine atoms is the work of Clyne and Stedman[128,129]. Cl_2 is partially dissociated in a 27 Mc discharge, and the resulting $Cl + Cl_2$ mixture can be mixed in a flow system with inert gases so as to control Cl, Cl_2, and inert gas concentrations independently. The observed emission is from the $^3\Pi_{0^+u}$ state, emission being observed from 5000–10,000 A corresponding to vibrational levels in the range $0 \leq v' \leq 14$. The lower levels are more populated. The temperature dependence of the emission (integrated over wavelength) corresponds to an activation energy of (-2.0 ± 0.5) kcal.mole^{-1}, essentially equal to that observed for the non-radiative recombination. This excludes the possibility that excited $Cl(^2P_{\frac{1}{2}})$, in equilibrium with $Cl(^2P_{\frac{3}{2}})$, is involved in the radiative recombination, since it is surely not involved in the non-radiative recombination. Since the $^3\Pi_{0^+u}$ state does not correlate with ground state atoms, one must assume it is populated by transfer from an intermediate state, probably $^1\Pi_{1u}$. Bader and Ogryzlo[130] suggest a spontaneous interaction, on the basis of an analysis of the absorption spectrum[131], but Clyne and Stedman[128,129] find that a third body is involved, Cl_2 being approximately seven times more effective than Ar. The excited molecules thus formed can be removed by (*i*) the reverse of the formation process, (*ii*) spontaneous radiation, (*iii*) electronic quenching, or (*iv*) vibrational deactivation in the excited electronic state. The kinetic analysis of Clyne and Stedman indicates that the only important electronic quenching step is

$$Cl + Cl_2(^3\Pi_{0^+u}) \rightarrow Cl_2 + Cl \tag{4}$$

which must be fast, proceeding at roughly one-tenth of the binary collision frequency. This is analogous to the fast reaction

$$N + N_2(^3\Sigma_u^+) \rightarrow N_2(^1\Sigma_g^+) + N \tag{5}$$

in nitrogen afterglows[132,133]. If the integrated emission intensity is expressed as

$$I \propto [Cl]^n \tag{6}$$

the effect of this reaction is to reduce n below the value of two[128,129] otherwise expected[130,134]. If an expression like (6) is written for each wavelength, n varies in the range $1 < n < 2$, approaching the limit $n = 2$ near the short wavelength limit, 5000 A.

Increase of third-body concentration (at constant [Cl]) shifts the emission to the red (emission from lower vibrational levels)[128–130] while increase in [Cl] at constant [M] shifts it to the blue (higher levels)[128,129]. This indicates vibrational relaxation by the third body. The blue shift with increasing [Cl] may be due to the fact that the shortened lifetime of $Cl_2(^3\Pi_{0+u})$, due to (4), allows less time for vibrational relaxation.

The radiative recombination of chlorine atoms in shock waves has also been studied (van Thiel et al.[135], Carabetta and Palmer[136]). This is an equilibrium system at temperatures in the neighborhood of 2000 °K, so there is no shortage of $Cl(^2P_{\frac{1}{2}})$. The temperature dependence of emission intensity at various wavelengths was measured and correlated with predictions based on potential curves[136]. These predictions assumed very simply that the observed continuous emission was due to vertical transitions between turning points for either the $^3\Pi_{0+u}-^1\Sigma_g^+$ or $^1\Pi_{1u}-^1\Sigma_g^+$ transitions. This simplest possible version of the Franck–Condon principle is justified by a great deal of use and a certain amount of success. It allows one to associate a given upper state energy with each emission wavelength. As applied to the emission from shock waves[136], the method indicated that most of the Cl_2 emission at wavelengths above 4700 A is due to the $^3\Pi_{0+u}$ state, while at shorter wavelengths, there may be a more important contribution from the $^1\Pi_{1u}$ state. It has recently been pointed out[129] that this work used an incorrect vibrational analysis, which would lead to an overestimate of the importance of the $^3\Pi_{0+u}$ state.

The potential curves for bromine (Fig. 4) are similar to those for chlorine, except that the excited atom, and the $^3\Pi_{0+u}$ state correlating with it, lie higher relative to the ground state Br + Br limit than in the case of chlorine. This means that the minimum of the $^3\Pi_{0+u}$ state is now very close to the dissociation limit of the ground state, rather than being considerably below it as in the case of Cl_2.

Palmer[137] has observed emission from bromine in the hot equilibrium gas behind a shock wave, in the temperature range 1300–2500 °K. The emission spectrum appeared to be a continuum, with several broad maxima in the range 4000–6100 A, though the resolution was presumably not adequate to distinguish a continuum from a dense band spectrum. Comparison was made of observed activation energies for emission at various wavelengths with predictions based on the simplest possible application of the Franck–Condon principle to the potential curves. The results clearly indicate that Br_2 is not involved as a third body in the radiative recombination process. The principal emitters are the $^3\Pi_{1u}$ and $^3\Pi_{0^+u}$ states of Br_2 and these are assigned to various features of the observed continuum. There is presumably an ample equilibrium population of $Br(^2P_{\frac{3}{2}})$ and no curve crossing need be postulated.

In connection with observations of emission from hot equilibrium gases, it should be pointed out that the observed emission spectrum must be simply the product of the black body intensity distribution, times the emissivity of the gas. Therefore the emission spectrum contains no more information about the excitation mechanism than does the absorption spectrum of the same gas at the same temperature and path length.

Radiative recombination of bromine atoms in room temperature afterglows has been studied by Gibbs and Ogryzlo[138], and by Clyne and Coxon[139]. They disagree on the most fundamental deduction, the identity of the emitter. The emission is discrete, showing well defined bands, but there is some overlapping of different band systems, and isotopic splitting causes further confusion. Gibbs and Ogryzlo[138] assigned the bands to the $^3\Pi_{0^+u}-^1\Sigma_g^+$ transition, but they observed a number of other bands which they were unable to assign. Clyne and Coxon[139] observed the same emission spectrum, though their observations extended somewhat farther into the infrared. They assign the bands to the $^3\Pi_{1u}-^1\Sigma_g^+$ transition, and this assignment is quite successful after a slight revision of a previous vibrational analysis of the infrared part of the system. The emission intensity is proportional to $[Br]^2[Br_2]$. The temperature dependence has not been measured, but it should correspond to a hump of a few hundred cm^{-1} if the emitter is $^3\Pi_{0^+u}$ while emission from $^3\Pi_{1u}$ should have zero or negative activation energy.

Clyne and Coxon[140] have studied radiative recombination of $Br + Cl$ and $I + Cl$, and have reviewed the results for other halogens. The emitters were identified as $BrCl(^3\Pi_{0^+})$, correlating with an excited $Cl(^2P_{\frac{1}{2}})$, and $ICl(^3\Pi_1)$, correlating with ground state atoms. In BrCl, the $^3\Pi_{0^+}$ state which normally correlates with an excited Cl, interacts with a repulsive state correlating with ground state atoms. The result is that the $^3\Pi_{0^+}$ state does, in fact, correlate over a hump with ground state atoms. The hump is estimated to have a height of a few hundred cm^{-1}. In ICl, the situation is probably similar, except that the hump in the $^3\Pi_{0^+}$ curve going over to ground state atoms is sufficiently high to strongly disfavor recombination on that curve at room temperature. Recombination into the $^3\Pi_1$ curve requires no activation energy. Unlike the curve-crossing mechanism, where vibrational levels are populated only up to the crossing point, in simple recombination on a single curve all vibrational levels up to the dissociation limit will be populated. Emission from high levels is often weak or hard to identify, but in ICl emission from levels up to $v = 18$, very near the limit, has been observed. Emission from $IBr(^3\Pi_1)$ has also been observed in a room temperature afterglow[139].

Vibrational intensity distributions have not been measured in any of the room temperature afterglow work, but it seems clear that they are concentrated toward low levels. In most cases the emission is quite weak, and one has to work in the torr pressure range where vibrational relaxation of the halogens may be significant.

This discussion of radiative recombination in the halogens clearly illustrates a recurring difficulty. A third body can act in three ways which are extremely difficult to disentangle. It can participate in the recombination step, it can electronically quench the excited molecule, and it can vibrationally (and rotationally) relax the excited molecule. In many cases vibrational relaxation will shift some of the emission spectrum into wavelength regions which are not observed, or are observed with low sensitivity, thus mimicking electronic quenching. All these processes appear to be significant in the halogen recombination reactions.

Radiative recombination of halogen atoms, and especially of hydrogen atoms with halogen atoms, has been observed in fuel-rich hydrogen+oxygen flames[141]. The intensity is proportional to [H][X] but a third body is probably involved.

3.1.3 Association reactions of oxygen atoms

The radiative recombination of oxygen atoms has been observed in the afterglow of a discharge through oxygen[142], and has been studied in more detail in nitrogen afterglows titrated with NO[143,144]. The processes observed are

$$O+O+M \rightarrow O_2(b\,^1\Sigma_g^+)+M \quad k_{N_2} = 6.10 \times 10^4 \ l^2.mole^{-2}.sec^{-1} \qquad (7)$$

emitting the atmospheric bands, and

$$O+O+M \rightarrow O_2(A\,^3\Sigma_u^+)+M \quad k = 1.4 \ l.mole^{-1}.sec^{-1}$$

$$\text{for} \quad N_2 > 1.7 \times 10^{-5} \ mole.l^{-1}$$

$$k_{N_2} = 7.6 \times 10^4 \ l^2.mole^{-2}.sec^{-1} \quad \text{for} \quad N_2 < 1.7 \times 10^{-5} \ mole.l^{-1} \qquad (8)$$

emitting the Herzberg bands. Both of these excited states of O_2 correlate with ground state atoms, see Fig. 5, so the mechanism is presumably simple stabilization of the collision by the third body, and no curve crossing is involved. Reaction (8) has also been observed on a nickel surface[145], where about one-tenth of recombining O atoms form $O_2(A\,^3\Sigma_u^+)$.

The reaction

$$O+H \rightarrow OH(A\,^2\Sigma^+)v = 0, N = 13 \qquad (9)$$

has been observed in a flow system containing O, H, H_2, and N_2[146]. The reactants are definitely ground state atoms, and since these do not correlate with the excited molecule, the interaction of two potential curves (inverse predissociation) must be involved. The situation is qualitatively similar to that discussed in connection with Fig. 2, but the interaction of the two potential curves is assumed to be strong enough to allow some radiative recombination. The region of the interaction may lie on the left limb of the $^2\Sigma^+$ curve. The identity of the second curve is unknown, but the energetics require that, for thermal collisions at room temperature, the only states which can be populated in two-body recombination are $v = 0$, $N = 13$ and $v = 1$, $N = 4$. The emission intensity is proportional to the product [O][H] and the constant of proportionality decreases with increasing pressure, indicating quenching of OH($^2\Sigma^+$). The OH($^2\Sigma^+$) emission spectrum is sufficiently open for

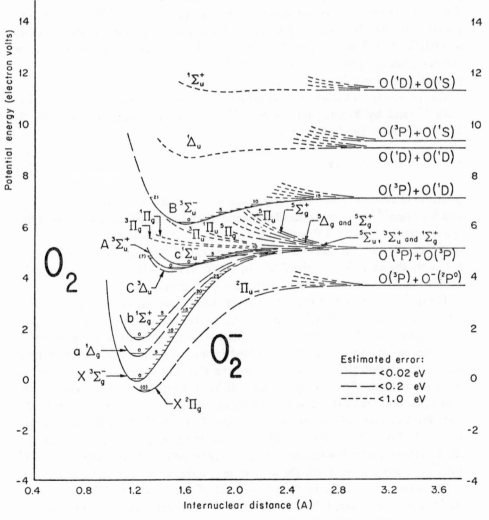

Fig. 5. Potential energy curves for O_2 adapted from Gilmore[329].

individual rotational lines to be resolved. The average lifetime of the upper state, determined by quenching collisions, is long enough to allow a lot of rotational relaxation. There is nevertheless a definite excess population at $N = 13$. This is one of the very few known cases in which the rotational distribution has been observed in a state populated by two-body radiative recombination. There is some indication of a corresponding underpopulation of this level relative to neighboring levels in $OH(^2\Sigma^+)$ emission from low pressure microwave discharges, as would be expected for predissociation.

A predissociation, which may or may not be related to the one just discussed, is observed in hot flames[147] and in cool atomic flames[148]. For rotationless states the predissociating curve appears to cross the bound $^2\Sigma^+$ state very near $v = 2$. The corresponding inverse predissociation has been proposed[149,150] as an explanation for the observed overpopulation of the first and second vibrational levels of $OH(^2\Sigma^+)$ in flames where there is a considerable excess population (over thermodynamic equilibrium) of O and H atoms. This process may produce a population inversion in nozzle expansion of a dissociated gas [151].

The two-body radiative recombination of N+O has been observed by Tanaka[152], and by Young and Sharpless[153] in nitrogen afterglows partially titrated with NO. The fast reaction

$$N+NO \rightarrow N_2+O \tag{10}$$

produces O atoms in amounts corresponding to the amount of NO added. Young and Sharpless[153] have made detailed kinetic studies of the NO^e emission from this system, particularly as a function of pressure, atom concentrations, and added quenchers. The δ, γ, β and Ogawa bands are emitted (see Fig. 6). We will discuss here only the excitation of the δ bands, $NO(C^2\Pi)-(X^2\Pi)$, since this is logically simpler, and has been confirmed by predissociation studies.

Tanaka[152] showed that the emission spectrum from the N+O afterglow cuts off sharply below 1915 A, corresponding to the dissociation energy of NO, 6.49 eV. The $v = 0$ level of the C state coincides to within 0.01 eV with the energy of separated atoms, so only $v = 0$ is excited in these systems. The excitation must involve interaction with another state, and the $a^4\Pi$ state is almost certainly the one involved, in spite of the change in multiplicity. If instead, a Σ state were the precursor, selection rules (for case b coupling in both states) would require that a Σ^+ state interact with only the Π^+ components of $^2\Pi$, or Σ^- with Π^-. In the first case, only P and R branches would be present in the emission due to the $C(^2\Pi)-A(^2\Sigma^+)$ transition; in the second case, only a Q branch would be present. Since all three branches are observed, the $a^4\Pi$ state is indicated as the perturber.

The emission intensity in the $(0, 0)$ band of the $C(^2\Pi)-X(^2\Pi)$ system is independent of pressure, $0.2 < P < 10$ torr[153], indicating that no third body is involved in the inverse predissociation

$$N+O \rightarrow NO(a\,^4\Pi) \rightarrow NO(C\,^2\Pi) \rightarrow NO(X\,^2\Pi)+h\nu \quad (\delta \text{ bands}) \qquad (11)$$

Furthermore, the emission is not strongly quenched by N_2O or CO_2[153], or by N_2 (half quenching pressure $= 22$ torr)[154] so this association reaction seems to be particularly simple (*cf.* the halogens, where a third body appears to be involved in three distinct ways).

Callear and Smith[154] have made a careful study of the fluorescence and pre-dissociation of the C state, and agree that the $a\,^4\Pi$ state is involved. NO ($C, v = 0$)

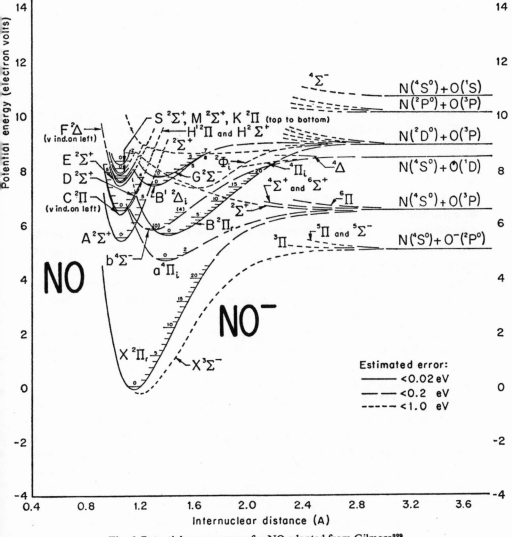

Fig. 6. Potential energy curves for NO adapted from Gilmore[329].

is essentially in equilibrium with $N+O$ and this is only slightly perturbed by spontaneous radiation to the ground state[153,154]. Only 3 % of the molecules in the $C, v = 0$ state radiate in competition with predissociation[154]. The spontaneous radiation rate coefficient for NO $(C, v = 0) \rightarrow (X, \text{all } v)$ is 2.2×10^7 sec^{-1} and the predissociation rate coefficient is 6.6×10^8 sec^{-1}. The rate coefficient for formation of NO $(C, v = 0)$ from $N+O$ is[154] $k = 3.1 \times 10^5$ l.mole^{-1}.sec^{-1}. The kinetics of emission from several other excited states in the $N+O$ system has been studied[153,155,156].

The temperature dependence of emission from the $N+O$ two-body recombination in a glow discharge shock tube has been measured[156a]. The rate coefficient for emission of the δ bands (from $v' = 0$) is 4×10^3 $(T/300)^{-0.35}$ l.mole^{-1}.sec^{-1}.

3.1.4 Nitrogen atom recombination

It seems inappropriate to review here the immense literature on active nitrogen. We will simply mention some of the recent work on the recombination process itself and the work cited will provide an introduction and access to the main body of the literature on the afterglow. The subject has recently been reviewed[157,332].

The principal emission from a gas containing a percent or so of N atoms in a few torr N_2 is from the B state (Fig. 7) but emission is also observed from the states B' and a. Since none of these states correlate with ground state atoms, they must be populated by interaction with a potential curve which does so correlate. The possibilities are X, A, and $^5\Sigma$. Bayes and Kistiakowsky[158] have proposed a mechanism in which the shallow $^5\Sigma$ state populates all the above states by collisional transfer. This is reasonably consistent with known potential curves (Fig. 7) and is further indicated by their observation that the B and B' states are populated only up to a limit which is 850 cm^{-1} below the energy of two ground state nitrogen atoms. This they interpret as the depth of the minimum in the $^5\Sigma$ potential well. The state is supposed to be essentially in equilibrium with atoms. The $^5\Sigma$ state has been described by Carroll[159] on the basis of observed predissociations, and Mulliken[160] has given a theoretical calculation. Carroll's potential curve is consistent with the demands of Bayes and Kistiakowsky[158].

Harteck et al.[161] have found that the afterglow intensity in N_2 is proportional to the total pressure to the power 2.5 as the pressure is varied in the range 5–30 μ, the composition being constant. They interpret this to indicate that an inverse predissociation mechanism is competing with collision induced transition. According to this, the interaction $^5\Sigma_g^+ - ^3\Pi_g$ is both spontaneous and produced by collisions. At higher pressures, in the range 1–10 torr, the B–A emission intensity is second order, and Thrush[162] gives $I(B–A) = 4 \times 10^{-18}[N]^2$cm^{-3}.sec^{-1}. Recent measurements by Brennen and Brown[163], using electron spin resonance to measure N atom concentrations, indicate that the B–A emission is simply proportional to $[N]^2$ over the entire range 0.040–70 torr. As with halogen recombination, the dependence of intensity on $[N]$ and $[N_2]$ is complicated by the many possible ways in which a third body may be involved.

Recently Campbell and Thrush[164] have pointed out the importance to the afterglow mechanism of quenching of the B state by N_2. Including this, they estimate that roughly half the $N + N$ recombinations go into the B state, though at

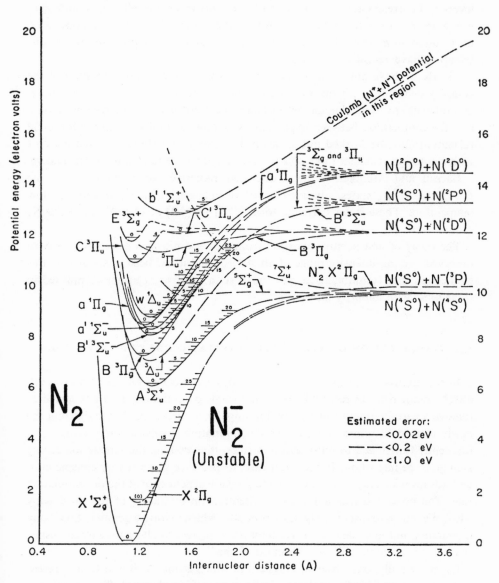

Fig. 7. Potential energy curves for N_2 adapted from Gilmore[329].

pressures above 1 torr most B state molecules are quenched rather than radiating. They feel that the shallow $^5\Sigma$ state, even if in equilibrium with $N + N$, will have insufficient population to account for the large rate of production of the B state. They therefore propose that the A state is the precursor of B, and interpret results of Kistiakowsky and Warneck[165] (comparing $^{15}N_2$ and $^{14}N_2$) as indicating the importance of energy matching between A and B at low vibrational levels in B. There will also be strong collisional interaction between A and B where the highest levels of A correspond with $v = 11$ or 12 in B, and again at levels corresponding to $v = 6$ in B, thus explaining qualitatively the observed vibrational population distribution in B. The effect of quenching and diluents on this vibrational distribution have recently been studied[166].

Clearly we have not yet at hand a rigorously convincing interpretation of the complex behavior of nitrogen afterglows as a function of pressure, temperature, and diluents. In particular, the importance of a third body in the recombination, and the competition between quenching, spontaneous radiation, and vibrational relaxation must be worked out in detail. It is likely that studies using apparatus with resolution sufficient to resolve the rotational structure would be informative.

Carroll and Mulliken[167] have discussed radiative recombination into the $C^3\Pi_u$ and $C'^3\Pi_u$ states, observed in active nitrogen at high pressure and low temperature. These states correlate with $N(^4S) + N(^2D)$ at 12.1 eV above the N_2 ground state.

The decay of first positive emission from a pulsed nitrogen afterglow has been observed over an intensity range of almost 10^6 and it has been postulated in this connection that recombination into the B state involves N atoms complexed in some sense with impurity molecules[168].

3.2 RECOMBINATION OF ATOMS WITH EXCITATION OF THE THIRD BODY

In the discussion up to now the third body, if present at all, has played a relatively minor role. In recombination on a single potential curve, it has served to remove a certain amount of energy and angular momentum, thus stabilizing the newly formed diatomic molecule. In recombination involving curve crossing, it frequently functioned to relax selection rules, thus making the interaction strong enough to be important. In the latter case, its presence was not very conspicuous, and was revealed only by its effect on the pressure dependence of the recombination rate. The present discussion focuses attention on excitation of the third body itself. We intend to discuss only those processes which involve a genuine three-body interaction, but in some cases it is necessary to digress to alternative mechanisms, since a unique mechanism is seldom established.

Experimentally, one observes a system in which atoms A, B and C are present and emission of light from electronically excited C is observed. To simplify the discussion, let A and B be the same kind of atom. Then a possible mechanism is

$$A+A+C \rightarrow A_2+C^e \tag{12}$$

However one can not obviously exclude the competing combination

$$A+C+C \rightarrow AC+C^e \tag{13}$$

unless the ratio of concentrations [A]/[C] is large. Apart from the uncertainty in mechanism due to the possible importance of reaction (13), there is the further possibility that the third body in the recombination may not be C at all, so that C is excited in a two step process, *viz.*

$$A+A+M \rightarrow A_2+M^* \tag{14}$$

$$M^*+C \rightarrow M+C^e \tag{15}$$

in which M is a relatively abundant third body and M* is relatively long-lived. If one thinks of reaction (12) as atom C colliding with the collision pair $A+A$, then (14) has the kinetic advantage that the energetic target, being M* instead of $A+A$, has a longer lifetime so that C can be excited at a greater rate. If M* is vibrationally excited, and subject to stepwise collisional deactivation, the C^e may well be formed with a wider distribution of energies than will be the case in (12). The mechanisms mentioned so far may be thought of as involving genuine three-body collisions, in which there is strong interaction between all three atoms simultaneously at some point in the collision. In some cases it may be appropriate to consider stepwise processes in which two of the three atoms interact strongly to maintain an equilibrium concentration of energetic dimers which then react with the third atom.

Theoretically, one usually tries to understand reactions of the type (12) in terms of potential energy *surfaces* for electronic states of the molecule A_2C. Whereas for the process

$$A+A+C \rightarrow A_2^e+C \tag{16}$$

it is generally quite sufficient to discuss the problem in terms of potential energy *curves*, with the role of C almost ignored, the situation is quite different here. We have a three-body problem, and must discuss potential *surfaces*. This increases the conceptual and computational difficulty by at least an order of magnitude, so that results are correspondingly few and uncertain.

The first group of these reactions to be considered is that in which an alkali metal atom is excited. These are especially favorable cases because the atoms have low-lying electronic states which do not require the entire heat of reaction for their excitation. They are also favorable because of the existence of low-lying ionic states which can provide paths between potential energy surfaces[29].

The prototype for the class of reactions (12) is surely

$$H + H + Na \rightarrow H_2 + Na^e(^2P) \tag{17}$$

This reaction has been postulated to explain emission from seeded flames and may also be important when sodium is added to hydrogen atoms pumped from an electric discharge tube. The emission from metal additives in flames has been studied particularly by Padley and Sugden[169,170] and reviewed by Sugden[171]. Excitation of sodium in dilute $H_2 + O_2 + N_2$ flames at atmospheric pressure is apparently due both to $H + H + Na$ and $H + OH + Na$.

A kinetic analysis of the results, based on (17) and its $O + OH$ analog, is in satisfactory agreement with observations on a wide variety of flames. These flames are relatively cool, and the concentrations of H and OH exceed their equilibrium values even in the burned gases, so that the observed sodium emission is definitely chemiluminescent. The third order rate coefficients for excitation by $H + H$ and $H + OH$ are estimated to be 8×10^9 and 2×10^{10} l^2.mole^{-2}.sec^{-1}, corresponding to an efficiency near unity per triple collision. The possible importance of mechanisms of the type (14, 15) has not been carefully studied.

Emission of the sodium D lines has also been observed in diffusion flames of sodium with hydrogen atoms by Polanyi et al.[172-174]. They were unable to decide between (17) and

$$H + Na + Na \rightarrow NaH + Na(^2P) \tag{18}$$

since the ratio [Na]/[H] was much larger than in the flame work. Polanyi and Sadowski[172] explained the rapid disappearance of sodium in their system by the interesting reaction

$$Na + H_2(v > 6) \rightarrow NaH + H \tag{19}$$

With reasonable estimates of rate coefficients, this does not require an unreasonable concentration of vibrationally excited H_2 in the gas flowing from the discharge.

A crude potential energy surface for linear $H + H + Na$ has been calculated and discussed by Magee and Ri[175]. They point out that $H + H + Na(^2S)$ can correlate directly with $H_2 + Na$ in either the ground ^2S state or the excited ^2P state, so that reaction (17) can take place adiabatically, without transitions between potential surfaces. With their calculated activation energy of 6.5 kcal.mole^{-1} for the chemiluminescent recombination, they estimate the ratio of rate coefficients for (18) and the corresponding reaction producing ground state sodium to be 0.015. Hence the energy released tends to go into vibration of the product H_2, or into translation, rather than into electronic excitation of Na.

Padley and Sugden[170] have studied chemiluminescence from a number of metal vapors in flames, in addition to sodium. Their results are consistent with excitation of the metal as third body in recombination of $H+H$ or $H+OH$. Thallium is principally excited by $H+H$, while lead is excited by $H+OH$. Zhit-kevich *et al.*[176,177] have also studied chemiluminescent emission from metals in flames, but the mechanism of excitation is at best speculative. Emission is often observed in carbon-containing flames from energy levels higher than the maximum of 4.4 eV obtainable from $O+H$. The source of this energy is probably a highly exothermic reaction in the mechanism of flame propagation; among the most exothermic are those forming CO. The energy may be passed on directly to the metal as a third body, or to an intermediate.

In atomic flames, where one might expect direct excitation of a metal atom as third body, an intermediate is often involved instead. For example, in studies of excitation of iron by nitrogen atoms in CO it has been postulated that Fe is excited by collision with excited N_2 or CO produced as third bodies in the atom recombination process[178]. Brennen and Kistiakowsky[55] studied the excitation of nickel, iron and other metals in active nitrogen and concluded that the metal atom is not excited as third body in the recombination process, but by interaction with the metastable $N_2(A^3\Sigma_u^+)$.

The reaction

$$O+O+O \rightarrow O_2+O(^1S)+21 \text{ kcal.mole}^{-1} \tag{20}$$

has been proposed to explain the observed emission in the upper atmosphere of the oxygen auroral line at 5577 A. There is evidence confirming this mechanism in laboratory work[143,179]. It has also been suggested[143,180] that the oxygen 5577 A line is excited in the upper atmosphere by

$$N+N+O \rightarrow N_2+O(^1S) \tag{21}$$

Up to now we have discussed only cases in which the third body being excited is an atom. There is little known for certain in cases where it is a species containing more than one atom. Processes of the type

$$H+OH+OH \rightarrow H_2O+OH(^2\Sigma^+) \tag{22}$$

have been discussed to account for non-thermal radiation from OH in flames (refs. [150,181,330]). However, other work[182,183] indicates that the excitation mechanism may be

$$H+O_2+H_2 \rightarrow H_2O+OH(^2\Sigma^+) \tag{23}$$

A process corresponding to (22) has been invoked to explain S_2 emission spectra observed[184] when SO_2 is added to hydrogen flames.

Garvin and Broida[185] have proposed the reactions

$$N+N+X \rightarrow N_2+X^* \tag{24}$$

to explain the formation of electronically excited NH, NO, and OH in the $H+N+O_3$ reaction. They also consider that N_2^* may be the source of excitation. Their paper and the discussion which follows it illustrate the difficulty of distinguishing between excitation in the reaction which initially forms a molecule, and subsequent excitation of a molecule originally produced in the ground state.

3.3 THE RADIATIVE COMBINATION OF AN ATOM WITH A DIATOMIC MOLECULE

Several reactions of the type

$$A+BC \rightarrow ABC^e \tag{25}$$

have been studied fairly extensively in recent years, so that the questions to be answered and the difficulties involved are rather well defined. The reaction leading to ground state ABC is by no means a trivial subject of investigation, and the reaction leading to an excited product presents a number of additional features concerning which only the most qualitative generalities are at present possible.

Experimentally, one is interested first of all in the order of the reaction, its absolute rate, and the temperature dependence of that rate. In addition to these primary data of chemical kinetics, one can observe the emission spectrum of ABC^e under various experimental conditions. From these observations one tries to infer information about the vibrational and electronic states involved, and their interactions, with or without collisions. Theoretically, interest centers on potential surfaces. Unfortunately, these are all too often thought of as potential curves, so that two of the three internal coordinates of the molecule are ignored. The experimental and theoretical difficulties are such that, even in terms of such an oversimplified model, it is seldom possible to arrive at a unique, widely agreed upon picture of the reaction process.

3.3.1 $O+NO = NO_2^e$

This is perhaps the most extensively studied radiative association of an atom with a diatomic molecule. We will discuss first those features of this reaction about which there is general agreement, and then continue to consider points which are more uncertain or speculative. For reviews see Refs. 186 and 187.

The reaction has been studied in flow systems in the torr pressure range[188] and in large spherical vessels at micron pressures[189,190]. Kaufman[188] has established that the intensity is proportional to the first power of the O atom concentration and to the first power of the NO concentration. This has been either assumed or checked in all subsequent work. At pressures in the torr range, the intensity does not depend on third-body concentration, though it does depend on the identity of the third body[191,192]. The luminous reaction, like many association reactions, has a negative temperature coefficient. Measurements using a flow system near room temperature[192] are not in good agreement with shock tube measurements[193,194] but if the rate is assumed to be proportional to T^{-n}, the range of exponents is $1.5 < n < 3$. At 276 °K the absolute rate coefficient [195] for emission in the wavelength range 0.4–$1.4\,\mu$ is 3.9×10^4 l.mole^{-1}.sec^{-1}. Hence light is emitted in roughly one collision in 10^6.

The emission spectrum is apparently continuous, with weak diffuse structure toward the short wavelength limit, 3980 A, set by the energy of $O + NO$ relative to ground state NO_2[196-198]. These emission bands correlate rather well with features in the absorption spectrum. The intensity maximum is near 6300 A[195], and emission has been observed into the near infrared. There is apparently very little intensity beyond $1.4\,\mu$[195], but emission has been observed, using interference spectroscopy, to wavelengths as long as $3.3\,\mu$[58]. There has been considerable uncertainty as to whether or not the emission is really continuous. The known absorption spectrum of NO_2 in the regions of interest is discrete, but extremely complex, so it is quite possible that lack of observed structure in the chemiluminescent emission may be due to insufficient instrumental resolving power, or even to the overlapping of lines lying closer together than their Doppler width. The emission spectrum, including transitions from many vibrational levels of the upper electronic state (or states) to many levels of the ground electronic state, is expected to be considerably more complex than the room temperature absorption spectrum, where almost all transitions occur from the ground vibrational state. We have observed that most of the vibrational structure in the room temperature absorption spectrum disappears on heating. This is due in part to increasing breadth of rotational distributions, as well as to the greater number of vibrational transitions. It is therefore not obvious that an emission spectrum which is in principle discrete (line separation much greater than natural width) would appear discrete, even with spectroscopic resolution adequate to resolve rotational structure in the room temperature absorption spectrum. Kaufman and Kelso[191] have reported that the emission looks continuous with this resolving power and there seems to be little more one can do spectroscopically.

We have just said that a discrete spectrum can look continuous. The converse also has a certain meaning. A spectrum due to emission from a continuous level above a potential well will have a good deal of structure[117]. Thus the problem of deciding, on the basis of the spectrum, whether or not the observed emission comes from NO_2^e which has been stabilized by a third body is far from simple, either experimentally or theoretically.

There has been no extensive laboratory study of the effects of temperature, pressure, and composition on the spectrum of the chemiluminescence, but most of the results available[191,195] show no effect of these variables over moderate ranges. There is evidence for a small shift to the blue in the spectrum at pressures below 0.1 torr[199]. There is also a small effect of water vapor on the intensity distribution[200].

An obvious point of inquiry concerns the identity of the electronic state or states from which emission is observed. The absorption spectrum of NO_2 in the visible and near infrared is extremely complex and has for the most part resisted analysis[33]. This in itself is an indication that perturbations are probably important, but the spectrum gives no obvious indication of more than one electronic transition. According to recent calculations[201] there are five electronic states located at energies above the ground state corresponding to absorption in the visible. For these the only fully allowed transitions are to the linear $^2B_1$ state and the somewhat higher bent $^2B_2$ state. The first of these correlates with ground state $O + NO$, since it and the ground state $^2A_1$ are the two surfaces which result from bending a linear $ONO(^2\Pi_u)$. The higher bent state $^2B_2$ correlates with $NO(^2\Pi) + O(^1D)$. A small part of the NO_2 absorption spectrum has recently been analyzed[198] in the region 3700–4600 A. It is found to be due to a $^2B_1$–$^2A_1$ transition, in agreement with theoretical predictions. It is estimated that the origin of the transition lies in the region 6500–8500 A. In view of these results it is most reasonable and simple to assume that the observed absorption and emission in NO_2 involves the single electronic transition $^2B_1$–$^2A_1$. The rotational structure in the absorption spectrum becomes diffuse below 3979 A, though the vibrational structure is unaffected. This probably indicated a unimolecular type dissociation in which energy is transferred from the bending mode initially excited to the asymmetric stretching mode which leads to dissociation[198].

We can most simply postulate that recombination occurs on the ground state potential surface and that radiation is then the result of vibronic interaction between the high vibrational levels so populated and corresponding levels in the $^2B_1$ state. The principal experimental argument for the importance of two excited electronic states in the visible emission of NO_2[196] has been the fact that the observed radiative lifetime[202,203] of 4.4×10^{-5} sec differs so greatly from the value 2.6×10^{-7} sec derived from the integrated absorption coefficient. It has been shown, however, that there are several ways of explaining this discrepancy without postulating a second electronically excited state[203]. The vibronic interaction with the ground state will, in effect, greatly increase the degeneracy of the upper state, so that the spontaneous radiative transition rate from an individual level is much decreased, producing the observed long lifetime. The absorption is not correspondingly weakened, and the lifetime derived from it is not physically meaningful.

Since a second excited state is not required to reconcile the observed lifetime with the integrated absorption coefficient, its retention must be justified on other grounds. The emission spectrum of laboratory afterglows is independent of temperature, composition, and pressure (0.1 to 5 torr)[191,195], but a quite different emission spectrum has been observed from NO released from a rocket in the upper atmosphere at an ambient pressure of 1×10^{-4} torr[204,209]. This was attributed to simple recombination into a second excited state[196] of NO_2 but the NO may be present in these high velocity expansions as a dimer or polymer of some sort[205]. Hence, it may not be necessary to invoke a second excited state to explain the rocket headglow spectrum, but it is certainly convenient. There does not seem to be any other evidence which indicates a second excited state so forcibly. It has not otherwise been observed spectroscopically.

There is no complete agreement on the fundamental question of whether a third body is essentially involved in the radiative recombination. The dependence[191] of the intensity in the torr pressure range on the identity of M suggests that it is. Further strong evidence that a third body is involved in the torr pressure range is the fact that electronic quenching must be important, as shown by Myers et al.[206]. Collisional relaxation of vibration in the upper electronic state is also kinetically important[207]. The observed second order behavior can then be explained only by the participation of a third body in the recombination. This predicts that at sufficiently low pressure, when quenching no longer competes successfully with spontaneous radiation, the recombination should become third order. Kaufman and Kelso[191] have determined the emission intensity divided by [O][NO] as a function of pressure over the range 0.05 to 0.5 torr and find that the values do indeed fall off sharply at pressures below about 0.15 torr, indicating third order behavior of the emission intensity. This necessarily indicates the participation of a third body. However, Reeves et al.[208] and Applebaum et al.[189] find second order dependence down to pressures as low as 3×10^{-3} torr, which they interpret as indicating two-body recombination. Doherty and Jonathan[190] find similar results in the range $(0.85–400) \times 10^{-3}$ torr, but they apply a three-body mechanism involving two excited electronic states. Recombination is supposed to go into one of these, followed by a radiationless transition into the state from which radiation is observed[196]. If the interaction between these two states is described in terms of a rate of transition, the rate coefficient must be $< 10^3$ sec^{-1}. This means that there will be a large population in the initially formed state, and radiation from it should be significant, but was not observed. Further evidence for two-body recombination at low pressures may come from the rocket experiments[204,209], where Spindler[204] has shown the emission to be second order in the range $(4–0.4) \times 10^{-4}$ torr. The weight of evidence thus seems to indicate second order recombination at sufficiently low pressures.

It is merely a truism that recombination must be two-body at sufficiently low pressure. The question of interest is whether this recombination will be fast enough to be observed. The rocket observations and corresponding wind tunnel experiments by Del Greco et al.[210] and van der Bliek et al.[211] show that the two-body recombination is indeed fast. It is about four orders of magnitude faster than would be predicted from the results in the torr pressure region. This startling result has not been fully explained, but supercooling and formation of polymers or crystallites in the rapid expansion of NO in these experiments is probably important[205,212]. A reaction such as

$$O + (NO)_n \rightarrow NO_2^e + (NO)_{n-1}$$

would need no third body and could easily have a collisional efficiency much greater than that for a simple recombination reaction.

Since our main interest in this chapter is the primary process producing electronic excitation and in view of the qualitative uncertainties remaining in the O+NO problem, it does not seem appropriate to review the detailed kinetic treatments which have been put forward. The kinetics has been thoroughly reviewed by Heicklen and Cohen[187], and good discussions are given by Clyne and Thrush[192] and Kaufman and Kelso[191].

3.3.2 $O + CO = CO_2^e$

This radiative recombination is observed in flames[181,213], electric arcs[214], shock tubes[194,215,216], and atomic flames[192,217-219]. In flames and arcs, the emission consists of diffuse bands on a continuum, whereas in atomic flames (the reaction of O with CO at low pressures and room temperature) there is no continuum and clear vibrational structure extends from below 3000 A out to at least 6000 A. The process has not been studied as thoroughly as O+NO, but enjoys the same complexities and uncertainties.

The reaction has been investigated in flow systems near room temperature in the torr pressure range in a 1 inch tube[192,217] and in a 5 litre bulb[219] with discordant results. Both the temperature dependence and the dependence on third-body concentration are substantially different. The results using the large bulb show strong quenching by O_2 at room temperature. Strong quenching is also indicated by the fact that the emission intensity is expressible in the form $I = I_0 [CO][O]$ in which I_0 depends on the nature of a third body but not on its concentration[192,217]. The room temperature studies agree on the activation energy only to the extent of saying that it is substantially positive. The shock tube studies agree on a value of 2.5 kcal.mole^{-1}. Unlike the atomic flames, the emission here presumably contains a good deal of continuum.

In interpreting the O+CO emission, the salient fact is that the ground state of

CO_2 correlates with ground state CO plus excited $O(^1D)^{29,192,217}$. Unlike the $O+NO$ case, correlation with ground states $CO(^1\Sigma^+)+O(^3P)$ is spin forbidden. Hence recombination of $CO+O(^3P)$ can not lead directly to the ground state. This presumably explains the high efficiency of the luminous reaction. The quantum yield, extrapolated to zero pressure, is 1/350 (quanta per molecule of CO_2 formed) at 298 °K, and increases with increasing temperature[219].

Recombination presumably begins on a triplet potential curve correlating with ground state fragments. Emission may then be a spin forbidden transition to the singlet ground state[219], or there may be a radiationless transition to an excited singlet state which then radiates[192,217]. The observed activation energy, or rather the algebraic difference between this and the negative activation energy for the $O+NO$ chemiluminescence, is presumed to correspond to a maximum in the triplet recombination path. Dixon[220] has studied the CO_2 afterglow spectrum with fairly high resolution (0.5 cm^{-1}, 5 day exposure) and interprets it in terms of an emitting state of species B_2, probably $^1B_2$. At some stage there must be a singlet–triplet transition, but it is not clear whether this is the radiation process itself, or preliminary to it. Mahan and Solo[219] have found that O_2 is a strong quencher of the emission, and the effect on the quantum yield is strongly temperature dependent. Understanding of the electronic and vibrational quenching by O_2 and other molecules in the system will be important in reconciling the many disagreements concerning the interpretation of the observations. As usual we are plagued by insufficient knowledge of the excited states of polyatomic molecules.

3.3.3 $O+SO = SO_2^e$

This radiative recombination has been observed in the photolysis of H_2S+O_2[221] and, in greater detail, in afterglows of a discharge in SO_2[222] and in systems such as $O+COS$ in which SO is produced[223,224]. The chemiluminescent reaction in the SO_2 afterglow is second order in the torr pressure range with a rate coefficient $k = 1.5 \times 10^5$ l.mole^{-1}.sec^{-1}. As with $O+NO$, there is the rather surprising observation that, as produced in the system $O+OCS$, it appears to be second order at pressures of only a few microns[225]. However Sharma et al.[224] interpret their results for this reaction in terms of third order association at pressures of 800 μ or greater. The absorption spectrum of SO_2 is somewhat better understood than that of NO_2[33], and this helps in interpreting the emission. Herman et al.[226] have discussed the correlation between the absorption and emission spectra, and Halstead and Thrush[227] have assigned regions in the emission spectrum (2240–5000 A) to several transitions known in absorption. Towards short wavelengths, the emission does not extend to the $O+SO$ energy limit, 2183 A, but cuts off at about 2240 A, an energy difference of about 1000 cm^{-1}. Near the short wavelength limit of the emission there are some bands of the $\tilde{C}$–$\tilde{X}$ and $\tilde{D}$–$\tilde{X}$ transitions[33,228]. However, most of the emission in the afterglow diluted with argon appears continuous, extending to about 5000 A. As with $O+NO$, one would expect a rather dense spectrum in emission, so it might well appear continuous even though actually discrete. Halstead and Thrush[227] have attributed the emission peaking at 2750 A to the $\tilde{C}-X$ and $\tilde{D}-X$ transitions, that in the 3500 A region to $\tilde{A}\,^1B_1-\tilde{X}$, and the component peaking at 4500 A to $\tilde{a}\,^3B_1-\tilde{X}$. The relative intensity of emission from these three regions is independent of addition of the quencher SO_2, which seems to indicate that they are not populated independently. Halstead and Thrush[227] suggest that recombination actually occurs on the ground state potential surface, and that the excited states are populated by interactions between the ground state and excited states.

Greenough and Duncan[229] have measured the radiative lifetime (τ) of the $\tilde{A}\,^1B_1$ state by a flash excited fluorescence method. Extrapolating to low pressure, they find $\tau = 4 \times 10^{-5}$ sec. The extrapolation is necessary because SO_2 is a strong quencher of its own fluorescence. The half quenching pressure is only 5×10^{-3} torr, so that the M (third-body) dependence of a three-body recombination step will be cancelled by that of the quenching down to a pressure of this order of magnitude. Halstead and Thrush[227] interpret their quenching experiments in this way, whereas Rolfes et al.[225] interpret their low pressure results in terms of a simple two-body radiative recombination.

In addition to electronic quenching of the $\tilde{A}\,^1B_1$ state, SO_2 also produces vibrational relaxation in the relatively long lived $\tilde{a}\,^3B_1$ state, as evidenced by the growth of banded emission at the expense of continuum in the 4500 A region when SO_2 is added. A partial analysis of absorption bands in the $\tilde{a}-\tilde{X}$ transition has been given[230]. This transition has also been observed in emission in shock heated SO_2[231, 231a].

As with NO_2, the observed radiative lifetime of the $\tilde{A}\,^1B_1$ state of SO_2 $(4 \times 10^{-5}$ sec) is much longer than the value calculated from the integrated absorption coefficient $(2 \times 10^{-7}$ sec)[229]. This is again presumably due to strong interaction with the electronic ground state[203].

3.3.4 H + NO = HNO[e]

Of all the radiative recombinations of type $A + BC \rightarrow ABC^e$, the reaction H + NO is the best from the spectroscopic point of view. Only one electronic transition is known, $\tilde{A}\,^1A''-\tilde{X}\,^1A'$. Unlike the cases discussed up to now, the emission is discrete and has been rather fully analyzed[33]. Though this is a tremendous advantage in principle, our understanding of this reaction is really not much deeper than that of the others already discussed.

Cashion and Polanyi[66] observed red and infrared emission from the reaction H + NO. This is the electronic transition $^1A''-^1A'$ in the region 0.6–1.4 μ, and vibrational transitions in the region 2.7–3.9 μ. The emission was studied under high resolution by Clement and Ramsay[232], who observed a predissociation in the rotational structure at energies above 17,000 cm^{-1} (48.6 kcal.mole^{-1}). This predissociation has also been observed in absorption[233], where it is weak.

Clyne and Thrush[68] determined that the emission is represented by $I = I_0$ [H][NO]. At room temperature with Ar as third body, $I_0 = 3.2 \times 10^2$ l.mole^{-1}. sec^{-1}, independent of total pressure. There is a negative temperature coefficient which corresponds to an activation energy of -1.4 ± 0.3 kcal.mole^{-1} or a rate proportional to $T^{-2.8 \pm 0.4}$. There is the usual question as to whether the observed second order light emission is due to a second order radiative recombination, or to a three-body recombination in which the M effect is cancelled by the M effect of quenching. Here there are no independent quenching experiments to provide rate coefficients for that step and also there are no experimental results in the micron pressure range. However, I_0 is found to depend on the nature, but not the concentration, of the third body in the torr pressure range, so it is likely that we are concerned with the same sort of three-body recombination and radiation dominated by quenching that is found in studies of this type on a number of other systems.

Since structure is observed in the absorption spectrum at energies higher than that at which predissociation sets in, the upper state $^1A''$ correlates with ground state H + NO only through interaction with a suitable repulsive state which does so correlate. This interaction presumably takes place at an energy somewhat above the energy of ground state fragments H + NO. If this is the case, the recombination must take place on a different potential surface, since the emission has a negative temperature coefficient. A good possibility is the $^3A''$ state, which should be suitably located and correlates with ground state fragments. It is also assumed to be responsible for the perturbations observed in the $^1A''$ state[233]. According to schematic potential curves[68] (Fig. 8) the $^3A''$ state lies below the emitting $^1A''$ state. If vibrational relaxation in the $^3A''$ state is important, molecules in this

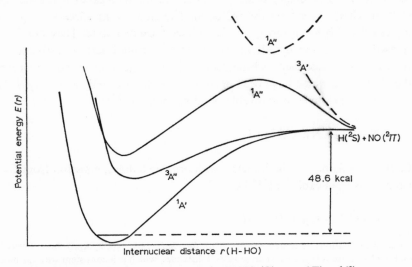

Fig. 8. Schematic potential energy curves for H–NO (Clyne and Thrush[68]).

state may be brought to levels below the lowest level of the $^1A''$ state, so that radiation is prevented. The importance of vibrational relaxation is suggested by the fact that most of the observed emission from $\tilde{A}$ comes from the lowest vibrational level, even though this is 3900 cm^{-1} (11 kcal.mole^{-1}) below the H+NO energy. H_2O is particularly effective in this vibrational relaxation[234].

Recombination of ground state H+NO directly into ground state HNO is possible for bent configurations (the equilibrium angle is 109°) but the ground state in the linear configuration correlates[31] with H+excited NO $^2\Delta$.†

3.3.5 H+OH = H₂O

Padley[235] has proposed that the blue continuum emitted in the region 2200–6000 A from hydrogen–oxygen flames is due to radiative recombination H+OH → $H_2O+h\nu$. The intensity is proportional to the product of the H and OH concentrations, determined by photometric methods (Sugden[171]). There has been no published speculation about what excited states may be involved.

3.4 ATOM TRANSFER REACTIONS

3.4.1 A+BC → AB^e+C

Examples of this type of reaction, where A, B, and C are atoms, are hard to find. Clear, well understood examples are particularly rare, and one must look instead in the uncertain field of elementary steps postulated as parts of complex mechanisms. A necessary condition for the reaction to occur is for the AB bond to be much stronger than the BC bond. The chances for success are presumably increased if AB has a low lying electronically excited state. They are further increased if formation of AB in the electronic ground state is forbidden by spin conservation. Since there is little detailed knowledge of even the few processes of the above type which have been proposed, we can give only a cursory discussion.

A perfect example of favorable bond strengths is [236,237]

$$N+NI \rightarrow N_2(A\,^3\Sigma_u^+)+I \tag{26}$$

In diffusion flames of alkali metals with organic halides, emission from C_2 (Swan bands) has been explained[238] by

$$C+CH \rightarrow C_2(A\,^3\Pi_g)+H \tag{27}$$

---

† Recent calculations by Krauss (*J. Res. Natl. Bur. Std.*, 73A, No. 2, April–June, 1969) indicate that the $^3A''$ state is repulsive at large H–NO distances, which is inconsistent with the mechanism just discussed.

in which certain vibrational levels of C_2 are preferentially populated. Both of these processes are suggested as steps in complex mechanisms.

3.4.2 $A + BC \rightarrow AB + C^e$

Reactions of this type are nearly as scarce as those in which the diatomic product is electronically excited, and presumably for the same reasons. Again they tend to be proposed as parts of complex mechanisms. A relatively simple possibility is

$$H + HI \rightarrow H_2 + I(^2P_{\frac{1}{2}}) \tag{28}$$

but recent work indicates that at most a few percent of the I atoms produced are excited[239]. Excited bromine atoms $(^2P_{\frac{1}{2}})$ are produced in the reaction of H with HBr and with Br_2, perhaps by

$$H + HBr \rightarrow H_2 + Br^e \tag{29}$$

which by analogy with (28) should have low yield. The mechanism is uncertain (refs. [240,241]). Another process of the type discussed in this section, though involving a vibrationally excited reactant, is

$$KBr^v + Na \rightarrow K(^2P) + NaBr \tag{30}$$

studied by Moulton and Herschbach[242] using a "triple beam" apparatus. They find that the cross section is at least 10 A^2. The qualitative treatment of potential surfaces for this reaction given many years ago[243] is still reasonable.

A number of reactions of type (3.4.2) have been discussed in connection with the excitation of sodium in the upper atmosphere[244], viz.

$$NaO + O \rightarrow Na(^2P) + O_2$$

$$NaH + O \rightarrow Na(^2P) + OH$$

$$NaH + H \rightarrow Na(^2P) + H_2$$

The last two are known to be exothermic. Tanaka and Ogawa[245] have suggested the importance of mechanisms involving Na_2 in producing $Na(^2P)$.

3.4.3 Transfer reactions in systems of more than three atoms

The only clear cut reaction of the formal type

$$A + BCD \rightarrow AB^e + CD$$

seems to be

$$Br + ClO_2 \rightarrow BrCl^e + O_2 \tag{31}$$

although it is presumably not a simple atom transfer reaction. The excited state is $^3\Pi_1$ and many bands in the transition to the ground state are observed in the range 5700–880 A[246]. No Cl_2 or Br_2 emission is observed, indicating that the excitation is directly produced in reaction (31), rather than by the radiative recombination $Br + Cl$. It is interesting that the vibrational distribution in this process is quite different from that observed in the $Cl + Br$ atom recombination emission. In this latter case the vibrational distribution is much more energetic than in the case of reaction (31).

The reaction

$$O_3 + NO \rightarrow O_2 + NO_2^e \tag{32}$$

is quite similar to the radiative recombination

$$O + NO \rightarrow NO_2^e \tag{33}$$

The emission from (32) is banded, but the same excited state is populated. The principal difference is that the NO_2^e is formed with < 52 kcal.mole^{-1} of excitation energy (the exothermicity plus the activation energy), whereas in the recombination, the upper limit is 72 kcal.mole^{-1}. The emission spectrum, observed in the range 0.6–2.6 micron, is quite similar to thermal emission from NO_2 at 1200 °K, and is much more peaked in the infrared than the emission from (33) or the fluorescence excited by 4358 A radiation[247,248]. About 10 % of reactive collisions lead to NO_2 in the postulated $^2B_1$ state, and only a few of the lowest vibronic levels of this state are populated[249]. At the energies available in the reaction with O_3, there is apparently little interaction between the $^2B_1$ state and the $^2A_1$ ground state. At the higher levels reached in (33), this interaction appears to be large. According to Clough and Thrush[249], the rate coefficient for the production of $NO_2(^2B_1)$ in the reaction with O_3 is $k = (7.6 \pm 1.5) \times 10^8 \exp(-4180 \pm 300/RT)$l.mole^{-1}.sec^{-1} and the rate coefficient for production of the ground electronic state is $k = (4.3 \pm 1.0) \times 10^8 \exp(-2330 \pm 150/RT)$l.mole^{-1}.sec^{-1}.

The reaction

$$O_3 + SO \rightarrow O_2 + SO_2 \tag{34}$$

is formally similar to the corresponding NO reaction, but there are a number of interesting differences, one of which is the larger exothermicity, 106.6 kcal.mole^{-1}. Perhaps the principal one is the fact that a transition from one potential surface to another is necessary to produce the ^1B state of SO_2[250]. In the case of NO, with its degenerate Π ground state, the reactants correlate with two potential surfaces, one of which leads to $NO_2(^2B_1)$. Since the ground state of SO is a Σ state, this degeneracy is lacking and the ^1B state can only be reached by interaction with another surface. Observations show that the emission reaction has a much higher activation energy than the reaction producing the ground state, indicating that this interaction occurs near the transition state[250]. If the reaction produced molecules in the electronic ground state with high vibrational energy, and the radiating ^1B state was populated by intersystem crossing from these, one would not expect such a difference in activation energy for the luminous and non-luminous reactions.

According to Halstead and Thrush[250], the rate coefficients for production of the three possible electronic states in the reaction of O_3 with SO are

$$k(\tilde{X}\ ^1A_1) = 1.5 \times 10^9 \exp\left(-2100/RT\right)$$

$$k(\tilde{A}\ ^1B_1) = 10^8 \exp\left(-4200/RT\right)$$

$$k(\tilde{a}\ ^3B_1) = 3 \times 10^7 \exp\left(-3900/RT\right)\text{l.mole}^{-1}.\text{sec}^{-1}$$

Because of the necessity for the interaction of potential curves, the pre-exponential factor for O_3 with SO producing the ^1B state of SO_2, is much less than that for the reaction producing the ground state. In the $O_3 + NO$ reaction these pre-exponential factors differ only by a factor of 2.

Emission from the $^3B_1$ state is also observed. It is probably populated by collisional quenching of the ^1B state. Because of the long lifetime of this triplet state, it is vibrationally relaxed, and its emission spectrum is very similar to that in the $O + SO$ emission.

3.5 COMPLEX CHEMILUMINESCENT SYSTEMS

3.5.1 Excitation of additives in active nitrogen

Several types of reaction are involved in these systems, and the kinetics is far from clear in most cases. It therefore seems better to present in one place a brief summary of results and interpretations rather than dissecting out each type reaction for discussion in its appropriate section of this chapter. The kinetics of these reactions has been reviewed[157,251].

Active nitrogen itself is hardly a simple system, and the addition of reactive compounds is unlikely to make it simpler. It is possible, however, that addition of such compounds may divert attention from subtle details which have only marginal importance for the overall kinetic behavior. The principal long-lived energetic species in active nitrogen are N, $N_2(A^3\Sigma_u^+)$, N_2^v, and perhaps $N_2(^5\Sigma_g^+)$.[†] It is presumably from these that the energy source is to be chosen in discussing excitation of additives such as C_2N_2, ClCN, $CHCl_3$, C_2H_2, C_3O_2, $Ni(CO)_4$, $(C_2H_5)_2Zn$, I_2, PbI_2, SF_6, and $SeCl_4$. Campbell and Thrush[255] have discussed the role of the A state in a number of these excitation processes, and Thrush[162] has estimated its concentration as $[N_2(A^3\Sigma_u^+)] \approx 6 \times 10^{-22}[N][N_2]$ in active nitrogen where wall removal is unimportant.

The most extensive studies are those of CN emission due to addition of halogenated hydrocarbons and molecules containing the CN bond. Rotational and vibrational distributions have been observed as a function of pressure and nature of the additive[256,257]. CH emission is also observed in some cases. Two electron-

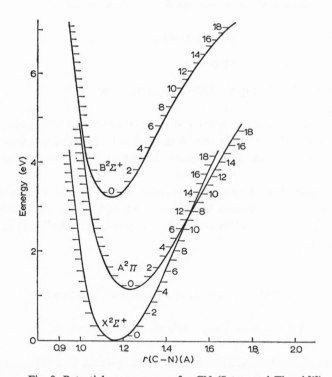

Fig. 9. Potential energy curves for CN (Setser and Thrush[262]).

[†] Čermák[252] has found evidence for a metastable state of N_2 near 11.5 eV, and Kenty[253] has frequently urged the importance of a metastable $^3\Delta_u$ state near 7.35 eV. Neither has been observed spectroscopically. Also the existence of electrons and ions in active nitrogen is well established[254].

ically excited states of CN are populated, the observed transitions are $B(^2\Sigma^+)$–$X(^2\Sigma^+)$, the violet system and $A(^2\Pi)$–$X(^2\Sigma^+)$, the red system (Fig. 9). Excitation mechanisms try to explain the relative excitation rates of the two electronic states, and the vibrational distributions within them. There is the further interesting complication that certain levels of the A and B states perturb each other rather strongly. When the excitation mechanism is such that the A state is predominantly produced, those transitions in the B–X system for which the upper level interacts with the A state will be strongly enhanced relative to other B–X transitions. Extra lines also appear, which are A–X transitions whose transition probabilities are much increased by mixing with B[256]. This mixing has been studied in considerable detail[258], including the effect of magnetic field, and rotational relaxation. Microwave pumping of appropriate A–B transitions has been used to enhance further the selective population of certain rotational levels in $v = 0$ of the B state, and for further study of rotational relaxation of these levels[259]. Mixing is also important in populating higher vibrational levels of the B state in cases where the chemical excitation mechanism principally produces the A state[260].

In the first extensive discussion of the mechanisms of production of electronically excited CN in active nitrogen containing organic additives, Bayes[261] pointed out that the vibrational distribution in the A state can be described as a superposition of two distributions. The distribution P_1 is peaked at $v = 0$ and falls to zero at $v = 4$. The second distribution, P_2, extends from $v = 2$ upward and peaks at $v = 6$ or 7. Bayes[261] suggested possible mechanisms for populating the A state with these distributions, and others[257,262,263] have elaborated the discussion. They observed that the P_1 distribution in the A state is associated with high vibrational levels in the B state, whereas P_2 in A is associated with level $v = 0$ in the B state. Here, as often, one has to distinguish between more or less physical processes which excite CN previously formed, and chemical processes which produce electronically excited CN directly. It seems probable that the P_1 vibrational distribution in the A state of CN, and the corresponding high vibration excitation in B, arise from the first of these mechanisms in which a previously existing CN bond is excited[257,261,262]. The P_2 distribution in A is probably produced chemically by a step in which a new CN bond is simultaneously formed and excited.

A third vibrational distribution has been observed[257] at pressures above 1 torr and with trace amounts of additive. Since its occurrence is independent of the nature of the additive, it is felt to arise from the process

$$N + N + CN \rightarrow N_2 + CN(A \text{ or } B) \tag{35}$$

or

$$N_2^e + CN \rightarrow N_2 + CN(A \text{ or } B) \tag{36}$$

In addition to organic cyanides and halides, metals have been added to active nitrogen, in the form of carbonyls[55] and alkyls[264]. In the carbonyl additions, rates of formation of Ni^e in successively higher excited states is a monotonically decreasing function of energy and apparently reflects the decreasing population of successively higher vibrational levels of $N_2(A)$ which supplies the energy. No CN emission was observed, although weak emission is observed when pure CO is added[256]. In interpreting their results with Zn, Al, B and Hg alkyls, March and Schiff[264] postulated $N_2(^5\Sigma)$ as the source of energy, thus avoiding the necessity of invoking high vibrational levels of the $N_2(A)$ state (see Fig. 7). The $^5\Sigma$ state is also postulated by Phillips[265,266] as the energy source for atomic line emission when metal halides are added to active nitrogen. However, some observed excited states require more energy than the $N+N$ limit if they are produced directly from the halide molecule. In these cases a stepwise process is proposed.

3.5.2 Reaction of oxygen atoms with acetylene

In the low temperature reaction of oxygen atoms with acetylene at pressures in the torr range, chemiluminescent emission is observed from one or more electronically excited states of CO, CH, C_2, OH and CHO. In none of these cases is the mechanism well established, so we will only mention briefly some recent work.

Emission from CO $A(^1\Pi)$, the fourth positive bands, has been observed in the vacuum ultraviolet[267,268] and emission from the triplet states $d(^3\Delta)$ and e ($^3\Sigma$) appears to arise from the same excitation process[268], possibly $O+C_2O \rightarrow CO^e+CO$.

The mechanism of chemi-ionization and of $CH(A^2\Delta)$ emission has been studied in considerable detail[269,331] but it was not possible to decide on a unique mechanism for the emission process. Observed rotational distributions in the $CH(A^2\Delta)$ state have been interpreted in terms of rotational relaxation[270] but the uncertainty in the excitation mechanism renders such an enterprise somewhat ambiguous.

Emission of the hydrocarbon flame bands, due to HCO^e, has recently been studied, along with chemi-ionization[271]. Again the mechanism is uncertain.

Rotational and vibrational distributions, as well as relative intensities of bands emitted by different molecules, have been surveyed by Krishnamachari and Broida[272]. They worked with $O+C_2H_2$ and also studied effects of added O_2.

4. Rotational excitation

A case in which the conservation laws demand rotational excitation of the product is the reaction[2, 273, 274, 328]

$$K+HBr \rightarrow KBr+H \tag{37}$$

Because of the relatively large mass of the K atom and the large cross section of the reaction $(34 A^2)$, the initial orbital angular momentum will be large, while the rotational angular momentum of the HBr will be small at room temperature, due to its small moment of inertia. The situation is just the reverse for the products. The departing H can have relatively little orbital angular momentum, so conservation requires that the total be concentrated in rotation of the KBr product. This is energetically possible because of the relatively large moment of inertia of KBr.

Highly energetic non-equilibrium rotational distributions are observed where they are not strictly required by the conservation laws, for example, in infrared chemiluminescence studies[4] in which special efforts are made to reduce rotational relaxation. In the reaction

$$Cl + HI \rightarrow HCl(v = 2) + I \tag{38}$$

almost all the available energy (in addition to that needed to excite $v = 2$) goes into rotation of the product. In the reaction

$$H + Cl_2 \rightarrow HCl(v = 2) + Cl \tag{39}$$

the fraction of energy going into rotation is smaller, but still substantial.

Highly energetic non-equilibrium rotational distributions are commonly observed in chemiluminescent reactions producing electronically excited molecules in ordinary flames[275-278, 329], atomic flames[148, 270, 272] and afterglows[279]. Because of the short lifetime of the electronic states involved, there is little time for rotational relaxation. In few cases is the excitation process known with any certainty. The only safe conclusion seems to be the qualitative one that non-equilibrium rotational excitation is quite common. There seems to be little reason why it should be more common among electronically excited products where it is relatively easy to detect, than among ground state products, where it is much harder to detect.

5. Chemical lasers

The absorption coefficient per unit length, averaged over line shape for a single transition is[280, 281]

$$\bar{k} \equiv -\frac{1}{I}\frac{dI}{dx} = \bar{\sigma}\left(n'' - \frac{g''n'}{g'}\right) \tag{40}$$

where $\bar{\sigma}$ is the absorption cross section (averaged over the line shape)

$$\bar{\sigma} = \frac{h\nu}{c\Delta\nu}B \tag{41}$$

Here B is the Einstein coefficient for absorption, Δv is the line width, and g'' and g' are the degeneracies of the lower and upper states. The absorption thus depends on the *difference* in population between the lower and upper states, $n'' - n'(g''/g')$. Mitchell and Zemansky[280] pointed out long ago that effects of stimulated emission, due to n', may be significant in electrical discharges. If there is a population inversion, meaning that $n'' < n'g''/g'$, we have negative absorption, or gain. If the gain is larger than the losses in an optical system, there is oscillation, *i.e.*, laser action.

Population inversions have been observed in a number of chemical and photochemical reactions. In a few of these cases, laser action has been produced in a suitable cavity. In most cases of molecular laser emission, there is only partial inversion[282] in which several vibration–rotation transitions are inverted even though the total population in the upper vibrational state does not exceed that in the lower. In this case there is laser action in P branch transitions only.

Perhaps it is most useful to define a chemical laser simply as one in which the population inversion is produced by a chemical reaction. This includes cases in which energetic precursors may be supplied externally or produced by external energy input, and in fact all known chemical lasers are initiated or driven by some external energy input. We exclude lasers in which the inversion is produced directly in the primary process of photodissociation. Possible chemical laser systems and mechanisms have been reviewed[282–286]. We will discuss here a few particular cases.

Most known chemical lasers oscillate on vibration–rotation transitions of a hydrogen halide. The first such laser was driven by the flash initiated explosion of $H_2 + Cl_2$ mixtures[287]. Here the flash dissociates the Cl_2 to start the chain decomposition, and the population inversion is due to the subsequent reactions

$$Cl + H_2 \rightarrow HCl + H$$

$$H + Cl_2 \rightarrow HCl^v + Cl \tag{42}$$

In this case, as with all other hydrogen halide lasers, only P branch transitions are observed, indicating that only partial inversion is attained. The vibrational transitions observed are $1 \rightarrow 0$ and $2 \rightarrow 1$. There is a definite threshold flash energy, below which no laser action is observed because the chain decomposition is not fast enough. The development in time of the emission spectrum was observed and discussed in terms of rotational relaxation.

In a somewhat similar approach [288], the flash initiated reaction of Cl_2 with HI gives pulsed laser action with inversion produced by the reaction

$$Cl + HI \rightarrow HCl^v + I \tag{43}$$

The laser emission occurs in a pulse during, but faster than the photolysis flash.
A number of P branch lines are observed in $\Delta v = 1$ transitions with $v' \leqq 3$.
The two reactions just discussed (42), (43), along with

$$H + Br_2 \rightarrow HBr^v + Br \tag{44}$$

have been studied with the continuous mixing of reactants, and the presence of
inversion in several vibration–rotation transitions has been demonstrated[4,46].
Build-up of population in the lower levels was prevented by rapid removal of
product molecules at a cold surface. In the case of reactions (42) and (43), the
occurrence of stimulated emission has been demonstrated by the appearance of
abnormal intensity ratios in different lines having different gain. In reaction (43)
intermittent oscillation was observed as emission spikes. In these experiments the
atoms taking part in the reaction which produces the inversion are produced outside the laser cavity in an electric discharge, so that this source of energy takes the
place of the flash lamp in the pulse lasers.

Hydrogen halide laser emission has also been produced in pulsed discharges
through mixtures of H_2 with halogens or their compounds[289–291]. The fact that
laser emission is not observed from a similar discharge through the hydrogen
halide itself is taken to indicate that the excitation is indeed the result of a chemical
reaction. Laser emission pulses have been observed on vibration–rotation transitions of HF, DF, HCl, DCl, HBr, and DBr. P branch lines of $\Delta v = 1$ transitions
are observed, in some cases with $v' \leqq 5$. Nothing is known about the mechanism,
but in the pulsed discharge there is no lack of energetic species. The absence of
hydrogen halide initially is presumably necessary to ensure a sufficiently low
population of the lower laser level. In pulsed discharges through $H_2 + CF_4$,
laser action was observed on pure rotation transitions in several of the lowest
vibrational states of HF^{289}.

Laser action is observed on vibration–rotation transitions of CO in the flash
photolysis of $CS_2 + O_2$ mixtures[61]. $\Delta v = 1$ transitions are observed with v' in the
range 6–14. Only P branch lines are observed, as usual. The excitation is presumably
chemical rather than by energy transfer to ground state CO. A suggested mechanism involves

$$SO + CS \rightarrow CO^v + S_2 \tag{45}$$

but there are several alternatives.

Non-equilibrium excitation in flames has been discussed from the point of view
of possible inversions[286]. The possibility of laser action on several transitions of
CN excited in active nitrogen has been discussed[292] in terms of relevant rate
equations and the threshold condition for oscillation. A chemical laser is of course
a physical phenomenon, the performance of which depends critically on the rate

processes involved. In principle then, measurements on a chemical laser, such as dependence of threshold and gain on composition, should be useful in estimating rate coefficients. Relatively little along these lines has been done with atomic lasers, and the situation with molecular lasers is less encouraging.

ACKNOWLEDGEMENT

R. L. Brown, L. Burnelle, M. A. A. Clyne, F. Kaufman and J. C. Polanyi have provided us with the results of their researches prior to publication. F. Kaufman and H. I. Schiff have aided with comments on parts of this paper. F. R. Gilmore, M. A. A. Clyne, E. A. Ogryzlo and B. A. Thrush have kindly provided figures. Professor Polanyi, the Chemical Kinetics Information Center (NBS), the Microwave Spectra Data Center (NBS) and the Diatomic Molecule Spectra and Energy Level Center (NBS) have supplied us with an abundance of reference material. The organization of the bibliography, tabular material and the typescript has been done by Mrs. M. C. Peter. To all of these we express our appreciation.

REFERENCES

1 B. S. Rabinovitch and M. C. Flowers, *Quart. Rev. (London)*, 18 (1964) 122.
2 D. R. Herschbach, *Advan. Chem. Phys.*, 10 (1966) 319.
3 J. C. Light, *Discussions Faraday Soc.*, 44 (1967) 14.
4 K. G. Anlauf, P. J. Kuntz, D. H. Maylotte, P. D. Pacey and J. C. Polanyi, *Discussions Faraday Soc.*, 44 (1967) 183.
5 T. Carrington, *Discussions Faraday Soc.*, 33 (1962) 44.
6 K. J. Laidler and J. C. Polanyi, *Progr. Reaction Kinetics*, 3 (1965) 1.
7 J. C. Polanyi, *Discussions Faraday Soc.*, 44 (1967) 293.
8 W. B. Miller, S. A. Safron and D. R. Herschbach, *Discussions Faraday Soc.*, 44 (1967) 108.
9 R. E. Minturn, S. Datz and R. L. Becker, *J. Chem. Phys.*, 44 (1966) 1149.
10 Z. Herman, J. Kerstetter, T. Rose and R. Wolfgang, *Discussions Faraday Soc.*, 44 (1967) 123.
10a W. R. Gentry, E. A. Gislason, Y-T. Lee, B. H. Mahan and C-W. Tsao, *Discussions Faraday Soc.*, 44 (1967) 137.
11 J. P. Simons, *Nature*, 186 (1960) 551.
12 P. J. Kuntz, E. M. Nemeth, J. C. Polanyi, S. D. Rosner and C. E. Young, *J. Chem. Phys.*, 44 (1966) 1168.
13 F. T. Smith, *J. Chem. Phys.*, 31 (1959) 1352.
14 M. Krauss, *Natl. Bur. Std. (U.S.), Tech. Note* 438 (1968).
15 D. L. Bunker and N. C. Blais, *J. Chem. Phys.*, 41 (1964) 2377.
16 R. N. Porter and M. Karplus, *J. Chem. Phys.*, 40 (1964) 1105.
17 M. Karplus and M. Godfrey, *J. Am. Chem. Soc.*, 88 (1966) 5332.
17a L. M. Raff and M. Karplus, *J. Chem. Phys.*, 44 (1966) 1212.
18 J. C. Polanyi, *J. Quant. Spectry. Radiative Transfer*, 3 (1963) 471.
19 I. W. M. Smith, *Discussions Faraday Soc.*, 44 (1967) 194.
20 L. M. Raff, *J. Chem. Phys.*, 44 (1966) 1202.
21 M. Karplus, R. N. Porter and R. D. Sharma, *J. Chem. Phys.*, 43 (1965) 3259.

22 M. KARPLUS AND K. T. TANG, *Discussions Faraday Soc.*, 44 (1967) 56.
23 R. J. SUPLINSKAS AND J. ROSS, *J. Chem. Phys.*, 47 (1967) 321.
24 M. S. CHILD, *Proc. Roy. Soc. (London)*, A292 (1966) 272.
25 R. A. MARCUS, *J. Chem. Phys.*, 45 (1966) 4493, 4500.
26 G. HERZBERG, *Molecular Spectra and Molecular Structure*, 2nd Ed., Vol. I, *Spectra of Diatomic Molecules*, Van Nostrand, New York, 1950.
27 D. MICHA, *J. Chem. Phys.*, 41 (1964) 1947.
28 D. R. BATES, H. C. JOHNSTON AND I. STEWART, *Proc. Phys. Soc. (London)*, 84 (1964) 517.
29 K. J. LAIDLER, *The Chemical Kinetics of Excited States*, Oxford University Press, 1955.
30 C. A. COULSON AND K. ZALEWSKI, *Proc. Roy. Soc. (London)*, A268 (1962) 437.
31 G. HERZBERG AND H. C. LONGUET-HIGGINS, *Discussions Faraday Soc.*, 35 (1963) 77.
32 K. E. SHULER, *J. Chem. Phys.*, 21 (1953) 624.
33 G. HERZBERG, *Molecular Spectra and Molecular Structure*, Vol. III, *Electronic Spectra and Electronic Structure of Polyatomic Molecules*, Van Nostrand, New York, 1966.
34 T. CARRINGTON, *J. Chem. Phys.*, 41 (1964) 2012.
35 W. D. MCGRATH AND R. G. W. NORRISH, *Proc. Roy. Soc. (London)*, A242 (1957) 265.
36 W. D. MCGRATH AND R. G. W. NORRISH, *Proc. Roy. Soc. (London)*, A254 (1960) 317.
37 N. BASCO AND R. G. W. NORRISH, *Can. J. Chem.*, 38 (1960) 1769.
38 N. BASCO AND R. G. W. NORRISH, *Proc. Roy. Soc. (London)*, A260 (1961) 293.
39 N. BASCO AND R. G. W. NORRISH, *Discussions Faraday Soc.*, 33 (1962) 99.
40 R. V. FITZSIMMONS AND R. J. BAIR, *J. Chem. Phys.*, 40 (1964) 451.
40a V. D. BAIAMONTE, D. R. SNELLING AND E. J. BAIR, *J. Chem. Phys.*, 44 (1966) 673.
40b D. R. SNELLING, V. D. BAIAMONTE AND E. J. BAIR, *J. Chem. Phys.*, 44 (1966) 4137.
41 W. M. JONES AND N. DAVIDSON, *J. Am. Chem. Soc.*, 84 (1962) 2868.
42 F. J. LIPSCOMB, R. G. W. NORRISH AND B. A. THRUSH, *Proc. Roy. Soc. (London)*, A233 (1956) 455.
43 N. BASCO, *Proc. Roy. Soc. (London)*, A283 (1965) 302.
43a J. R. AIREY, F. D. FINDLAY AND J. C. POLANYI, *Can. J. Chem.*, 42 (1964) 2193.
44 L. F. PHILLIPS AND H. I. SCHIFF, *J. Chem. Phys.*, 42 (1965) 3171.
45 R. E. MARCH, S. G. FURNIVAL AND H. I. SCHIFF, *Photochem. Photobiol.*, 4 (1965) 971.
46 K. G. ANLAUF, D. H. MAYLOTTE, P. D. PACEY AND J. C. POLANYI, *Phys. Letters*, 24A (1967) 208.
47 A. W. REED, *Progr. Reaction Kinetics*, 3 (1965) 205.
48 A. I. OSIPOV, *Russ. J. Phys. Chem.*, 36 (1962) 972.
49 J. C. POLANYI, *J. Chem. Phys.*, 31 (1959) 1338.
50 C. E. H. BAWN, *Ann. Rept. Progr. Chem. (Chem. Soc. London)*, 39 (1942) 36.
51 E. WARHURST, *Quart. Rev. (London)*, 5 (1951) 44.
52 A. F. TROTMAN-DICKENSON, *Gas Kinetics*, Butterworth, London, 1955, p. 212.
53 A. F. TROTMAN-DICKENSON, *Tables of Bimolecular Gas Reactions* (National Standard Reference Data Series–National Bureau of Standards, No. 9, Washington D.C., 1967, pp. 28–37). This reference summarizes the available kinetic information.
54 M. POLANYI, *Atomic Reactions*, Williams and Norgate, London, 1932.
55 W. R. BRENNEN AND G. B. KISTIAKOWSKY, *J. Chem. Phys.*, 44 (1966) 2695.
56 R. ENGLEMAN, JR., *J. Am. Chem. Soc.*, 87 (1965) 4193.
57 N. BASCO AND R. G. W. NORRISH, *Proc. Roy. Soc. (London)*, A268 (1962) 291.
58 A. T. STAIR, JR. AND J. P. KENNEALY, *J. Chim. Phys.*, 64 (1967) 124.
59 A. B. MEINEL, *Astrophys. J.*, 111 (1950) 207, 433, 555.
60 H. S. HEAPS AND G. HERZBERG, *Z. Physik*, 133 (1952) 48.
61 M. A. POLLACK, *Appl. Phys. Letters*, 8 (1966) 237.
62 W. D. MCGRATH, *J. Chem. Phys.*, 33 (1960) 297.
63 E. HIROTA, *Bull. Chem. Soc. Japan*, 31 (1958) 130.
64 R. L. KUCZKOWSKI, *J. Am. Chem. Soc.*, 86 (1964) 3617.
65 J. K. CASHION AND J. C. POLANYI, *Proc. Roy. Soc. (London)*, A258 (1960) 529.
66 J. K. CASHION AND J. C. POLANYI, *J. Chem. Phys.*, 30 (1959) 317.
67 M. A. A. CLYNE AND B. A. THRUSH, *Trans. Faraday Soc.*, 57 (1961) 1305.

68 M. A. A. CLYNE AND B. A. THRUSH, *Discussions Faraday Soc.*, 33 (1962) 139.
69 R. SIMONAITIS, *J. Phys. Chem.*, 67 (1963) 2227.
70 O. P. STRAUSZ AND H. E. GUNNING, *Trans. Faraday Soc.*, 60 (1964) 347.
71 D. RAPP AND H. S. JOHNSTON, *J. Chem. Phys.*, 33 (1960) 695.
72 P. J. BODDY AND J. C. ROBB, *Proc. Roy. Soc. (London)*, A249 (1959) 518.
73 J. H. CURRENT AND B. S. RABINOVITCH, *J. Chem. Phys.*, 38 (1963) 783.
74 J. H. CURRENT AND B. S. RABINOVITCH, *J. Chem. Phys.*, 38 (1963) 1967.
75 B. S. RABINOVITCH, D. H. DILLS, W. H. MCLAIN AND J. H. CURRENT, *J. Chem. Phys.*, 32 (1960) 493.
76 W. E. FALCONER, B. S. RABINOVITCH AND R. J. CVETANOVIC, *J. Chem. Phys.*, 39 (1963) 40.
77 B. S. RABINOVITCH AND R. W. DIESEN, *J. Chem. Phys.*, 30 (1959) 735.
78 J. W. SIMONS, D. W. SETSER AND B. S. RABINOVITCH, *J. Am. Chem. Soc.*, 84 (1962) 1758.
79 B. S. RABINOVITCH AND M. J. PEARSON, *J. Chem. Phys.*, 41 (1964) 280.
80 C. A. HELLER AND A. S. GORDON, *J. Chem. Phys.*, 36 (1962) 2648.
81 J. H. CURRENT, B. S. RABINOVITCH, C. A. HELLER AND A. S. GORDON, *J. Chem. Phys.*, 39 (1963) 3535.
82 B. S. RABINOVITCH AND D. W. SETSER, *Advan. Photochemistry*, 3 (1964) 1.
83 B. S. RABINOVITCH, R. F. KUBIN AND R. E. HARRINGTON, *J. Chem. Phys.*, 38 (1963) 405.
84 W. BRAUN AND M. LENZI, *Discussions Faraday Soc.*, 44 (1967) 252.
85 J. V. MICHAEL AND R. E. WESTON, JR., *J. Chem. Phys.*, 45 (1966) 3632.
86 G. HERZBERG AND J. SHOOSMITH, *Nature*, 183 (1959) 1801.
87 H. M. FREY, *Progr. Reaction Kinetics*, 2 (1964) 131.
88 H. M. FREY, in *Carbene Chemistry*, W. KIRMSE (Ed.), *Organic Chemistry*, Vol. 1, Academic Press, New York, 1964, p. 217.
89 J. A. BELL, *Progr. Phys. Org. Chem.*, 2 (1964) 1.
90 H. M. FREY AND G. B. KISTIAKOWSKY, *J. Am. Chem. Soc.*, 79 (1957) 6373.
91 G. O. PRITCHARD, J. T. BRYANT AND R. L. THOMMARSON, *J. Phys. Chem.*, 69 (1965) 2804.
92 D. W. SETSER AND B. S. RABINOVITCH, *Can. J. Chem.*, 40 (1962) 1425.
93 M. C. FLOWERS AND H. M. FREY, *J. Chem. Soc.*, (1962) 1157.
94 H. M. FREY, *Chem. Commun.*, (1965) 260.
95 R. W. CARR, JR. AND G. B. KISTIAKOWSKY, *J. Phys. Chem.*, 70 (1966) 118.
96 G. B. KISTIAKOWSKY AND P. H. KYDD, *J. Am. Chem. Soc.*, 79 (1957) 4825.
97 F. H. DORER AND B. S. RABINOVITCH, *J. Phys. Chem.*, 69 (1965) 1952.
98 J. A. BELL AND G. B. KISTIAKOWSKY, *J. Am. Chem. Soc.*, 84 (1962) 3417.
99 W. J. DUNNING AND C. C. MCCAIN, *J. Chem. Soc. (B)*, (1966) 68.
100 G. Z. WHITTEN AND B. S. RABINOVITCH, *J. Phys. Chem.*, 69 (1965) 4348.
101 J. N. BUTLER AND G. B. KISTIAKOWSKY, *J. Am. Chem. Soc.*, 82 (1960) 759.
102 J. N. BUTLER AND G. B. KISTIAKOWSKY, *J. Chem. Phys.*, 83 (1961) 1324.
103 R. GOMER AND G. B. KISTIAKOWSKY, *J. Chem. Phys.*, 19 (1951) 85.
104 G. B. KISTIAKOWSKY AND E. K. ROBERTS, *J. Chem. Phys.*, 21 (1953) 1637.
105 A. SHEPP, *J. Chem. Phys.*, 24 (1956) 939.
106 W. G. ALCOCK AND E. WHITTLE, *Trans. Faraday Soc.*, 61 (1965) 244.
107 R. D. GILES AND E. WHITTLE, *Trans. Faraday Soc.*, 61 (1965) 1425.
108 J. C. HASSLER, D. W. SETSER AND R. L. JOHNSON, *J. Chem. Phys.*, 45 (1966) 3231.
109 J. C. HASSLER AND D. W. SETSER, *J. Chem. Phys.*, 45 (1966) 3237.
110 J. C. HASSLER AND D. W. SETSER, *J. Chem. Phys.*, 45 (1966) 3246.
111 J. A. KERR, A. W. KIRK, B. V. O'GRADY, D. C. PHILLIPS AND A. F. TROTMAN-DICKENSON, *Discussions Faraday Soc.*, 44 (1967) 263.
112 B. S. RABINOVITCH AND D. W. SETSER, *J. Am. Chem. Soc.*, 83 (1961) 750.
113 H. M. FREY AND I. D. R. STEVENS, *J. Chem. Soc.*, (1962) 3865.
114 H. M. FREY AND I. D. R. STEVENS, *J. Am. Chem. Soc.*, 84 (1962) 2647.
115 B. S. RABINOVITCH, K. W. WATKINS AND D. F. RING, *J. Am. Chem. Soc.*, 87 (1965) 4960.
116 J. AKRICHE, L. HERMAN AND H. GRENAT, *J. Phys. Radium*, 19 (1958) 649.
117 F. H. MIES AND A. L. SMITH, *J. Chem. Phys.*, 45 (1966) 994.
118 C. A. BARTH, *Ann. Geophys.*, 20 (1964) 182.

119 H. B. PALMER AND R. A. CARABETTA, *J. Chem. Phys.*, 46 (1967) 1538.

120 H. B. PALMER, *J. Chem. Phys.*, 47 (1967) 2116.

121 A. L. SMITH, *J. Chem. Phys.*, 48 (1968) 4817.

122 Y. TANAKA AND K. YOSHINO, *J. Chem. Phys.*, 39 (1963) 3081.

123 D. C. ALLISON, J. C. BROWNE AND A. DALGARNO, *Proc. Phys. Soc. (London)*, 89 (1966) 41.

124 D. VILLAREJO, R. R. HERM AND M. G. INGRAHAM, *J. Opt. Soc. Am.*, 56 (1966) 1574.

125 R. A. GERBER, G. F. SAUTER AND H. J. OSKAM, *Physica*, 32 (1966) 2173.

126 M. P. TETER AND W. W. ROBERTSON, *J. Chem. Phys.*, 45 (1966) 661.

127 W. B. HURT AND C. B. COLLINS, *J. Chem. Phys.*, 45 (1966) 295.

128 M. A. A. CLYNE AND D. H. STEDMAN, *Chem. Phys. Letters*, 1 (1967) 36.

129 M. A. A. CLYNE AND D. H. STEDMAN, *Trans. Faraday Soc.*, 64 (1968) 1816.

130 L. W. BADER AND E. A. OGRYZLO, *J. Chem. Phys.*, 41 (1964) 2926.

131 A. E. DOUGLAS, CHR. KN. MØLLER AND B. P. STOICHEFF, *Can. J. Phys.*, 41 (1963) 1174.

132 C. H. DUGAN, *J. Chem. Phys.*, 47 (1967) 1512.

133 R. A. YOUNG AND G. A. ST. JOHN, *J. Chem. Phys.*, 48 (1968) 895.

134 E. HUTTON AND M. WRIGHT, *Trans. Faraday Soc.*, 61 (1965) 78.

135 M. VAN THIEL, D. J. SEERY AND D. BRITTON, *J. Phys. Chem.*, 69 (1965) 834.

136 R. A. CARABETTA AND H. B. PALMER, *J. Chem. Phys.*, 46 (1967) 1325.

137 H. B. PALMER, *J. Chem. Phys.*, 26 (1957) 648.

138 D. B. GIBBS AND A. E. OGRYZLO, *Can. J. Chem.*, 43 (1965) 1905.

139 M. A. A. CLYNE AND J. A. COXON, *J. Mol. Spectry.*, 23 (1967) 258.

140 M. A. A. CLYNE AND J. A. COXON, *Proc. Roy. Soc. (London)*, A298 (1967) 424.

141 L. F. PHILLIPS AND T. M. SUGDEN, *Can. J. Chem.*, 38 (1960) 1804.

142 H. P. BROIDA AND A. G. GAYDON, *Proc. Roy. Soc. (London)*, A222 (1954) 181.

143 R. A. YOUNG AND G. BLACK, *J. Chem. Phys.*, 44 (1966) 3741.

144 R. A. YOUNG AND R. L. SHARPLESS, *J. Chem. Phys.*, 39 (1963) 1071.

145 P. HARTECK AND R. R. REEVES, JR., *Discussions Faraday Soc.*, 37 (1964) 82.

146 S. TICKTIN, G. B. SPINDLER AND H. I. SCHIFF, *Discussions Faraday Soc.*, 44 (1967) 218.

147 A. G. GAYDON AND H. G. WOLFHARD, *Proc. Roy. Soc. (London)*, A208 (1951) 63.

148 D. W. NAEGELI AND H. B. PALMER, *J. Mol. Spectry.*, 23 (1967) 44.

149 M. CHARTON AND A. G. GAYDON, *Proc. Roy. Soc. (London)*, A245 (1958) 84.

150 W. E. KASKAN, *J. Chem. Phys.*, 31 (1959) 944.

151 I. R. HURLE AND A. HERTZBERG, *Phys. Fluids*, 8 (1965) 1601.

152 Y. TANAKA, *J. Chem. Phys.*, 22 (1954) 2045.

153 R. A. YOUNG AND R. L. SHARPLESS, *Discussions Faraday Soc.*, 33 (1962) 228.

154 A. B. CALLEAR AND I. W. M. SMITH, *Discussions Faraday Soc.*, 37 (1964) 96.

155 C. A. BARTH, W. J. SCHADE AND J. KAPLAN, *J. Chem. Phys.*, 30 (1959) 347.

156 I. M. CAMPBELL AND B. A. THRUSH, *Proc. Roy. Soc. (London)*, A296 (1967) 222.

156a R. W. F. GROSS AND N. COHEN, *J. Chem. Phys.*, 48 (1968) 2582.

157 B. BROCKLEHURST AND K. R. JENNINGS, *Progr. Reaction Kinetics*, 4 (1967) 1.

158 K. D. BAYES AND G. B. KISTIAKOWSKY, *J. Chem. Phys.*, 32 (1960) 992.

159 P. K. CARROLL, *J. Chem. Phys.*, 37 (1962) 805.

160 R. S. MULLIKEN, *J. Chem. Phys.*, 37 (1962) 809.

161 P. HARTECK, R. R. REEVES AND D. APPLEBAUM, *Symposium on Chemiluminescence*, U.S. Army Research Office, Durham, N.C., April 1965, p. 91.

162 B. A. THRUSH, *J. Chem. Phys.*, 47 (1967) 3691.

163 W. BRENNEN AND R. L. BROWN, unpublished.

164 I. M. CAMPBELL AND B. A. THRUSH, *Proc. Roy. Soc. (London)*, A296 (1967) 201.

165 G. B. KISTIAKOWSKY AND P. WARNECK, *J. Chem. Phys.*, 27 (1957) 1417.

166 R. L. BROWN AND S. DITTMANN, *Chem. Commun.*, (1967) 1144.

167 P. K. CARROLL AND R. S. MULLIKEN, *J. Chem. Phys.*, 43 (1965) 2170.

168 J. M. ANDERSON, *Proc. Phys. Soc. (London)*, 87 (1966) 299.

169 P. J. PADLEY AND T. M. SUGDEN, *Proc. Roy. Soc. (London)*, A248 (1958) 248.

170 P. J. PADLEY AND T. M. SUGDEN, *Symp. Combustion 7th*, (1959) 235.

171 T. M. SUGDEN, *Ann. Rev. Phys. Chem.*, 13 (1962) 369.

172 J. C. POLANYI AND C. M. SADOWSKI, *J. Chem. Phys.*, 36 (1962) 2239.
173 E. M. NEMETH, J. C. POLANYI AND C. M. SADOWSKI, *J. Chem. Phys.*, 40 (1964) 2054.
174 J. D. MCKINLEY, JR. AND J. C. POLANYI, *Can. J. Chem.*, 36 (1958) 107.
175 J. L. MAGEE AND T. RI, *J. Chem. Phys.*, 9 (1941) 638.
176 V. F. ZHITKEVICH, A. I. LYUTYI, N. A. NESTERKO, V. S. ROSSIKHIN AND I. L. TSIKORA, *Opt. Spectry.*, 14 (1963) 180.
177 V. F. ZHITKEVICH, A. I. LYUTYI, V. S. ROSSIKHIN AND I. L. TSIKORA, *Opt. Spectry.*, 15 (1963) 217.
178 H. P. BROIDA AND K. E. SHULER, *J. Chem. Phys.*, 27 (1957) 933.
179 R. A. YOUNG AND G. BLACK, *Planetary Space Sci.*, 14 (1966) 113.
180 J. KAPLAN, W. J. SCHADE, C. A. BARTH AND A. F. HILDENBRANDT, *Can. J. Chem.*, 38 (1960) 1688.
181 A. G. GAYDON, *The Spectroscopy of Flames*, Wiley, New York, 1957.
182 F. E. BELLES AND M. R. LAUVER, *J. Chem. Phys.*, 40 (1964) 415.
183 V. N. KONDRATIEV AND M. ZISKIN, *Acta Physicochim. URSS*, 7 (1937) 65.
184 T. M. SUGDEN AND A. DEMERDACHE, *Nature*, 195 (1962) 596.
185 D. GARVIN AND H. P. BROIDA, *Symp. Combustion, 9th*, (1963) 678.
186 H. I. SCHIFF, *Ann. Geophys.*, 20 (1964) 115.
187 J. HEICKLEN AND N. COHEN, *The Role of Nitric Oxide in Photochemistry*, Aerospace Report No. TR-1001(2250-40)-4, Aerospace Corporation, El Segundo, California, 1966.
188 F. KAUFMAN, *Proc. Roy. Soc. (London)*, A247 (1958) 123.
189 D. APPLEBAUM, P. HARTECK AND R. R. REEVES, *Photochem. Photobiol.*, 4 (1965) 1003.
190 G. DOHERTY AND N. JONATHAN, *Discussions Faraday Soc.*, 37 (1965) 73.
191 F. KAUFMAN AND J. R. KELSO, *Symposium on Chemiluminescence*, U.S. Army Research Office, Durham, N.C., April 1965, p. 65.
192 M. A. A. CLYNE AND B. A. THRUSH, *Proc. Roy. Soc. (London)*, A269 (1962) 404.
193 B. P. LEVITT, *J. Chem. Phys.*, 42 (1965) 1038.
194 R. A. HARTUNIAN, W. P. THOMPSON AND E. W. HEWITT, *J. Chem. Phys.*, 44 (1966) 1765.
195 A. FONTIJN, C. B. MEYER AND H. I. SCHIFF, *J. Chem. Phys.*, 40 (1964) 64.
196 H. P. BROIDA, H. I. SCHIFF AND T. M. SUGDEN, *Trans. Faraday Soc.*, 57 (1961) 259.
197 D. E. PAULSEN, W. F. SHERIDAN AND R. E. HUFFMAN, *Bull. Am. Phys. Soc.*, 11 (1966) 746.
198 A. E. DOUGLAS AND K. P. HUBER, *Can. J. Phys.*, 43 (1965) 74.
199 E. FREEDMAN AND J. R. KELSO, *Bull. Am. Phys. Soc.*, 11 (1966) 453.
200 D. B. HARTLEY AND B. A. THRUSH, *Discussions Faraday Soc.*, 37 (1964) 220.
201 L. BURNELLE, A. M. MAY AND R. A. GANGI, unpublished results.
202 D. NEUBERGER AND A. B. F. DUNCAN, *J. Chem. Phys.*, 22 (1954) 1693.
203 A. E. DOUGLAS, *J. Chem. Phys.*, 45 (1966) 1007.
204 G. B. SPINDLER, *Planetary Space Sci.*, 14 (1966) 53.
205 A. FONTIJN AND D. E. ROSNER, *J. Chem. Phys.*, 46 (1967) 3275.
206 G. H. MYERS, D. M. SILVER AND F. KAUFMAN, *J. Chem. Phys.*, 44 (1966) 718.
207 F. KAUFMAN, private communication.
208 R. R. REEVES, P. HARTECK AND W. H. CHACE, *J. Chem. Phys.*, 41 (1964) 764.
209 D. GOLOMB, N. W. ROSENBERG, C. AHARONIAN, J. A. F. HILL AND H. L. ALDEN, *J. Geophys. Res.*, 70 (1965) 1155.
210 F. P. DEL GRECO, D. GOLOMB, J. A. VAN DER BLIEK, R. A. CASSANOVA AND R. E. GOOD, *J. Chem. Phys.*, 44 (1966) 4349.
211 J. A. VAN DER BLIEK, R. A. CASSANOVA, D. GOLOMB, F. P. DEL GRECO, J. A. F. HILL AND R. E. GOOD, *Advances in Applied Mechanics*, Vol. II, Academic Press, New York, 1967, p. 1543.
212 T. A. MILNE AND F. T. GREENE, *J. Chem. Phys.*, 47 (1967) 3668.
213 W. E. KASKAN, *Combust. Flame*, 3 (1959) 39.
214 M. W. FEAST, *Proc. Phys. Soc. (London)*, A63 (1950) 772.
215 B. F. MYERS AND E. R. BARTLE, *J. Chem. Phys.*, 47 (1967) 1783.
216 T. A. BRABBS AND F. E. BELLES, *Symp. Combustion, 11th*, (1967) 125.
217 M. A. A. CLYNE AND B. A. THRUSH, *Symp. Combustion, 9th*, (1963) 177.
218 H. P. BROIDA AND A. G. GAYDON, *Trans. Faraday Soc.*, 49 (1953) 1190.

219 B. H. MAHAN AND R. B. SOLO, *J. Chem. Phys.*, 37 (1962) 2669.
220 R. N. DIXON, *Discussions Faraday Soc.*, 35 (1963) 105.
221 R. G. W. NORRISH AND A. P. ZEELENBERG, *Proc. Roy. Soc. (London)*, A240 (1957) 293.
222 M. A. A. CLYNE, C. J. HALSTEAD AND B. A. THRUSH, *Proc. Roy. Soc. (London)*, A295 (1966) 355.
223 A. SHARMA, J. P. PADUR AND P. WARNECK, *J. Chem. Phys.*, 43 (1965) 2155.
224 A. SHARMA, J. P. PADUR AND P. WARNECK, *J. Phys. Chem.*, 71 (1967) 1602.
225 R. T. ROLFES, R. R. REEVES, JR. AND P. HARTECK, *J. Phys. Chem.*, 69 (1965) 849.
226 L. HERMAN, J. AKRICHE AND H. GRENAT, *J. Quant. Spectry. Radiative Transfer*, 2 (1962) 215.
227 C. J. HALSTEAD AND B. A. THRUSH, *Proc. Roy. Soc. (London)*, A295 (1966) 363.
228 I. DUBOIS AND B. ROSEN, *Discussions Faraday Soc.*, 35 (1963) 124.
229 K. F. GREENOUGH AND A. B. F. DUNCAN, *J. Am. Chem. Soc.*, 83 (1961) 555.
230 A. J. MERER, *Discussions Faraday Soc.*, 35 (1963) 127.
231 B. P. LEVITT AND D. B. SHEEN, *J. Chem. Phys.*, 41 (1964) 584.
231a B. P. LEVITT AND D. B. SHEEN, *Trans. Faraday Soc.*, 63 (1967) 540.
232 M. J. Y. CLEMENT AND D. A. RAMSAY, *Can. J. Phys.*, 39 (1961) 205.
233 J. L. BANCROFT, J. M. HOLLAS AND D. A. RAMSAY, *Can. J. Phys.*, 40 (1962) 322.
234 D. B. HARTLEY AND B. A. THRUSH, *Proc. Roy. Soc. (London)*, A297 (1967) 520.
235 P. J. PADLEY, *Trans. Faraday Soc.*, 56 (1960) 449.
236 D. I. WALTON, M. C. McEWAN AND L. F. PHILLIPS, *Can. J. Chem.*, 43 (1965) 3095.
237 L. F. PHILLIPS, *Can. J. Chem.*, 43 (1965) 369.
238 W. J. MILLER AND H. B. PALMER, *J. Chem. Phys.*, 40 (1964) 3701;
   D. W. NAEGLI AND H. B. PALMER, *J. Chem. Phys.*, 48 (1968) 2372.
239 P. CADMAN AND J. C. POLANYI, *J. Phys. Chem.*, 72 (1968) 3715.
240 J. K. CASHION AND J. C. POLANYI, *Proc. Roy. Soc. (London)*, A258 (1960) 570.
241 J. R. AIREY, P. D. PACEY AND J. C. POLANYI, *Symp. Combustion, 11th*, (1967) 85.
242 M. C. MOULTON AND D. R. HERSCHBACH, *J. Chem. Phys.*, 44 (1966) 3010.
243 J. L. MAGEE, *J. Chem. Phys.*, 8 (1940) 687.
244 J. W. CHAMBERLAIN, *Physics of the Aurora and Airglow*, Academic Press, New York, 1961, p. 565.
245 Y. TANAKA AND M. OGAWA, in *The Airglow and the Aurorae*, E. B. ARMSTRONG AND A. DALGARNO (Eds.), Pergamon, London, 1955, p. 270.
246 M. A. A. CLYNE AND J. A. COXON, *Chem. Commun.*, (1966) 285.
247 J. C. GREAVES AND D. GARVIN, *J. Chem. Phys.*, 30 (1959) 348.
248 M. A. A. CLYNE, B. A. THRUSH AND R. P. WAYNE, *Trans. Faraday Soc.*, 60 (1964) 359.
249 P. N. CLOUGH AND B. A. THRUSH, *Trans. Faraday Soc.*, 63 (1967) 915.
250 C. J. HALSTEAD AND B. A. THRUSH, *Proc. Roy. Soc. (London)*, A295 (1966) 380.
251 H. G. V. EVANS, G. R. FREEMAN AND C. A. WINKLER, *Can. J. Chem.*, 34 (1956) 1271.
252 V. ČERMÁK, *J. Chem. Phys.*, 44 (1966) 1318.
253 C. KENTY, *J. Chem. Phys.*, 47 (1967) 2545.
254 H. P. BROIDA AND I. TANAKA, *J. Chem. Phys.*, 36 (1962) 236.
255 I. M. CAMPBELL AND B. A. THRUSH, *Chem. Commun.*, (1967) 932.
256 N. H. KIESS AND H. P. BROIDA, *Symp. Combustion, 7th*, (1959) 207.
257 T. IWAI, M. I. SAVADATTI AND H. P. BROIDA, *J. Chem. Phys.*, 47 (1967) 3861.
258 H. E. RADFORD AND H. P. BROIDA, *J. Chem. Phys.*, 38 (1963) 644.
259 K. M. EVENSON AND H. P. BROIDA, *J. Chem. Phys.*, 44 (1966) 1637.
260 R. L. BROWN AND H. P. BROIDA, *J. Chem. Phys.*, 41 (1964) 2053.
261 K. D. BAYES, *Can. J. Chem.*, 39 (1961) 1074.
262 D. W. SETSER AND B. A. THRUSH, *Proc. Roy. Soc. (London)*, A288 (1965) 256.
263 I. M. CAMPBELL AND B. A. THRUSH, *Proc. Chem. Soc.*, (1964) 410.
264 R. E. MARCH AND H. I. SCHIFF, *Can. J. Chem.*, 45 (1967) 1891.
265 L. F. PHILLIPS, *Can. J. Chem.*, 41 (1963) 732.
266 L. F. PHILLIPS, *Can. J. Chem.*, 41 (1963) 2060.
267 N. JONATHAN, F. F. MARMO AND J. P. PADUR, *J. Chem. Phys.*, 42 (1965) 1463.
268 K. H. BECKER AND K. D. BAYES, *J. Chem. Phys.*, 45 (1966) 396.

269 C. A. ARRINGTON, W. BRENNEN, G. P. GLASS, J. V. MICHAEL AND H. NIKI, *J. Chem. Phys.*, 43 (1965) 1489.
270 W. BRENNEN AND T. CARRINGTON, *J. Chem. Phys.*, 46 (1967) 7.
271 A. FONTIJN, *J. Chem. Phys.*, 44 (1966) 1702.
272 S. L. N. G. KRISHNAMACHARI AND H. P. BROIDA, *J. Chem. Phys.*, 34 (1961) 1709.
273 D. BECK, E. F. GREENE AND J. ROSS, *J. Chem. Phys.*, 37 (1962) 2895.
274 E. F. GREENE, A. L. MOURSUND AND J. ROSS, *Advan. Chem. Phys.*, 10 (1966) 135.
275 H. P. BROIDA AND D. F. HEATH, *J. Chem. Phys.*, 26 (1957) 223.
276 H. P. BROIDA AND H. J. KOSTKOWSKI, *J. Chem. Phys.*, 23 (1955) 754.
277 R. BLEEKRODE, *J. Chem. Phys.*, 45 (1966) 3153.
278 R. BLEEKRODE AND W. C. NIEUPOORT, *J. Chem. Phys.*, 43 (1965) 3680.
279 W. BRENNEN, *J. Chem. Phys.*, 44 (1966) 1793.
280 A. C. G. MITCHELL AND M. W. ZEMANSKY, *Resonance Radiation and Excited Atoms*, Cambridge University Press, 1934 (reprinted 1961).
281 A. YARIV AND J. P. GORDON, *Proc. IEEE*, 51 (1963) 1.
282 J. C. POLANYI, *Appl. Opt. Suppl.*, No. 2 (1965) 109.
283 R. A. YOUNG, *J. Chem. Phys.*, 40 (1964) 1848.
284 K. E. SHULER, T. CARRINGTON AND J. C. LIGHT, *Appl. Opt. Suppl.*, No. 2 (1965) 81.
285 H. P. BROIDA, *Appl. Opt. Suppl.*, No. 2 (1965) 105.
286 R. BLEEKRODE, *J. Chim. Phys.*, 64 (1967) 141.
287 J. V. V. KASPER AND G. C. PIMENTEL, *Phys. Rev. Letters*, 14 (1965) 352.
288 J. R. AIREY, *IEEE J. Quantum Electron.*, QE3 (1967) 208.
289 T. F. DEUTSCH, *Appl. Phys. Letters*, 11 (1967) 18.
290 T. F. DEUTSCH, *IEEE J. Quantum Electron.*, QE3 (1967) 419.
291 T. F. DEUTSCH, *Appl. Phys. Letters*, 10 (1967) 234.
292 T. T. KIKUCHI AND H. P. BROIDA, *Appl. Opt. Suppl.*, No. 2 (1965) 171.
293 J. R. AIREY, R. R. GETTY, J. C. POLANYI AND D. R. SNELLING, *J. Chem. Phys.*, 41 (1964) 3255.
294 J. K. CASHION AND J. C. POLANYI, *J. Chem. Phys.*, 29 (1958) 455.
295 J. K. CASHION AND J. C. POLANYI, *J. Chem. Phys.*, 30 (1959) 1097.
296 P. E. CHARTERS AND J. C. POLANYI, *Discussions Faraday Soc.*, 33 (1962) 107.
297 F. CABRE AND L. HENRY, *J. Chim. Phys.*, 64 (1967) 119.
298 J. K. CASHION AND J. C. POLANYI, *J. Chem. Phys.*, 35 (1961) 600.
299 P. E. CHARTERS, B. N. KHARE AND J. C. POLANYI, *Discussions Faraday Soc.*, 33 (1962) 276; *Nature*, 193 (1962) 367.
300 B. A. THRUSH, *Symp. Combustion, 7th*, (1959) 243.
301 P. E. CHARTERS AND J. C. POLANYI, *Can. J. Chem.*, 38 (1960) 1742.
302 J. K. CASHION AND J. C. POLANYI, *J. Chem. Phys.*, 30 (1959) 316.
303 R. N. DIXON AND B. F. MASON, *Nature*, 197 (1963) 1198.
304 H. P. BROIDA, *J. Chem. Phys.*, 36 (1962) 444.
305 T. M. CAWTHON AND J. D. MCKINLEY, JR., *J. Chem. Phys.*, 25 (1956) 585.
306 D. GARVIN, *J. Am. Chem. Soc.*, 81 (1959) 3173.
307 D. GARVIN, H. P. BROIDA AND H. J. KOSTKOWSKI, *J. Chem. Phys.*, 32 (1960) 880.
308 D. GARVIN, H. P. BROIDA AND H. J. KOSTKOWSKI, *J. Chem. Phys.*, 37 (1962) 193.
309 F. KAUFMAN, *Ann. Geophys.*, 20 (1964) 106.
310 F. KRAUS, *Z. Naturforsch.*, 12 (1957) 479.
311 J. D. MCKINLEY, JR., D. GARVIN AND M. J. BOUDART, *J. Chem. Phys.*, 23 (1955) 784.
312 A. T. STAIR, JR., J. P. KENNEALY AND S. P. STEWART, *Planetary Space Sci.*, 13 (1965) 1005.
313 L. F. PHILLIPS AND H. I. SCHIFF, *J. Chem. Phys.*, 37 (1962) 1233.
314 M. A. A. CLYNE AND B. A. THRUSH, *Trans. Faraday Soc.*, 57 (1961) 2176.
315 F. P. DEL GRECO AND F. KAUFMAN, *Discussions Faraday Soc.*, 33 (1962) 128.
316 M. A. A. CLYNE AND B. A. THRUSH, *Discussions Faraday Soc.*, 33 (1962) 286.
317 A. M. BASS AND D. GARVIN, *J. Chem. Phys.*, 40 (1964) 1772.
318 R. A. KANE, J. J. MCGARVEY AND W. D. MCGRATH, *J. Chem. Phys.*, 39 (1963) 840.
319 W. D. MCGRATH AND R. G. W. NORRISH, *Nature*, 182 (1958) 235.
320 W. D. MCGRATH AND R. G. W. NORRISH, *Z. Physik. Chem.*, 15 (1958) 245.

321 K. Dressler, *J. Chem. Phys.*, 30 (1959) 1621.
322 L. F. Phillips and H. I. Schiff, *J. Chem. Phys.*, 36 (1962) 3283.
323 F. Kaufman and J. R. Kelso, *J. Chem. Phys.*, 28 (1958) 510.
324 J. E. Morgan, L. F. Phillips and H. I. Schiff, *Discussions Faraday Soc.*, 33 (1962) 118.
325 P. N. Clough and B. A. Thrush, *Discussions Faraday Soc.*, 44 (1967) 205.
326 K. G. Anlauf, R. G. MacDonald and J. C. Polanyi, *Chem. Phys. Letters*, 1 (1968) 619.
327 E. D. Kaufman and J. F. Reed, *J. Phys. Chem.*, 67 (1963) 896.
328 C. Maltz and D. R. Herschbach, *Discussions Faraday Soc.*, 44 (1967) 176.
329 F. R. Gilmore, *J. Quant. Spectry. Radiative Transfer*, 5 (1965) 369.
330 P. J. Th. Zeegers and C. Th. J. Alkemade, *Symp. Combustion*, 10*th*, (1965) 33.
331 A. Fontijn, W. J. Miller and J. M. Hogan, *Symp. Combustion*, 10*th*, (1965) 545.
332 A. N. Wright and C. A. Winkler, *Active Nitrogen*, Academic Press, New York, 1968.

Chapter 4

The Transfer of Energy between Chemical Species

A. B. CALLEAR

AND

J. D. LAMBERT

1. Introduction

Molecular systems exist in discrete quantum states, the study of which lies in the realm of molecular structure and wave mechanics. Transitions between quantum states occur either by absorption or emission of radiation (spectroscopy) or by collisional processes. There are two main types of collisional transitions which are important in chemical physics; these are first, reactive processes in which chemical rearrangement takes place (reaction kinetics), and secondly collisions in which the energy distribution is changed without overall chemical reaction. It may therefore be concluded that the energy transfer processes discussed here are of fundamental importance in all molecular systems, and that the subject, like molecular structure, is enormously varied and complex.

In 1925, Pierce[1] discovered that the velocity of sound in CO_2 is a function of the frequency of the sound waves. Shortly afterwards it was realised that this is due to the 'metastability' of vibrational energy with respect to interconversion with translational energy. The velocity of sound depends on the specific heat of a medium, and at high frequencies the internal energy of molecules may be unable to respond to the oscillatory temperature change, and will no longer contribute to the specific heat. By the 1930's it was recognised that the interconversion of translational and vibrational energy is slow, and especially if the vibrational frequency is high, a very large number of gas kinetic collisions are required for deactivation[2]. For example, for relaxation of carbon monoxide excited to the first vibrational level, on average about 10^{10} collisions are required with carbon monoxide molecules with zero point vibrational energy at room temperature. The classical Landau–Teller[3] theory correctly interpreted the dependence of relaxation times on the reduced mass of the collision partners, on the vibrational frequency, and on the temperature. In 1931, Zener[4] published an elegant wave mechanical solution for vibrational energy transfer in gases, which forms the basis of modern theory. However, only in the 1950's was Zener's theory reduced to a form suitable for numerical comparison with experiment, under the stimulus of shock-tube measurements of the vibration–translation relaxation rates of simple molecules. Vibrational relaxation rates of a large number of molecules have now been measured by the two main methods, ultrasonic absorption and dispersion, and the shock tube, and the results are in general accord with theory.

Also by the 1930's, it was realised that electronically excited molecules may require a large number of gas kinetic collisions for deactivation, particularly with the inert gases[5-7]. Excited species were produced photochemically, *e.g.* $Na(3\,^2P)$ and $Hg(6\,^3P)$, or by electric discharge, *e.g.* $He(2\,^1S_0)$, and were studied by stationary and non-stationary methods. These highly excited species (compared to the low energy vibrational levels) usually show evidence of chemical affinity in the collision complex. Even if there is no overall chemical change, a substantial electronic rearrangement may occur in the transition state. Such systems are difficult to treat theoretically and indeed also present considerable difficulties from the experimental viewpoint. Consequently, compared to vibrational energy transfer, our knowledge of electronic energy transfer is still in its infancy. In presenting and discussing this topic, we have endeavoured to draw a distinction between the case of parallel potential curves where there is little electronic rearrangement (*e.g.* spin–orbit relaxation of a low lying state), and the other extreme case of crossed potential curves. It is hoped that these ideas will at least aid the formulation of an outline of the subject.

Considering the four types of energy, translational, rotational, vibrational and electronic, it is convenient to classify ten types of molecular energy transfer. In some, the 'type' of energy is preserved, for example the E–E process which produces population inversion in the He/Ne laser[8]

$$He(2\,^3S_1) + Ne(2\,^1S_0) \rightarrow He(1\,^1S_0) + Ne(2s)$$

In other processes, the energy is converted from one form into another, for example the spin–orbit relaxation of $Hg(6\,^3P_1)$ by N_2[9]

$$Hg(6\,^3P_1) + N_2\,(v = 0) \rightarrow Hg(6\,^3P_0) + N_2\,(v = 1)$$

As yet, only V–T and V–V transfer are understood in any detail. Processes involving R–R and R–E transfer are practically unknown. In the sections which follow, each of the different classes is examined in relation to experimental results, and simple theoretical discussion. The chapter is concluded with a brief survey of the relationships between energy transfer and reaction kinetics[10].

Supposing $P_{1,0}$ is the transition probability per collision for a change in internal quantum number from $1 \rightarrow 0$, the average number of collisions required for a molecule to lose one quantum of this particular form of energy will be $Z_{1,0} = 1/P_{1,0}$. If Z is the number of collisions one molecule suffers per second, a *relaxation time*, β, for the appropriate kind of internal energy may be defined by the equation

$$Z_{1,0} = Z\beta \tag{1}$$

Z is proportional to the gas pressure, and, since $Z_{1,0}$, the *collision number* for energy transfer, is constant for a particular transition, the actual value of β is inversely proportional to the pressure. For convenience relaxation times are usually referred to a pressure of 1 atm. Equation (1) is an approximation, and requires modification to take into account the reversibility between quantum states 0 and 1. For example, the correct equation for vibrational relaxation of a simple harmonic oscillator of fundamental frequency, v, is

$$Z_{1,0} = Z\beta(1+e^{-hv/kT}) \tag{2}$$

which approximates closely to (1) for large values of hv/kT.

It is usually convenient to express the efficiency of electronic energy transfer in terms of a cross-section (Q), defined by

$$\text{Rate} = n_1 n_2 Q \left(\frac{8kT}{\pi\mu}\right)^{\frac{1}{2}}$$

where μ is the reduced mass and n_1 and n_2 are the concentrations of the two colliding species.

2. Experimental measurement of relaxation times

2.1 ACOUSTIC METHODS

These methods provide an accurate means of investigating translation–vibration and translation–rotation transfer. The passage of a sound wave through a gas involves rapidly alternating adiabatic compression and rarefaction. The adiabatic compressibility of a gas is a function of γ, the ratio of the specific heats, and the classical expression for the velocity, V, of sound in a perfect gas is

$$V^2 = \gamma RT/M = RT/M(1+R/C_v) \tag{3}$$

For a monatomic gas, where the heat capacity involves only translational energy, V is independent of sound oscillation frequency (except at ultra-high frequencies, where a classical visco-thermal dispersion sets in). For a relaxing polyatomic gas this is no longer so. At sound frequencies, where the period of the oscillation becomes comparable with the relaxation time for one of the forms of internal energy, the internal temperature lags behind the translational temperature through-out the compression–rarefaction cycle, and the effective values of C_v and V in equation (3) become frequency dependent. This phenomenon occurs at medium ultrasonic frequencies, and is known as *ultrasonic dispersion*. It is accompanied by

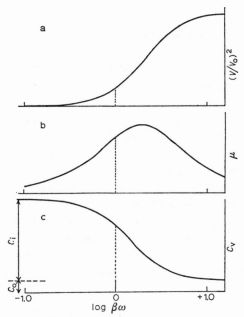

Fig. 1. Ultrasonic dispersion and absorption curves.

a non-classical *absorption* of sound, because the internal temperature lag causes dissipation of energy.

The variation of sound velocity with cyclic frequency, ω ($= 2\pi f$), is given by

$$V^2 = RT/M \left[1 + \frac{R(C_{v_0} + C_\infty \omega^2 \beta^2)}{C_{v_0}^2 + C_\infty^2 \omega^2 \beta^2} \right] \qquad (4)$$

where C_{v_0} is the static specific heat of the gas, which involves the total heat capacity, and corresponds to V_0, the value of V at low frequencies. $C_\infty = (C_{v_0} - C_i)$, where C_i is the contribution due to the relaxing form of internal energy, so that C_∞ corresponds to V_∞, the value of V at high frequencies. A theoretical plot of $(V/V_0)^2$ against $\log \beta\omega$ is shown in Fig. 1(a). The relaxation time may be determined by fitting measurements of sound velocity at varying frequencies to a theoretical curve. An alternative presentation is to plot the 'effective' value of C_v calculated from equation (3) against $\log \beta\omega$. This gives the dispersion curve shown in Fig. 1 (c), and has the advantage that C_i, the heat content of the relaxing form of internal energy, may be read off directly from the graph, making easier the assignment of individual relaxation times to particular modes of internal energy. Corrections for gas imperfection are necessary, and require an accurate knowledge of the second virial coefficient over a range of temperatures.

The variation with frequency of the sound absorption coefficient, μ, which rises to a maximum in the centre of the dispersion zone, is given by

$$\mu = \frac{2\pi\omega C_{\infty}\beta}{C_{v_0}} \left[\frac{V^2/V_0^2 - 1}{(V^2/V_0^2)(C_{\infty}\omega\beta/C_{v_0})^2 + 1} \right] \tag{5}$$

This gives the curve shown in Fig. 1 (b), and relaxation times may also be estimated by fitting sound absorption measurements, corrected for classical absorption, to the appropriate theoretical curve.

Measurements of sound velocity at ultrasonic frequencies are usually made by an acoustic interferometer. An example of this apparatus[11] is shown in Fig. 2. An optically flat piezo-quartz crystal is set into oscillation by an appropriate electrical circuit, which is coupled to an accurate means of measuring electrical power consumption. A reflector, consisting of a bronze piston with an optically flat head parallel to the oscillating face of the quartz, is moved slowly towards or away from the quartz by a micrometer screw. The electrical power consumption shows successive fluctuations as the distance between quartz and reflector varies between positions of resonance and non-resonance of the gas column. Measurement of the distance between resonance positions gives a value for $\lambda/2$, and if f

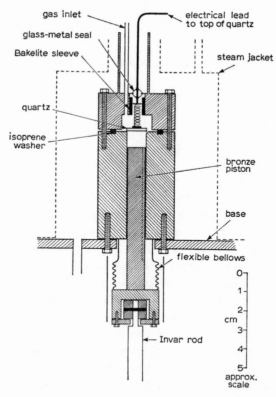

Fig. 2. 4 Mc.sec^{-1} acoustic interferometer. (Reproduced from Lambert and Rowlinson[11] by kind permission of the Council of the Royal Society)

is the frequency of the quartz, $V = f/\lambda$, can be calculated. Frequencies ranging between 100 kc.sec^{-1} and 10 Mc.sec^{-1} are used, giving wavelengths ranging between 0.5 and 0.005 cm. As the wavelength is always very small compared to the interferometer dimensions, 'tube effects' are unimportant, and wavelengths can usually be determined to 1 in 1000. It is most convenient to drive the quartz at its natural frequency, and to scan the dispersion zone by varying the gas pressure. Since binary collisions are responsible for energy transfer, halving the pressure (and hence the gas-kinetic collision rate) is equivalent to doubling the frequency in the context of equations (1) and (4), and f/p may be taken as the variable in equation (4) instead of f. A large part of the available data on relaxation times has been obtained by acoustic interferometry, and the method is reviewed in detail by Cottrell and McCoubrey[12]. It is the most accurate technique available, and has the advantage that it requires only a small sample of gas, whose purity (the critical importance of this will be stressed below) can be easily controlled. Its disadvantage is that measurements at temperatures above 500 °C are difficult, as quartz loses its piezo-electric properties at higher temperatures, and no other suitable transducers are available for high temperature operation. Its use is therefore restricted to the investigation of translation–rotation and translation–vibration transfers involving low-lying energy states.

Measurements of acoustic absorption can be made with instruments similar to the interferometer described above, but with a receiving transducer substituted for the reflector. It is best to use a pulse technique, and quartz crystals are sometimes replaced by condenser transducers, where the oscillating surface is a thin, aluminized Melinex sheet, stretched over a flat metal backing-plate[13]. This is capable of operating effectively at very low gas pressures, and correspondingly high f/p values. This type of more sophisticated apparatus may also be used for dispersion measurements. Observations of reverberation time in a resonant cavity provide another method of measuring absorption[14]; this technique has been used at temperatures as high as 1186 °C by employing a quartz transducer external to the hot cavity as the source of sound[15].

2.2 SHOCK-TUBE METHODS

The shock-tube technique, described in Volume 1 of this series as a tool for investigating high temperature chemical reactions, may also be employed for measuring translation–rotation and translation–vibration relaxation times at more moderate temperatures. When a shock wave travels through a gas, a thin layer behind the shock front undergoes extremely rapid adiabatic compression, resulting in an increase of density and temperature. If there is relaxation of internal energy, the translational energy alone is affected directly behind the front, resulting in an initial rise in temperature and density corresponding to an effective specific

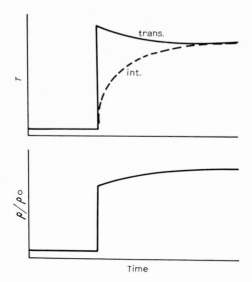

Fig. 3. Relaxation of internal energy behind a shock front.

heat, $(C_p - C_i)$. As the energy subsequently "leaks" into the internal degrees of freedom, the translational temperature decreases and the density increases, until they reach values corresponding to the static value of C_p a short distance behind the front. The theory of this process in terms of relaxation times has been worked out by Bethe and Teller[16]. Fig. 3 shows diagrammatically the time variation of temperature and density behind the shock front. The density profile may be followed by a Mach–Zehnder optical interferometer or the variation in the population of the vibrational states of the gas may be followed spectroscopically[17]. This technique enables relaxation times to be measured at temperatures up to *ca.* 3000 °K for gases which remain chemically stable under these conditions, and is particularly useful for investigating vibrational relaxation of diatomic gases, whose vibrational modes are only active at high temperatures. Its disadvantage is that the shock tube requires a comparatively large amount of gas, and that it is not easy to maintain a high degree of purity.

Rapid aerodynamic flow past obstacles involves adiabatic compressions and rarefactions, and is influenced by relaxation of internal degrees of freedom in a way similar to shock phenomena. This effect has been quantitatively treated by Kantrowitz[18], who developed a method for obtaining relaxation times by measuring the pressure developed in a small Pitot tube which forms an obstacle in a rapid gas stream. This 'impact tube' is not a very accurate technique, and requires a very large amount of gas; it has been used to obtain a vibrational relaxation time for steam.

2.3 SPECTROSCOPIC METHODS

A number of interesting problems on vibrational and electronic energy transfer have been attacked by means of flash photolysis (Volume 1, p. 118). The spectroscopic record of the individual quantum states permits direct observation of the relaxation steps. However the method is limited in application to simple mol-

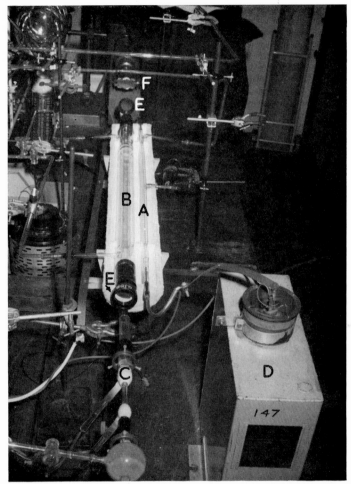

Fig. 4. Flash photolysis apparatus. A, photolysis flash lamp; B, reaction vessel; C, spectroscopic flash lamp; D, condenser; E, lenses; F, spectrograph.

ecules with banded electronic spectra, and thus far has been applied successfully only to diatomic molecules.

Fig. 4 provides an illustration of the simplest type of equipment that is normally employed for energy transfer studies. The experiment is basically simple, and consists of two operations:

(*a*) the production of a high intensity light pulse to be absorbed in the reaction vessel (photolysis flash duration $\sim 4 \times 10^{-5}$ sec), and

(*b*) the firing of a second flash (spectroscopic flash duration $\sim 10^{-5}$ sec) which is focused through the reaction vessel into a spectrograph.

Spectra are usually recorded photographically, and for kinetic measurements, a set of exposures at different delay times can be displayed on a single plate. After each strip has been exposed, the reaction vessel is refilled, the condensers recharged and the plateholder shifted according to the length of the spectrograph slit.

Fig. 5 is a set of absorption spectra taken during and following flash photolysis of nitric oxide, which is excited to the first vibrational level. The decay of $NO(v = 1)$ is clearly visible, and the relaxation rate could be roughly estimated by eye. Normally the change with time is automatically pen-recorded on graph paper by measuring the plate-density with a microdensitometer. In this experiment, the increased rate of decay caused by addition of foreign polyatomic gases permits direct measurement of V–V transfer rates. Several other examples of the flash photolysis method are included in the sections which follow.

Some other experimental methods also require brief discussion here. The technique of 'microwave-pulse flash-spectroscopy' is similar to that of flash photolysis, except that excitation is achieved by means of a powerful single pulse of microwave radiation from a magnetron[19]. The gas is contained in a quartz reaction vessel placed along the axis of a cylindrical cavity, tuned to the frequency

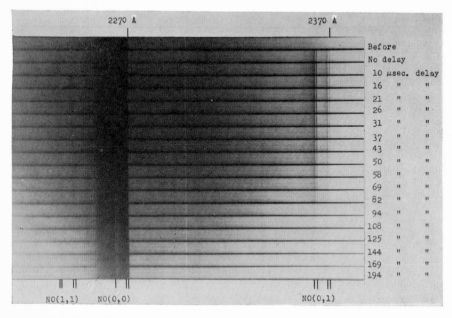

Fig. 5. Decay of NO $^2\Pi(v = 1)$ with time. Pressure of NO $= 5$ torr; pressure of $N_2 = 600$ torr. Flash energy $= 1600$ Joules.

of the magnetron. In a microwave field, electrons acquire high translational energies which cannot be dissipated to molecules as translational energy because of the disparity of the masses. However, electronic and especially vibrational excitation occur efficiently, and the method has been employed to excite nitric oxide to the first and second vibrational levels, and to study energy transfer to triatomic hydrides.

Millikan[20] has described an ingenious fluorescence technique for measuring relaxation rates of the $CO(v = 1)$ molecule. In a flow tube at 5–20 cm.sec^{-1}, CO is excited to $(v = 1)$ at the inlet with infrared emission from the CO fundamental (2143 cm^{-1}), a suitably intense source being a CH_4(rich)/O_2 flame. There are two competing processes by which the vibrational excitation can decay

(1) radiation $CO^* \rightarrow CO + h\nu$

(2) collision $CO^* + M \rightarrow CO + M +$ kinetic energy

At 300 °K and 1 atm pressure the radiative lifetime corresponding to (1), 0.033 sec, is much shorter than the collisional lifetime corresponding to (2), where M is CO or Ar, which is more than 1 sec. If deactivation by collision with the wall of the containing tube is eliminated by surrounding the streaming CO with an annular stream of Ar, the intensity of fluorescence down-stream measures the rate of (1). Many foreign gases are much more efficient collision partners in process (2), and, if sufficient is added to half-quench the fluorescence, the rates of the radiative and collisional processes, (1) and (2), must be equal. The rate of (1) is known, and hence the collisional efficiency of the foreign gas in the role of M has been determined.

Analysis of the fluorescence from electronically excited molecules in a conventional static gas system[21] provides a way of investigating vibrational relaxation of such molecules, and is also a means of studying selection rules for rotational relaxation[22]. It is now well established that multiple quantum rotational jumps can occur with high probability (see Section 6).

Another ingenious technique which has been suggested is the *spectrophone*[23]. Here an infrared-active molecular vibrational mode is excited by absorption of *pulsed* infrared radiation of the appropriate wavelength. The vibrational energy produced is degraded by intermolecular collisions to translational energy, producing a sound wave whose intensity depends on the relation between the pulse frequency and the vibrational relaxation time. This is, in principle, an ideal method for following the relaxation of a single vibrational mode, but the experimental difficulties are formidable, and only a few semiquantitative results have been obtained[24].

Bradley Moore andc o-workers[167] have ingeniously applied vibrational lasers to the study of V–T and especially V–V energy transfer. The principle is to expose a gas sample to a Q-switched laser pulse and to monitor the infrared fluorescence from the afterglow. A large number of measurements has been reported for relaxation of $CO_2(0, 0°, 1)$ excited with an N_2/CO_2 source. The microwave double resonance method, for rotational relaxation, will be described in section 6.

3. Theoretical considerations

3.1 THE LANDAU–TELLER THEORY

The probability of the occurrence of vibrational energy transfer during a molecular collision can be calculated in a very simple and elegant manner by wave mechanics, using the unperturbed harmonic-oscillator approximation. In general, the theoretical rates are in good agreement with experiment, and the theory reveals the molecular properties which are important in controlling the rate of transfer. However, before embarking on the development of the wave-mechanical method, it is instructive briefly to consider the classical theory of Landau and Teller[3], which incorporates some of the correct physical principles.

In Fig. 6, an atom A collides head on with the diatomic oscillator BC. This orientation is obviously the most efficient for coupling the vibrational coordinate (X) with that of translation (x); the real situation of random orientation is discussed later. It is convenient to fix the centre of gravity of the oscillator at the origin and consider the single translational coordinate separating the two molecules. The gas phase collision is then correctly described by averaging the energy in the single coordinate, according to the Boltzmann distribution.

The 'metastability' of vibrational energy results from two factors: (a) the amplitude of vibration is usually small compared to the range of the coordinate x for which the molecules experience each others repulsive forces, and (b) the period of vibration is usually short compared to the duration of the interaction during collision. Landau and Teller[3] pointed out that the probability of energy transfer depends on and increases with the ratio of the period of vibration to the duration of the collision. If the repulsive part of the interaction potential is shallow, the forces during the collision tend to act on the centres of gravity, rather than on

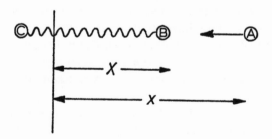

Fig. 6. Collision of an atom, A, with a diatomic oscillator, BC.

particular atoms, and energy transfer has low probability. It is the steep repulsive force which is important in vibrational energy transfer, because only this force (considering the classical analogue) can provide sufficient impulse to act on a particular atom in a molecule, and thereby either set it vibrating, or deactivate it. The weak long range attractive forces play only a minor role by increasing slightly the relative speed of the collision. With this classical model, Landau and Teller showed that the probability of relaxation is proportional to $\exp\left[-(l/v)/(1/2\pi v)\right]$, where v is the vibrational frequency in $\sec^{-1}$ units, v is the relative velocity, and with a model interaction potential of the form

$$V = C \exp\left(-x/l\right)$$

The quantity l is usually called the characteristic length of the interaction potential, and is also employed in the more realistic wave-mechanical treatment. Many authors employ the symbol, α, which is equal to l^{-1}. The form of the molecular interaction potential can be determined in terms of a suitable model, from experimental measurements of the temperature dependence of the viscosity of a gas, whence the characteristic length can be estimated.

3.2 GENERAL PRINCIPLES OF WAVE-MECHANICAL TREATMENT

The wave-mechanical solution for vibrational relaxation was first described by Zener[4] in 1931, and is given here following the refinements of Jackson and Mott[25]. Comparison with experimental values was first achieved by Schwartz et al.[26,27]. Although we retain the form of Fig. 6 to define the variables, the model is effectively an oscillating plate on which a beam of particles rebounds. Such a model is easier to deal with mathematically than that of a single wave packet. A very large number of collision or scattering problems in physics have been solved with this technique of employing a uniform beam, which in the absence of a field is correctly described by

$$\psi = A[\cos\left(kx-\omega t\right)+i\sin\left(kx-\omega t\right)] = A\mathrm{e}^{i(kx-\omega t)}$$

This is clearly a wave motion moving from left to right (Fig. 6). It is complex, and the number of particles per unit volume $|\psi\psi^*|$, usually written simply $|\psi^2|$, is equal to $|A^2|$. Absence of dependence on x and t occurs because the real and complex parts are π out of phase. A wave moving in the other direction is represented by

$$\psi = A\mathrm{e}^{i(-kx-\omega t)}$$

For the problem of vibrational relaxation, the frequency of the beam is not affected when the speed is changed by application of a field (*i.e.* intermolecular force), and for brevity the time factor is omitted. The wavelength is $2\pi/k$, and because of the de Broglie relationship $mv\lambda = h$

$$k = mv/\hbar$$

The above wave functions are solutions of the wave equation

$$\left\{\frac{\partial^2}{\partial x^2} + k^2\right\} \psi = 0 \quad \text{or} \quad \left\{\frac{\partial^2}{\partial x^2} + 2m \frac{(\frac{1}{2}mv^2)}{\hbar^2}\right\} \psi = 0$$

If the beam enters a retarding field due to a potential $V(x)$, the wavelength at all points is given by the well known equation

$$\left\{\frac{\partial^2}{\partial x^2} + \frac{2m}{\hbar^2}[W - V(x)]\right\} \psi = 0$$

where W is the kinetic energy at zero potential. As the beam is slowed down by a repulsive field, the wavelength and the amplitude must increase, because the number of particles crossing unit area, $k\hbar|A^2|/m$ (*i.e.* $v|A^2|$), must be independent of x.

One of the simplest of scattering problems, which can easily be solved by describing a beam as

$$\psi = Ae^{\pm ikx}$$

and which is included in most text books, is the calculation of the percentage transmission of light at an interface. If an incident wave is written e^{-ikx}, the reflected wave as Ae^{kx}, and the transmitted wave as $Be^{-ik'x}$ (the change of k corresponds to a change in light velocity or in refractive index, μ), the requirements that both ψ and $\partial\psi/\partial x$ are continuous at the interface yields the equations

$$1 + A = B$$

and

$$k(1 - A) = Bk'$$

The fraction of light reflected is therefore

$$|A^2| = \left(\frac{k - k'}{k + k'}\right)^2 = \left(\frac{\mu - \mu'}{\mu + \mu'}\right)^2$$

which is about 4 % at a glass/air interface.

The energy transfer problem is solved in a similar manner, by considering the boundary conditions and ensuring that the waves have the correct form at $x \rightarrow + \infty$, and vanish at $x = -\infty$. However, we have also to include the possibility of inelastic collisions.

3.3 VIBRATIONAL EXCITATION

For the problem of vibrational excitation (Fig. 1), we require a solution which at $x \rightarrow \infty$ has the form

$$f_0(x) = e^{-ik_0 x} + A_0 e^{ik_0 x}$$
$$f_1(x) = A_1 e^{ik_1 x}$$

where $f_0(x)$ is the sum of the incident and elastically scattered translational waves, and $f_1(x)$ is the inelastically scattered translational wave. The difference between k_0 and k_1 corresponds to the quantum of vibrational energy.

Considering the variables x and X, the complete wave equation is

$$\left\{ \frac{\hbar^2}{2\mu_{BC}} \frac{\partial^2}{\partial X^2} + \frac{\hbar^2}{2\mu} \frac{\partial^2}{\partial x^2} + E - V(X) - V(X, x) \right\} \Psi = 0$$

where $V(X)$ is a Hooke's Law or Morse potential *etc.* for the isolated vibrator, $V(X, x)$ is the interaction potential, E is the total energy, μ_{BC} is the reduced mass of the oscillator, and μ is the reduced mass of the system A–BC. ($|d\Psi^2|$ may be considered to express the combined probability of finding the oscillator in the range X to $X + dX$, and the translational coordinate in the range x to $x + dx$.) It is usually assumed that Ψ can be expanded in the form

$$\Psi = \sum_n \psi_n(X) f_n(x)$$

where $\psi_n(X)$ are normalised eigenfunctions of the isolated vibrator, and $f_n(x)$ is the associated translational wave function. This separation appears to hold precisely only if the amplitude of vibration is small compared to the range of the repulsive forces. The distortion of the oscillator during collision may be significant for hydrides. If we now substitute for Ψ in the above wave equation, we obtain

$$\sum_n \left(\frac{\partial^2}{\partial x^2} + k_n^2 - \frac{2\mu}{\hbar^2} V(X, x) \right) \psi_n(X) f_n(x) = 0 \tag{6}$$

because we have chosen $\psi_n(X)$ to be eigenfunctions of the isolated vibrator. Clearly k_n corresponds to the translational energy at $x \rightarrow \infty$.

To investigate the probability of transitions, we now multiply eqn. (6) by $\psi_i(X)$ and integrate with respect to X from $-\infty$ to $+\infty$. The oscillator wave functions are orthogonal, and therefore

$$\left(\frac{\partial^2}{\partial x^2} + k_i^2\right) f_i(x) = \frac{2\mu}{\hbar^2} \sum_n f_n(x) \int V(X, x)\psi_n(X)\psi_i(X)\mathrm{d}X$$

It is shown later that unless $i = n$ or $n \pm 1$, the terms on the right hand side can be neglected. If we consider the specific case of excitation from $v = 0$ to $v = 1$, the following two equations are required

$$\left(\frac{\partial^2}{\partial x^2} + k_0^2\right) f_0(x) = \frac{2\mu}{\hbar^2} f_0(x) \int V(X, x)\psi_0(X)\psi_0(X)\mathrm{d}X$$

$$+ \frac{2\mu}{\hbar^2} f_1(x) \int V(X, x)\psi_0(X)\psi_1(X)\mathrm{d}X \quad (7)$$

$$\left(\frac{\partial^2}{\partial x^2} + k_1^2\right) f_1(x) = \frac{2\mu}{\hbar^2} f_0(x) \int V(X, x)\psi_0(X)\psi_1(X)\mathrm{d}X$$

$$+ \frac{2\mu}{\hbar^2} f_1(x) \int V(X, x)\psi_1(X)\psi_1(X)\mathrm{d}X \quad (8)$$

Provided the probability of vibrational excitation is small, and this must apply at all velocities which make a significant contribution to the total transition probability, the second term on the right hand side of equation (7) can be neglected compared to the first term. Solution of (7) then yields $f_0(x)$, which may be substituted in (8) to solve for $f_1(x)$. This is called the 'distorted wave approximation'. The distortion refers to the change in wavelength due to the interaction potential, which in many other physical problems, such as the scattering of high energy particles, can be neglected (Born approximation). The remaining problem is purely mathematical.

3.4 THE INTERACTION POTENTIAL

Jackson and Mott[25] solved equation (7), with the interaction potential

$$V(X, x) = C \exp\left[-(x-X)/l\right] = U(x) \exp\left(X/l\right) \quad (9)$$

The occurrence of a closed solution for this particular function is of great importance, and results in a remarkably simple set of final equations.

From equation (9)

$$\int V(X, x)\psi_i(X)\psi_n(X)dX = U(x)\int \exp(X/l)\psi_i(X)\psi_n(X)dX = U(x)V_{in}$$

In Section 3.6 below, it is shown that provided l is very large compared to the vibrational amplitude (which usually holds), the diagonal elements $V_{nn} \cong 1$.[†] Equations (7) and (8) then take the forms

$$\left(\frac{\partial^2}{\partial x^2} + k_0^2 - \frac{2\mu}{\hbar^2} U(x)\right) f_0(x) = 0 \tag{10}$$

$$\left(\frac{\partial^2}{\partial x^2} + k_1^2 - \frac{2\mu}{\hbar^2} U(x)\right) f_1(x) = \frac{2\mu}{\hbar^2} U(x)V_{0,1} f_0(x) \tag{11}$$

Equation (11) can then be solved by letting $F_n(x)$ denote the solution of the equation

$$\left(\frac{\partial^2}{\partial x^2} + k_n^2 - \frac{2\mu}{\hbar^2} U(x)\right) F_n(x) = 0 \tag{12}$$

normalised to have the asymptotic form

$$F_n(x) = \cos(k_n x + \eta_n)$$

at $x \to \infty$. If the probability of energy transfer is small, the incident and elastically scattered waves have the same amplitude

$$f_0(x) = 2F_0(x)\,[††]$$

Writing $f_1(x) = y\,F_1(x)$, where y is a function of x, and substituting in (9), it may be seen after rearranging that

$$\frac{d}{dx}\left(F_1(x)^2 \frac{dy}{dx}\right) = \frac{4\mu}{\hbar^2} V_{0,1} U(x)F_0(x)F_1(x)$$

and

$$F_1(x)^2 \frac{dy}{dx} = \frac{4\mu}{\hbar^2} V_{0,1} \int_{-\infty}^{x} U(x)F_1(x)F_0(x)dx$$

[†] Strictly the V_{nn} are all equal when l is large but all V_{nn} are not equal to unity if the individual $\psi_n(X)$ are normalised. The effect of taking $V_{nn}=1$ is simply to change the constant C which does not affect the rate as shown in Section 3.5.

[††] $f_0(x)$ is now represented as a standing wave with the cosine form. An equally valid choice would be $2i\sin(k_n x + \eta_n)$ leading to the same, final equation.

References pp. 269–273

the integral being taken from $-\infty$ where $F_n(x)$ must vanish. At large x

$$\frac{dy}{dx} = \frac{4\mu}{\hbar^2} V_{0,1} T \sec^2 (k_1 x + \eta_1)$$

where

$$T = \int_{-\infty}^{+\infty} U(x) F_0(x) F_1(x) dx$$

becomes independent of x. Therefore

$$y = \frac{4\mu V_{0,1}}{\hbar^2 k_1} T[\tan(k_1 x + \eta_1) + \mathrm{const}]$$

Since as $x \to \infty$, $f_1(x)$ must have the form

$$f_1(x) = y F_1(x) = A_1 \exp\left[-i(k_1 x + \eta_1)\right]$$

the constant of integration is $-i$.

$$\therefore \quad f_1(x) = -\frac{4i\mu V_{0,1} T}{\hbar^2 k_1} \exp\left(-i k_1 x\right)$$

and

$$|A_1| = \frac{4\mu V_{0,1} T}{\hbar^2 k_1}$$

The probability of vibration excitation per collision, for the particular velocity corresponding to k_0, is therefore

$$P_{0,1} = |A_1^2|v_1/|A_0^2|v_0 = \frac{16\mu^2 V_{0,1}^2 T^2}{\hbar^4 k_0 k_1} \tag{13}$$

The delightful simplicity of this result reveals a transition rate which will be controlled largely by the factor T, 'the overlap of the translational wave functions'.

The matrix element $V_{0,1}$ is derived below for the harmonic oscillator, but evaluation of T involves a lengthy integration of equation (12) to obtain $F_n(x)$, and we accept the result of Jackson and Mott.

3.5 TRANSLATIONAL OVERLAP

The object in this section is to provide some simple and non-mathematical illustrations of how the transition probability is controlled by translational overlap

$$T = \int_{-\infty}^{+\infty} C \exp\left(-x/l\right) F_0(x) F_1(x) dx$$

In the absence of a field, two beams of different wavelength must cancel when integrated over a large distance; $F_0(x)$ and $F_1(x)$ obviously can only contribute to T where the interaction potential is finite, and the main contribution is close to the classical turning points.

Jackson and Mott[25] obtained a solution of equation (10) in terms of Bessel functions, from which T was derived in exponential functions. Following a simplification similar to that noted by Schwartz, Slawsky and Herzfeld[26] (SSH theory)

$$\frac{T^2}{k_0 k_1} = \pi^2 \Delta E^2 l^4 \exp\left[-2\pi l(k_0 - k_1)\right] = \pi^2 \Delta E^2 l^4 \exp\left[-\frac{4\pi l}{\hbar}\left(\frac{\Delta E}{v_0 + v_1}\right)\right]$$

(14)

where $\Delta E = h\nu$ for the case of vibrational excitation. This approximation deteriorates for systems of small reduced mass. The variations of T with l, v and ΔE are similar to the classical Landau–Teller[3] equations. The form of T, and the restriction of translational overlap, appear to be common to several different kinds of energy transfer, and this is discussed in various sections below.

Some simple molecular properties which can influence translational overlap in vibrational excitation are shown in Fig. 7. Case (a) illustrates for hard spheres the substantial overlap near the classical turning points. In example (b) with l large, the transition probability is very much smaller, because the final wave function overlaps the initial wave where the latter is executing very nearly pure SHM, even in the finite region of $U(x)$ which can contribute to the integrand. Cases (a) and (b) are examples of parallel potential curves, one of which may be derived from the other by vertical displacement with a fixed energy $h\nu$. If the potential curves (or surfaces) are parallel, it implies that the change of the state of excitation of the molecule B–C does not alter its mean dimensions with respect to the colliding species A. Cases (c) and (d) illustrate the effect of the surfaces becoming non-parallel. In case (c), the molecule has expanded in the excited state (divergent potential curves) and the overlap is reduced. The mean internuclear separation of a vibrating molecule increases slightly with increase of vibrational quantum number because of anharmonicity, and Mies[28] has shown that this can lower the tran-

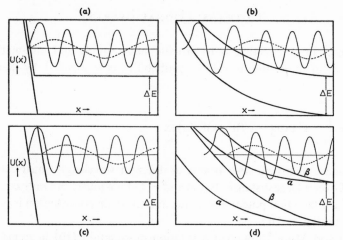

Fig. 7. Some simple molecular properties which can influence translational overlap in vibrational excitation (see text). Full lines represent initial, and dotted lines the final translational wave functions.

sition probability by 10–100 fold. Case (d) illustrates the increase of transition probability which can occur if the potential curves converge or cross. According to Nikitin[29], when two nitric oxide molecules ($X^2\Pi$) collide, the interaction potential splits to give the curves α and β, corresponding to a triplet and singlet collision complex. There is substantial overlap between β with zero point vibrational energy, and curve α of the vibrationally excited molecule. The theory can thereby account for the abnormally fast self-relaxation of nitric oxide.

It is interesting to note that the translational overlap is independent of the constant C, increase of which simply shifts all the potential curves and related wave functions to greater x. It is essentially because of the lack of dependence on C that the 'double normalisation' is permitted.

3.6 VIBRATIONAL MATRIX ELEMENTS

The evaluation of the vibrational matrix elements is quite straightforward; however, the problem has been formulated in terms of the amplitude of vibration with respect to the centre of gravity of the oscillator, so that a transformation of the usual oscillator wave functions is required. For a diatomic harmonic-oscillator

$$\psi_0(r) = (\alpha/\pi)^{\frac{1}{4}} \exp\left(-\tfrac{1}{2}\alpha r^2\right)$$
$$\psi_1(r) = (\alpha/\pi)^{\frac{1}{4}}\sqrt{2\alpha}\, r \exp\left(-\tfrac{1}{2}\alpha r^2\right) \tag{15}$$
$$\psi_2(r) = (\alpha/\pi)^{\frac{1}{4}}(1/\sqrt{2})(2\alpha r^2 - 1) \exp\left(-\tfrac{1}{2}\alpha r^2\right)$$

where r is the displacement of the internuclear separation from its equilibrium value[†], and $\alpha = 2\pi v\, \mu_{BC}/\hbar$. The vibrational amplitude is proportional to $\alpha^{-\frac{1}{2}}$. (The form of the eigenfunctions for polyatomic harmonic-oscillators is essentially identical, being the same functions of the displaced normal coordinate. Therefore the derivation for a diatomic molecule, which follows, is generally applicable.)

In the X coordinate with B vibrating, the amplitude is reduced such that

$$r m_C/(m_B+m_C) = X-X_e$$

where $X-X_e$ is the displacement of X from the equilibrium value X_e. It is necessary to retain the same normalised functions (15), and simply substitute $\alpha = 2\pi v\, \mu_{BC}(m_C+m_B)^2/\hbar(m_C)^2$. A factor $\exp(X_e/l)$ is common to all matrix elements and will drop out. Substituting in the equation

$$V_{i,\,n} = \int_{-\infty}^{+\infty} \exp(X/l)\psi_i(X)\psi_n(X)\mathrm{d}X$$

we have

$$V_{0,\,0} = 1$$
$$V_{0,\,1} = \sqrt{2\gamma}$$
$$V_{0,\,2} = \sqrt{2\gamma}$$
$$V_{1,\,1} = 1+2\gamma \simeq 1$$
$$V_{n,\,n+1} = \sqrt{2(n+1)\gamma} \tag{16}$$

where $\gamma = (4\alpha l^2)^{-1}$ and a common factor $\exp(\gamma)$ has been omitted.

Generally $\gamma \ll 1$, and the approximation $V_{1,1} = 1$ is found to be satisfactory. For example, for N_2, $\gamma = 0.01$. However, hydrides are abnormal because the amplitude of vibration is comparable to the range of the repulsive forces, and $V_{0,0}/V_{1,1}$ is considerably below unity for low frequency vibrations. Mies[28] has taken similar effects into account, using the more precise Morse functions. It should be emphasised that it is not the linear error due to $1+2\gamma$ being significantly greater than unity, e.g. in eq. (13), which is serious. In any case this correction could easily be made. A serious error arises because the terms $(2\mu/\hbar^2)U(x)$ on the LHS of equations (10) and (11) are multiplied by the diagonal matrix elements, a disparity in which has the effect of a relative shift of the potential curves which may substantially change the translational overlap. Then equation (14) does not hold.

In the general case for the harmonic oscillator

[†] The theory approximates the amplitude of vibration of the nuclear separation with that at the periphery of the molecule.

$$(V_{0,1})^2_{CB} = 2\gamma = \hbar m_C / 4\pi v l^2 m_B (m_B + m_C)$$

the probability is averaged over both ends of the molecule, which are equally accessible for collision, by inversion of B and C, addition of the two equations, and division by two

$$(V_{0,1})^2 = \frac{(V_{0,1})^2_{BC} + (V_{0,1})^2_{CB}}{2} = \frac{m_C^2 + m_B^2}{m_B m_C (m_B + m_C)} \frac{\hbar^2}{4l^2 \Delta E} \tag{17}$$

The mass function tends to be large for hydrides; for example for N_2 and HI it is respectively 0.07 and 1. This is because the large amplitude of vibration of a light atom favours efficient coupling with translation, which is also obvious on classical considerations. The amplitude factor gives rise to the special behaviour of hydrides in the Lambert–Salter[30] correlation (see Section 4.2).

The relative probability of a double quantum transition depends on

$$(V_{0,2}/V_{0,1})^2 = \gamma$$

which is $\ll 1$ except for hydrides. However, for V–T processes $T_{0,2}/T_{0,1}$ will usually prohibit a double quantum jump because ΔE occurs in the exponent. (The harmonic oscillator rules, and indeed the entire theory given, will not hold if there is chemical affinity between A and BC. The latter problem is much more complex.)

3.7 TRANSITION PROBABILITY

In this section, in the simplest possible manner, it is shown that because of random orientation of BC, the probability of a transition is lowered by 1/3.

Considering Fig. 6, if BC is orientated at an angle θ to the line joining the centres of gravity, the potential will be

$$V(X, x) = C \exp\left[-\frac{x - X \cos \theta}{l}\right] = U(x) \exp\left(\frac{X \cos \theta}{l}\right)$$

Thus the squared matrix element is

$$(V_{0,1})^2 = 2\gamma \cos^2 \theta$$

Since B–C has equal probability of all orientations over a hemisphere, we simply integrate $\cos^2 \theta$, weighted with the area of the hemisphere, to derive the orientation factor

$$\frac{\int_0^{\pi/2} \cos^2 \theta \, 2\pi X^2 \sin \theta \, d\theta}{2\pi X^2} = \tfrac{1}{3}$$

It may be verified that for a double quantum jump the steric factor is $\tfrac{1}{5}$.

A precise treatment must consider simultaneous rotational transitions, because the impulse during collision depends not only on θ but also on the manner in which θ changes. However, the major part of the overall transition probability occurs where θ is small, and coupling with rotation is then at its lowest efficiency. A detailed discussion of this complex problem is given by Herzfeld and Litovitz[32]. By integration of equation (14) through the velocity distribution, equation (13) takes the form

$$P_{0,1} = \frac{16\mu^2 (V_{0,1})^2 (T^2)_{Av}}{\hbar^4 k_0 k_1} = \frac{1}{3} \frac{16\mu^2}{\hbar^4} (V_{0,1})^2 \pi^2 l^4 \Delta E^2 \left(\frac{2\pi}{3}\right)^{\frac{1}{2}}$$

$$\times A^{\frac{1}{2}} \exp\left(-\frac{3}{2}A - \frac{\Delta E}{2kT} + \frac{\varepsilon}{kT}\right) \quad (18)$$

where

$$A^3 = \frac{4\pi^2 \mu l^2 \Delta E^2}{\hbar^2 kT}$$

(Due to approximations in the derivation, (18) ceases to hold as $\Delta E \to 0$.) The energy ε refers to the depth of the minimum of the interaction potential, and may be considered either as an acceleration effect, or as an enhancement of the collision frequency. Inclusion of the term usually increases the transition probability by a factor of about two fold for weak van der Waals interaction. The first term in the exponent of equation (18), $-\tfrac{3}{2}A$, is generally the dominant term, and it should be noted that the theory therefore predicts a linear dependence of $\log P_{0,1}$ on $T^{\frac{1}{3}}$.

Equation (18) may be derived in a manner similar to that described above for vibration–vibration energy transfer, with the difference that the product of the squared matrix elements of the two vibrations concerned, appears in the expression for transition probability. Since both molecules require suitable orientation for vibrational exchange, for two diatomic molecules the steric factor is usually taken to be $(\tfrac{1}{3})^2$. If $\Delta E \to 0$ (exact resonance), eqn. (18) is no longer valid and the following equation may be employed

$$P_{0-1}^{1-0} = \frac{1}{3^2} (V_{0,1})_1^2 (V_{0,1})_2^2 \frac{8\mu l^2 kT}{\hbar^2} \exp\left(\varepsilon/kT\right) \quad (19)$$

This differs by a factor of 2 from the equation of Herzfeld and Litovitz[32].

Schwartz et al.[26] deduced values of l from the Lennard-Jones parameters ε and r_0, which have been derived from viscosity measurements, and analysed and tabulated by Hirschfelder et al.[31]. A detailed account of the required calculation has been given by Herzfeld and Litovitz[32]. The problem is to obtain a satisfactory fit of the exponential interaction potential, with the repulsive region of the Lennard-Jones function

$$V = 4\varepsilon \left\{ \left[\frac{r_0}{r}\right]^{12} - \left[\frac{r_0}{r}\right]^{6} \right\}$$

Two fitting methods have been discussed. In method A, the two functions are made to coincide in slope and magnitude at the most favourable velocity for transitions, with the exponential function at $x \to \infty$ corresponding to the Lennard-Jones minimum. The most probable velocity (energy E_m) can be obtained from eq. (14), by multiplying by the velocity distribution, $(\mu v_0/kT) \exp(-\mu v_0^2/2kT)dv_0$, and deriving the maximum of the product function. The identity of slope and magnitude gives two equations and permits the elimination of C. r_0/l is then given by

$$r_0/l = 12 \left\{ \frac{1}{2} \left[1 + \left(\frac{E_m}{\varepsilon} + 1\right)^{\frac{1}{2}} \right] \right\}^{1/6} \left[1 + \left(\frac{E_m}{\varepsilon} + 1\right)^{-\frac{1}{2}} \right]$$

In method B, which seems to be devoid of theoretical support, the two functions are fitted in magnitude at the most probable velocity and also at r_0, the point of zero potential on the Lennard-Jones model. A comparison with experiment is given in Table 1. Z is the reciprocal probability or number of gas kinetic collisions required for deactivation. The calculated values are derived from equations (17) and (18).

It is seen that although the order of magnitude of agreement of Z_{exp} and Z_{calc} is reasonably satisfactory, Z_{calc} is very sensitive to the magnitude of l. The importance of the steepness of the repulsive potential is evident in its control of translational overlap according to equation (14). The uncertainty in l seems to be the main difficulty in the precise application of the theory at the present time. The derivation from Lennard-Jones parameters involves a considerable extrapolation to the high velocities required for energy transfer. The complications are worse for polar or polyatomic molecules. In a careful comparison of Z_{calc} and Z_{exp} for polyatomic gases, Stretton[33] concluded that $l \simeq 0.18$ A generally gives satisfactory agreement, which indicates that at the velocities required for energy transfer, the steepness of the potential is independent of the molecular system. Mies[34] has calculated relaxation rates for H_2–He collisions with a Hartree–Fock potential.

It may be noted that in the above analysis a substantial van der Waals interaction (for example due to dipole–dipole interaction or to hydrogen bonding)

TABLE 1

COLLISION NUMBERS FOR VIBRATIONAL RELAXATION[32, 32a]

System	$\nu(\text{cm}^{-1})$	Temperature (°K)	Z_{exp}	Z_A	Z_B	$l_A(\text{A})$	$l_B(\text{A})$
O_2	1556	288	1.6×10^7	6.8×10^8	2.0×10^7	0.198	0.179
O_2	1556	1372	3.6×10^4	8.2×10^4	0.7×10^4	0.199	0.183
N_2	2330	550	1.1×10^8	1.2×10^8	4.3×10^8	0.212	0.196
N_2	2330	3640	2.4×10^4	2.8×10^4	0.6×10^4	0.203	0.199
Cl_2	557	288	4.6×10^4	5.8×10^5	3.3×10^4	0.225	0.20
Cl_2	557	1000	550	4.9×10^3	360	0.235	0.209

has the effect of considerably steepening the repulsive potential for a particular incident velocity, because the zero of the exponential function is fitted to the Lennard-Jones potential at its minimum.

3.8 TANCZOS' THEORY FOR POLYATOMIC MOLECULES

Tanczos[35] has extended the theory (for V–T and V–V transfer) to polyatomic molecules, and a detailed comparison with experiment was recently given by Stretton[33]. Considering each surface atom, energy transfer depends on how the intermolecular potential varies with the oscillation of the atom. In deriving the result for the diatomic molecule from the harmonic-oscillator wave functions, we substituted

$$\alpha = 2\pi\nu\mu_{BC}(m_C + m_B)^2 / \hbar(m_C)^2$$

to correct the amplitude to the centre of gravity of the oscillator, which is the origin of both X and x. For polyatomic molecules, precisely the same method is applied by calculating the amplitude for each of the surface atoms and suitably adjusting α in the wave functions of the normal coordinate under consideration. A somewhat similar procedure was adopted in deriving the orientation factor of $\frac{1}{3}$; the amplitude of vibration in the X coordinate was written $X \cos \theta$. The model for polyatomic molecules may therefore be regarded as a sphere 'breathing' with the full amplitude of the Cartesian coordinate, which of course may or may not be in the direction of the line of centres of A and BC. (In the latter case, simultaneous rotational transitions may be specially important.) The derivation of the steric factor is essentially identical to that for a diatomic system.

It could not be considered appropriate to give a full description of the Tanczos theory here. However, it should be useful to outline its application with some simple examples. For a diatomic molecule

$$(V_{0,\,1})^2 = \frac{\hbar}{4\pi l^2 v} \frac{1}{2} \left\{ \frac{m_C}{m_B(m_B + m_C)} + \frac{m_B}{m_C(m_B + m_C)} \right\}$$

The amplitude of vibration is proportional to $\alpha^{-\frac{1}{2}}$, and therefore the squared amplitude depends on

$$\alpha^{-1} = \frac{\hbar}{2\pi v \mu_{BC}} \left\{ \frac{m_C^2}{(m_C + m_B)^2} \right\} = \frac{\hbar}{2\pi v} \left\{ \frac{m_C}{m_B(m_C + m_B)} \right\}$$

Thus for a given vibrational frequency, the squared amplitude is proportional to the term $m_C/m_B(m_B + m_C)$ and the mean square amplitude is proportional to

$$\frac{1}{2} \left\{ \frac{m_C}{m_B(m_B + m_C)} + \frac{m_B}{m_C(m_B + m_C)} \right\}$$

Evidently the squared matrix element is proportional to the mean square amplitude, and for polyatomic molecules Tanczos formalises

$$(V_{0,\,1})^2 = \frac{\hbar}{4\pi l^2 v} \left(\frac{1}{N_s} \sum_s \frac{A_s^2}{m_s} \right)$$

N_s is the number of surface atoms, A_s is the amplitude of vibration of the sth atom of mass m_s. The expression inside the brackets is called an amplitude factor, written by Stretton[33] as $(\overline{A^2})$. In general the orientation factor may be taken as $N_s/6$, since for each atom we need to consider orientation over a complete sphere.

From the above equations, it is seen that the amplitude factor for a homonuclear diatomic molecule is $(2m)^{-1}$, which for H_2 in atomic mass units would be 0.5. For H_2O, the amplitude of vibration in the bending mode would be very nearly the same as that of a hypothetical H_2 molecule with the same frequency. The amplitude factor is therefore again 0.5. Considering v_3 in CH_4, in which opposite pairs of H atoms are pinched together, the potential energy is shared between the two pairs. Considering Hooke's law, clearly the squared amplitude is lower by a factor of 2 compared to the hypothetical H_2 of the same frequency, and therefore the amplitude factor of $CH_4(v_3)$ is 0.25. (Note that for CH_4, H_2O and H_2, the product of the amplitude factor and orientation factor is constant.) Stretton[33] has compiled a useful list of amplitude factors for substituted methanes.

For polyatomic molecules equation (18) is employed with the vibrational matrix elements modified as described above. For vibrational exchange, in equation (18) the single vibrational matrix element is replaced by the product of the squares of the matrix elements for each molecule. In general, the theory leads to collision probabilities which are in good agreement with experiment.

3.9 ORIENTATION AND LOW-TEMPERATURE EFFECTS

At temperatures where the potential minimum, ε, is comparable to kT, equation (18) is not valid because the method of treating the thermal velocity distribution is no longer appropriate. When ε becomes comparable to kT, orientation effects are believed to become important because in general the depth of the potential minimum is dependent on the configuration of both molecules. The probability of a particular orientation can be derived from Boltzmann's equation.

The importance of orientation in molecular energy transfer has been postulated by Lambert and co-workers[30, 43, 65, 81, 82] on the basis of experimental evidence, and several cases have been developed theoretically by Shin. If the configuration of the molecules corresponds to a deep potential minimum, for a particular v_0 the repulsive potential will be steepened at the classical turning point. If this orientation is favourable for energy transfer (θ small) it may result in a negative temperature coefficient in the low-temperature range. Long range attraction becomes dominant if $\varepsilon \gg kT$, and order of magnitude calculations can easily be attempted by the direct application of equation (14), with $\frac{1}{2}\mu v_0^2 = \varepsilon$ and $\frac{1}{2}\mu v_1^2 = \varepsilon + h\nu$. However the reader may refer to the explicit solutions for a Morse intermolecular potential including allowance for the temperature dependent orientation[169].

Nitric oxide exhibits a negative temperature coefficient for vibrational relaxation in self-collisions, below about 700° K. It has been suggested[170] that this effect arises because the potential energy of the point of resonance, postulated by Nikitin, is strongly orientation dependent. (In this case the maximum depth of the potential minimum can be no greater than about 3 kcal. mole^{-1} which will not steepen the potential sufficiently to account for the observed relaxation rate, with $l = 0.18$ A.)

3.10 EXACT QUANTUM MECHANICAL TREATMENT

Secrest and Johnson[171] have computed an exact quantum mechanical solution for the case of a harmonic oscillator undergoing a colinear collision on an exponential repulsive potential. The method does not require the unperturbed oscillator approximation and all the possible channels for inelastic scattering were left open. Secrest and Johnson calculated the exact transition probabilities with various initial velocities and reduced masses. There are two aspects of this computation which are of special interest here. First comparison of numbers derived from equation (13) showed that the distorted wave–unperturbed harmonic oscillator approximation, in the form given above, does overestimate the transition rate. With m_B/μ large, however, the agreement is very nearly exact. With m_B/μ small, discrepancies of the order of ~ 5 fold were found; in the energy region where $\mu v_0^2/2 \leqslant 2h\nu$, which is frequently appropriate to room temperature conditions, the discrepancies are generally quite small for all mass parameters. Secrest and Johnson did not identify the deficiency in the Jackson–Mott formulation. It has been pointed out by Roberts[182] that at least part of the error in the Jackson–Mott equations arises from the assumption of equality of the diagonal matrix elements. If the potential surface is split, according to the ratio of the matrix elements (equation 16), the exact solution is close to that predicted by equation (13), modified such that $F_0(x)$ and $F_1(x)$ belong respectively to $V_{0,0} U(x)$ and $V_{1,1} U(x)$.

Secondly, Secrest and Johnson showed that for the problem of vibrational energy transfer with soft repulsive forces, it is valid not only to close channels which are forbidden classically (negative kinetic energy) but even channels which are open may be disregarded in a particular calculation. In setting up equations (10) and (11) to examine the problem of excitation from $v = 0 \to v = 1$ all channels involving transitions (real or virtual) to higher quantum states were ignored and the exact calculation justifies this procedure. It may easily be verified with the same spirit that

$$P_{0,2} = \frac{16\mu^2 V^2_{0,2} T^2_{0,2}}{\hbar^4 k_0 k_1}$$

provided $A_1/A_0 \ll V_{0,2}/V_{1,2}$.

3.11 CONCLUSION

Schwartz, Slawsky and Herzfeld[26] developed many of the refinements which were necessary in order to be able to compare the theory with experiment. The 'SSH' theory is remarkably successful in accounting, perhaps semiquantitatively, for all the principal features of vibrational energy transfer. It correctly predicts that a molecule with a high vibrational frequency will have a lower probability of relaxation per collision, than a molecule with a low vibrational frequency. It correctly accounts for the temperature dependence. It predicts that a molecule with a large vibrational amplitude, such as a hydride, will undergo relaxation comparatively easily. The theory predicts a reduced mass effect; light molecules are more efficient deactivators than are heavy molecules. It correctly points out that provided an intermediate energy level is available in the deactivating molecule, V–V transfer must occur.

There are presently two main difficulties which handicap attempts at exact calculation. The first concerns the intermolecular potential, and the hazards of extrapolation from models derived from viscosity measurements have been discussed. Furthermore, such a method is of dubious validity for polyatomic molecules, because the intermolecular repulsive potential will generally appear to become progressively shallower with increasing molecular dimensions if the viscosity data are cast, for example, in the Lennard-Jones form. Energy transfer depends on the repulsion between the deactivator and specific surface atoms, which should not be specially affected by gross molecular dimensions. Stretton[33] points out that, using $l = 0.18$ A, the theory leads to values which agree well with those found experimentally, and certainly the employment of such a potential does maximise the probability of achieving realistic probabilities.

The second main difficulty concerns simultaneous rotational excitation, which does play a role, though probably a minor one, in the vibrational relaxation of hydrides. Presently, there is no satisfactory theory, and little direct experimental evidence, to show that rotational excitation is strongly coupled to vibrational relaxation.

A number of applications of the theory may be found in the various sections which follow.

4. Vibration–translation transfer

4.1 DIATOMIC MOLECULES

Vibration–translation transfer for diatomic molecules presents theoretically the simplest and best understood relaxation phenomenon. Only one fundamental frequency is involved, and the corresponding quantum is commonly so large that only $1 \rightleftharpoons 0$ quantum transitions need be considered. The necessary intermolecular potentials and vibrational matrix elements are comparatively simple so that the SSH treatment, described in Section 3 above, can be easily applied. Experimental investigation is less satisfactory, as the relatively high frequency vibrations concerned make only a very small specific heat contribution at temperatures where the more accurate acoustic techniques may be employed, and shock-tube measurements at higher temperatures may be subject to serious inaccuracies due to the presence of impurities. More recent experimental data, where sufficient care has been taken, do show fairly good agreement with theoretical prediction. Dickens and Ripamonti[36] have compared a large number of relaxation times, calculated by the SSH method, with the more reliable experimental values available. These are plotted in Fig. 8, and the relevant data and sources are listed in Table 2. (These include some polyatomic molecules, which will be discussed below: here the agreement is less good.) The predicted temperature dependence, giving an approximately linear plot of log $Z_{1,0}$ against $T^{-\frac{1}{3}}$ is also found experimentally[36]. There are a few notable exceptions, where experimental relaxation times are much shorter than theoretically predicted. The abnormally fast vibrational relaxation of NO has already been mentioned in Section 3[46–48, 170]. A possible explanation was suggested in terms of converging potential curves. Borrell[49] has recently mentioned abnormally low relaxation times for HCl, HBr, HI, H_2 and D_2. For these molecules both the 'hydrogen effect' mentioned in Section 3 and the effect of dipole–dipole interaction (for the three polar molecules) may be important factors.

So far only $1 \rightleftharpoons 0$ transfers have been considered. For a molecule excited to a higher vibrational level there is a possibility of deactivation (or activation) by a multiple quantum jump, *e.g.*

$$N_2(v = 5) + M \rightleftharpoons N_2(v = 5 - \delta v) + M + K.E.$$

where δv might be 1, 2, 3, 4 or 5. The theoretical treatment, given in Section 3 above, predicts that jumps involving $\delta v > 1$ would have comparatively low probability. This has been confirmed by calculations made by Bauer and Cummings[50], who apply the modified wave number (MWN) approximation, due to Takayanagi[51], to the collisional deactivation of $N_2(v = 5)$ and $NO(v = 5)$, and show that for both cases the probability of a transfer for which $\delta v > 1$ is several orders of magnitude less than for $\delta v = 1$. Experimental investigation of this requires special care, since deactivation of highly excited molecules by homomolecular collisions usually occurs by the more rapid resonant vibration–vibration transfer (see below) instead of by vibration–translation transfer. The selection rules and dependence on v have not yet been adequately explored experimentally. Hooker and Millikan[52] observed the variation with time of the fundamental $(v = 1 \rightarrow 0)$ and overtone $(v = 2 \rightarrow 0)$ emission in shock wave heated CO, and showed that the former increased linearly, whereas the latter showed an induction period. Decius[53] has shown that on the multistate harmonic oscillator model of Montroll and Shuler[54] with the selection rule $\Delta v = 1$, $\log [1 - (I/I_\infty)]$ against time should be linear for the fundamental with a slope equal to the reciprocal relaxation time,

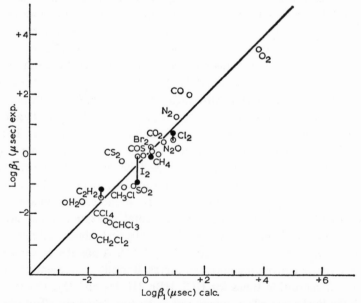

Fig. 8. Comparison of experimental and calculated relaxation times. (Reproduced from Dickens and Ripamonti[36] by kind permission of the Council of the Faraday Society)

TABLE 2

EXPERIMENTAL AND CALCULATED VIBRATIONAL RELAXATION TIMES

Molecule	$T(°K)$	$\beta_1 (\mu sec)$ calc.	$\beta_1 (\mu sec)$ expt.	References
O_2	288	5840	3180	37
N_2	3480	10	19	38
CO	2200	26.7	100	39
Cl_2	290	7.88	3.4	40
			5.0	41
Br_2	331	1.42	1.6	40
			0.81	41
I_2	453	0.495	0.85	40
			0.104	41
C_2H_2	298	0.0285	0.041	30
			0.067	14
N_2O	273	1.53	1.18	32
COS	288	0.827	0.902	32
CS_2	288	0.44	0.64	32
CO_2	300	3.80	2.53	32
H_2O	585	0.00158	0.03	42
SO_2[a]	373	0.36	0.089	43
CH_2Cl_2[a]	303	0.0158	0.00195	44
CH_4	303	2.70	1.06	45
CH_3Cl	303	0.147	0.078	44
$CHCl_3$	303	0.0510	0.0060	44
CCl_4	303	0.0420	0.0062	44

[a] Shorter relaxation time.
(This Table is reproduced from Dickens and Ripamonti[36] by kind permission of the Council of the Faraday Society)

whereas for the overtone, $[1-(I/I_\infty)^{\frac{1}{2}}]$ should be linear with the same slope. I/I_∞ is the ratio of the intensity of emission to the equilibrium intensity. Thus Hooker and Millikan obtained relaxation times of 172 μsec and 190 μsec for $v = 1$ and $v = 2$ respectively, which may be evidence for excitation of $v = 2$ by the step

$$CO(v = 1)+CO(v = 0) \rightarrow CO(v = 2)+CO(v = 0)$$

However, the relaxation time for the V–V exchange

$$2CO(v = 1) \rightarrow CO(v = 2)+CO(v = 0)$$

is very short at the pressures that they employed, and would appear to "Boltzmannise" the vibrational distribution irrespective of the detail of the mechanism.

There is, in fact, very little experimental support, as yet, either for the $\delta v = 1$ rule, or the linear variation of relaxation rate with v predicted by the ratio of the squared matrix elements. Flash photolysis offers a number of opportunities of

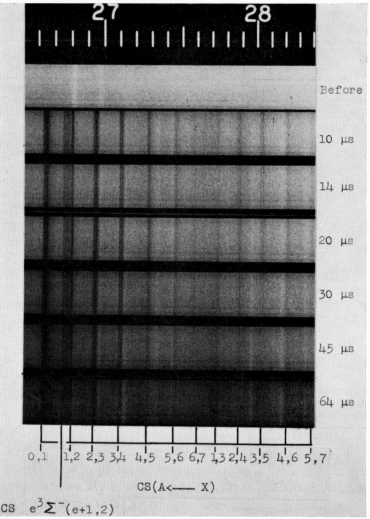

Fig. 9. Enlargement of the (0, 1) and (0, 2) sequences of CS ($A \leftarrow X$). 0.5 torr CS_2, 5 torr O_2 with 450 torr N_2. 1.5 m path length, 2000 J flash energy.

studying the $v > 1$ quantum states[55,56], though as yet few quantitative measurements have been made. Accurate vibrational distributions have recently been measured for vibrationally excited O_2 produced in the photolysis of ozone[57], and relaxation was shown to be consistent with single quantum transitions. However, the ozone photolysis produces O_2^* with 20 or more vibrational quanta, and the general form of the decay is rather insensitive to Δv. The radical CS, produced by flash photolysis of CS_2, was reported to relax by multiple vibrational jumps[58], and it was suggested that the abnormal relaxation occurs when CS collides with atomic sulphur. Similar conclusions were later drawn from the relaxation of CSe produced in flashed CSe_2[59]. Formation and relaxation of excited CS is shown in Fig. 9. A strong attractive interaction between the diatomic molecule and the atom may be the cause of the high probability of relaxation per collision. For strong collisions (l small) between particles of high energy, Shuler and Zwanzig[60] have shown theoretically that multiple quantum jumps have high probability.

The possibility of deactivation of vibrationally excited molecules by spontaneous *radiation* is always present for infrared-active vibrational modes, but this is usually much slower than collisional deactivation and plays no significant role; (this is obviously not the case for infrared gas lasers). CO is a particular exception in possessing an infrared-active vibration of high frequency (2144 cm^{-1}). The probability of spontaneous emission depends on the cube of the frequency, so that the radiative life *decreases* as the third power of the frequency, and is, of course, independent of both pressure and temperature; the collisional life, in contrast, *increases* exponentially with the frequency. Reference to the vibrational relaxation times given in Table 2, where CO has the highest vibrational frequency and shortest radiative lifetime of the polar molecules listed, shows that most vibrational relaxation times are much shorter than the 3×10^4 μsec radiative lifetime of CO. For CO itself radiative deactivation only becomes important at lower temperatures, where collisional deactivation is very slow indeed, and the specific heat contribution of vibrational energy is infinitesimal. Radiative processes do play an important role in reactions in the upper atmosphere, where collision rates are extremely slow.

4.2 POLYATOMIC MOLECULES

Polyatomic molecules may possess several vibrational modes which make a significant contribution to the specific heat, and might be expected to show highly complicated vibrational–relaxation phenomena. This does not happen in the vast majority of cases owing to rapid intramolecular vibration–vibration transfer, which maintains continuous equilibrium of vibrational energy between the active fundamental modes (see Section 5, p. 220, for a full discussion). The whole vibrational heat content of the molecule thus relaxes in a single vibration–translation transfer *via* the lowest mode, which will have a higher collisional transfer probability than upper modes. This mechanism is characterised by a single overall relaxation time, β, which may be shown to be related to β_1, the actual relaxation time of the lowest mode, by the equation $\beta_1 = (C_1/C_s)\beta$, where C_s is the total vibrational specific heat, and C_1 the contribution due to the lowest mode. A great many polyatomic molecules show vibrational relaxation under conditions where accurate acoustic techniques can be used, and it remains to discuss how far the experimental values of β_1 are in accord with the predictions of the SSH theory.

It will be seen from Fig. 8 that polyatomic molecules tend to have relatively small relaxation times, corresponding to the low frequency of their vibrational modes. Agreement between theory and experiment is markedly less good than for diatomic molecules, and becomes progressively worse as the relaxation times decrease. This may be due to various factors. In the first place there is much greater difficulty in finding and fitting an adequate intermolecular potential for polyatomic molecules, especially when they are polar. In addition, large molecules are likely to be appreciably deformed during the collision, so that first-order perturbation theory is unsatisfactory; it is in any case unreliable where high transfer probabilities are involved.

A simple empirical relation which correlates most of the available experimental relaxation times available at temperatures in the neighbourhood of 300 °K is the Lambert–Salter plot[30], which is shown in Fig. 10. Molecules fall into two classes, differentiated by the presence or absence of hydrogen atoms, each class showing a linear relation between $\log Z_{1,0}$ and v_{min}. It is difficult to see any clear theoretical explanation of this striking correlation between vibrational frequency and transition probability which neglects entirely the influence of both mass and inter-

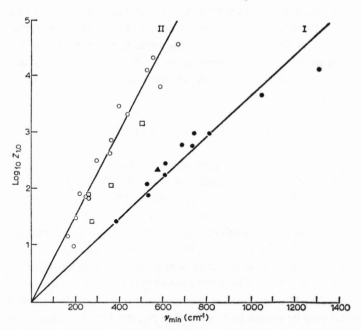

Fig. 10. The Lambert–Salter plot. Relation between $Z_{1,0}$ and v_{min} for simple polyatomic molecules at 300 °K. Values listed in order from the bottom upwards. $\bigcirc$, molecules containing no hydrogen atom: C_2F_4; CF_2Br_2; CF_2BrCl; CF_2Cl_2; $CFCl_3$; CCl_4; CF_3Br; CF_3Cl; SF_6; CF_4; CS_2; N_2O; COS; Cl_2; CO_2. $\square$, molecules containing one hydrogen atom: $CHCl_2F$; $CHCl_3$; $CHClF_2$; CHF_3. $\bullet$, molecules containing two or more hydrogen atoms: CH_2ClF; CH_3I; CH_2F_2; CH_3Br; C_2H_2; CH_3Cl; C_2H_4O; C_2H_4; $cyclo\text{-}C_3H_6$; CH_3F; CH_4. $\blacktriangle$, deuterated molecule: CD_3Br.

molecular potential, though the paramount importance of translational overlap has been stressed. The marked effect of the presence of hydrogen atoms in facilitating energy transfer can be interpreted in terms of ssh theory as due to the very large amplitude factor for vibrations involving hydrogen atoms (see Section 3). This factor is effective for molecules such as CH_3Cl, where the C–Cl bond and not the C–H bonds are concerned in the lowest fundamental mode, as well as for vibrations involving hydrogen atoms. It is possible that vibrational relaxation of methyl halides does not occur *via* the lowest mode, but *via* the next lowest, which has a higher frequency, but a much larger amplitude factor, and which *does* involve hydrogen atoms[33]. An alternative explanation of the 'hydrogen effect' in terms of vibration–rotation transfer has also been put forward. This possibility will be discussed in detail in Section 7.

Molecules, both diatomic and polyatomic, with lowest fundamental frequencies above 1000 cm^{-1}, do not follow the Lambert–Salter plot: their log $Z_{1,0}$ values fall consistently below the lines in Fig. 10 (extended to higher frequencies). For these the ssh treatment agrees better than the Lambert–Salter plot. H_2O, with lowest frequency 1595 cm^{-1}, shows the surprisingly low value of $Z_{1,0} \sim 70$ at 323 °K[61]. Here the very strong dipole–dipole interaction between colliding molecules produces a deep minimum in the intermolecular potential, corresponding to a numerically large value of ε in eqn. (18) and to a steeper repulsion potential (Sec. 3.9). A few polar molecules with frequencies below 1000 cm^{-1} also show transfer probabilities much higher than those predicted by the Lambert–Salter correlation. Thus NH_3 ($v_{min} = 950$ cm^{-1}) and CH_3CN ($v_{min} = 380$ cm^{-1}) both show no ultrasonic dispersion up to the highest f/p values at which measurements have been made, so that $Z_{1,0}$ must be lower than 10 collisions[30]. Dipole–dipole interaction is again very strong for both and is probably responsible, though an alternative possibility for NH_3 is that the inversion vibrational mode may play a special role in energy transfer by splitting the potential surface.

The torsional modes, due to restricted internal rotation in the more complex molecules, behave in exactly the same way with respect to relaxation as other fundamental modes[30]. Most molecules with internal rotation relax the whole of their vibrational energy *via* the torsional mode, which shows highly efficient vibration–translation transfer owing to its low frequency. The one exception, ethane, is discussed in Section 5 below. Holmes *et al.*[62] have recently shown by ultrasonic methods that the straight-chain hydrocarbons, *n*-pentane and *n*-hexane, whose torsional frequencies lie below 100 cm^{-1}, undergo vibration–translation transfer at approximately every collision. Other flexible organic molecules may be expected to show the same high transfer efficiency.

It is interesting that vibration–translation relaxation phenomena in *liquids*, where the molecules can be regarded as in continual close association, show the same general features as for the corresponding gases; energy transfer would appear to occur in binary collisions with the same transfer probability per collision as in the gas[63]. This is also the case with very highly compressed CO_2 at temperatures above the critical point, where the relaxation time remains inversely proportional to the density up to a pressure of 250 atm[64].

The transfer probabilities for the majority of polyatomic molecules, while their absolute values may not always be successfully predicted by the SSH theory, do follow the predicted temperature dependence, giving a linear plot of log $Z_{1,0}$ against $T^{-\frac{1}{3}}$. A few strongly polar molecules, however, show a reversed temperature dependence at lower temperatures, so that the transfer probability first falls to a minimum with increasing temperature and then rises in the usual way[65]. Fig. 11 shows the log $Z_{1,0}$ plot for CH_3F, CH_3Br and SO_2, with CF_4 as a normal case for comparison. (The measurements recorded for CH_3Cl by Corran et al.[65] were later shown to be in error owing to the presence of impurity.) The explanation for the reversal is probably the same as discussed above for the unexpected high transfer probabilities shown by the polar molecules H_2O, NH_3 and CH_3CN. Dipole–dipole interaction (Section 3.9) may dominate the other exponential terms at lower temperatures, causing *both* abnormally high values of $P_{1,0}$ *and* a reversal of temperature dependence. At higher temperatures this factor diminishes in relative importance and the normal temperature dependence takes over. There is some evidence that H_2O also shows, as would be expected, a reversed temperature dependence[168]. Roesler and Sahm[61] found $Z_{1,0} \sim 73$ at 323 °K by ultrasonic absorption measurements. Huber and Kantrowitz[42], using the impact-tube method, found that at 486 °K, $Z_{1,0} \sim 400$; at 585 °K, $Z_{1,0} \sim 300$; at 706 °K, $Z_{1,0} \sim 190$. Measurements by Fujii et al.[66], using an ultrasonic-pulse method in the inter-

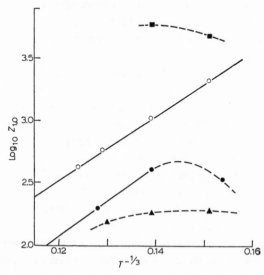

Fig. 11. Temperature dependence of $Z_{1,0}$ for polar and non-polar molecules: ■, CH_3F; ○, CF_4; ●, SO_2 (519 cm^{-1}); ▲, CH_3Br. (Data from Corran et al.[65])

mediate temperature range, 315 to 450 °K, show a considerable scatter, with $Z_{1,0}$ values varying between two and three times that of Roesler and Sahm. For NH_3 and CH_3CN no data are yet available for comparison.

4.3 MIXTURES

If a relaxing gas, A, is mixed with a non-relaxing gas, B, such as helium, there are two collision processes by which vibration–translation energy transfer may occur

(1) $A^* + A \rightarrow A + A$ (vib → trans)

(2) $A^* + B \rightarrow A + B$ (vib→ trans).

Since (1) and (2) will have different collisional efficiencies, the result will be a composite relaxation time for A*, given by

$$\frac{1}{\beta_A} = \frac{1-x}{\beta_{AA}} + \frac{x}{\beta_{AB}} \tag{20}$$

where x is the mole fraction of B in the mixture; β_{AA} is characteristic of process (1), and β_{AB} of process (2). This may be represented graphically by a *linear* plot of reciprocal relaxation time, $1/\beta_A$, against mole fraction, x. Such a plot for experimental measurements made with $CF_4 + He$ mixtures[67] is shown in Fig. 12. The intercept on the $x = 1.0$ axis gives a value for β_{AB}, from which Z_{AB} may be calculated.

The probability of vibration–translation transfer in heteromolecular collisions would be expected to follow the same theoretical pattern as for homomolecular collisions. Application of the SSH theory is less easy owing to uncertainty in the intermolecular potentials and other collision parameters. Olson and Legvold[67] have made measurements extending over the whole concentration range, as in Fig. 12, for mixtures of inert gases with CF_4, CHF_3 and CCl_2F_2. These are shown in Table 3. Qualitatively the expected variation with mass is found, helium being the most efficient partner and collisions with Ar being less efficient than self-collisions. Quantitatively the very fair agreement shown in the table was obtained using SSH calculations, with a Lennard-Jones 6 : 12 intermolecular potential for the inert gases, and the 7 : 28 potential, which was proposed by Hamann and Lambert[68] for 'spherical shell' molecules, for the halomethanes. The combining rules given by Hirschfelder *et al.*[31] were used for obtaining the necessary heteromolecular collision parameters.

A great many experimental studies have been made on the effect of adding *very small* quantities (1 to 2 %) of various additives, B, to a relaxing gas, A.

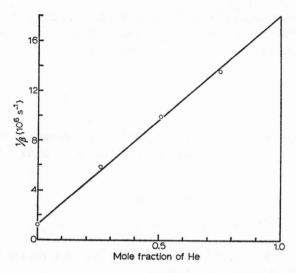

Fig. 12. Reciprocal relaxation times for $CF_4 + He$ mixtures. (Data from Olson and Legvold[67])

Values of β_{AB} and Z_{AB} have been calculated by applying equation (20), and often show striking "catalytic" effects of the additive; i.e. $Z_{AB} \ll Z_{AA}$. Thus a very small amount of additive produces reductions in β_A which are far greater than can be explained in terms of the SSH theory for vibration–translation transfer. Now that vibration–vibration transfer is better understood (see Section 5 below), it has become quite clear that many of these cases, where B is a polyatomic molecule, can be explained in terms of rapid *vibration–vibration* transfer from

TABLE 3

COLLISION NUMBERS FOR VIBRATION–TRANSLATION TRANSFER IN MIXTURES

Mixture	Z_{AB}(expt.)	Z_{AB}(calc.)
$CF_4 + CF_4$	2220	
$CF_4 + Ar$	3730	3370
$CF_4 + Ne$	859	656
$CF_4 + He$	223	230
$CHF_3 + CHF_3$	2180	
$CHF_3 + Ar$	7430	10600
$CHF_3 + Ne$	2430	1960
$CHF_3 + He$	298	580
$CCl_2F_2 + CCl_2F_2$	109	
$CCl_2F_2 + Ar$	238	296
$CCl_2F_2 + Ne$	112	153
$CCl_2F_2 + He$	72	228

(Data from Olson and Legvold[67])

TABLE 4

TEMPERATURE DEPENDENCE OF COLLISION NUMBERS FOR VIBRATION–TRANSLATION TRANSFER

Temp. (°K)	293	373	473	573	673
Collision partners	Z_{AB}	Z_{AB}	Z_{AB}	Z_{AB}	Z_{AB}
$N_2O + N_2O$	7500	4500	3300	2500	2100
$N_2O + H_2O$	105	190	210	–	–
$CO_2 + CO_2$	57000	29000	17000	10000	8000
$CO_2 + H_2O$	105	65	102	150	250

	Fundamental vibrational modes (cm^{-1})		
N_2O	582	1285	2223
CO_2	672	1351	2396
H_2O	1595	3652	3756

(Data from Eucken and Nümann[69])

A to B. If this occurs, equation (20) does not necessarily apply and the calculated values of Z_{AB} may be quantitatively meaningless; measurements extending to much higher concentrations of additive are necessary if meaningful quantitative conclusions are to be drawn. The vitally important experimental fact remains, that less than 1 % of a "catalytic" impurity may diminish the observed value of β_A by a factor of four or five. This is why scrupulous attention to chemical purity of the gases used is necessary when making relaxation measurements; a great many discrepancies between early observations by different experimenters were due to neglect of this factor.

Water vapour is a particularly effective 'catalyst' for vibrational relaxation, as illustrated by the data given in Tables 4 and 7. For $H_2O + O_2$ (Table 7, p. 229) near-resonant vibration–vibration transfer between the fundamental frequency of O_2, $v = 1554$ cm^{-1}, and the bending frequency of H_2O, $v = 1595$ cm^{-1}, accounts satisfactorily for the low value of Z_{AB}. This is discussed in more detail in Section 5 below. For the $H_2O + CO_2$ and $H_2O + N_2O$ mixtures there are no near-resonant frequencies, as may be seen from Table 4; neither N_2O nor CO_2 is polar, so that dipole–dipole interaction cannot be responsible, and some other explanation must be sought. It has been suggested both by Eucken and Küchler[70] and by Widom and Bauer[71] that for $CO_2 + H_2O$, because of 'incipient' chemical combination to H_2CO_3, there will be a transition state with particularly strong interaction, giving a very deep potential well for molecules in this preferred orientation. This picture is closely similar to the effect of orientation-dependent dipole–dipole interaction discussed in Sec. 4.2 and 3.9 above, and Table 4 shows that the same reversed temperature dependence is found for $CO_2 + H_2O$ mixtures between 373 and 673 °K. It is difficult to find a simple physical explanation for the opposite dependence occurring between 293 and 373 °K, shown in Table 4, and it is likely that the single value of Z_{AB} for 293 °K is in error. More recent measurements by Lewis and Lee[72] show a steady rise of Z_{AB} between 297 and 468 °K, though their absolute values are consistently lower than those of Eucken and Nümann. Similar considerations may apply to the $N_2O + H_2O$ mixtures, which show the same reversed temperature dependence over the whole range. Recent measurements on the vibrational relaxation of $CO_2 + H_2$ and $CO_2 + D_2$ mixtures[73] also show consistent reversed temperature dependence in the range 296 to 427 °K. For these pairs it is difficult to visualise any specific "chemical" interaction, but efficient energy transfer may again depend critically on the relative orientation of the approaching molecules. This factor appears to be generally more important for heteromolecular than for homomolecular collisions.

5. Vibration–vibration transfer

5.1 INTRAMOLECULAR TRANSFER OF VIBRATIONAL ENERGY

Experimental observation on the rate of vibrational relaxation of polyatomic molecules, which is in itself a vibration–translation transfer, can furnish indirect information about the rate of intramolecular vibration–vibration transfer between different modes. For lower quantum states of simple molecules, intramolecular transfer can only occur in collision: the energy in the different modes is quantized, and, except in rare cases where there is exact resonance between harmonics, the energy discrepancy can only be made up by translational energy. For a molecule with two active vibrational modes of frequency v_1 and v_2, there are three possible vibrational transitions, which are illustrated on the energy level diagram in Fig. 13. (a) Transfer of translational energy to $0 \rightarrow 1$ excitation of the mode 1 (frequency v_1), with relaxation time β_1. (b) Transfer of translational energy to $0 \rightarrow 1$ excitation of the mode 2 (frequency v_2), with relaxation time β_2. (c) The *complex* transfer of a quantum of vibrational energy from mode 1, plus the necessary increment of translational energy to give $0 \rightarrow 1$ excitation of the mode 2, with relaxation time $\beta_{1,2}$.

It was pointed out in Section 4.2 that most polyatomic molecules show only a single relaxation process, owing to rapid intramolecular vibration–vibration transfer between modes. This corresponds to a state of affairs where $\beta_2 \gg \beta_1 \gg \beta_{12}$. Vibrational energy enters the molecule *via* process (a), which is rate-determining,

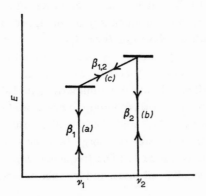

Fig. 13. Energy level diagram showing transitions for a molecule with two active vibrational modes.

and rapidly flows in complex collisions *via* process (c) to the second mode (and to any higher modes). The reverse occurs in deactivation. Process (b) is too slow to play any role. The relation of the observed overall relaxation time, β, to β_1 was discussed in Section 4.2. The general picture is that rapid vibration–vibration transfer maintains continuous equilibrium of vibrational energy between the various fundamental modes of the molecule, and that the whole of this energy relaxes in a single vibration–translation transfer process *via* the lowest mode[30]. The only conclusion that can be drawn about the rate of vibration–vibration transfer in these cases is that it is faster than vibration–translation transfer from the lowest mode.

For a few polyatomic molecules, where there is a large difference between v_1 and v_2, the rate of the complex process (c) is much slower, and the condition $\beta_2 \gg \beta_{12} > \beta_1$ applies. This gives rise to a double relaxation phenomenon. Process (b) is again too slow to play any role, but process (a) is now faster than process (c). The vibrational energy of mode 2 (and any upper modes) relaxes

TABLE 5

EXPERIMENTAL COLLISION NUMBERS FOR INTRAMOLECULAR VIBRATIONAL ENERGY TRANSFER AT 300 °K

Substance	v_1 (cm^{-1})	v_2 (cm^{-1})	i	Δv (cm^{-1})	$Z_{1,2}$	$Z_{1,0}$
SO_2	519	1151	2	110	2390	390
CH_2Cl_2	283	704	2	140	460	30
C_2H_6	290	820	3	50	74	20

$Z_{1,2}$ is the collision number for vibration–vibration transfer between i quanta of mode 1 (frequency v_1) and one quantum of mode 2 (frequency v_2).

(Table reproduced from Lambert[86] by kind permission of the Council of the Chemical Society)

via complex process (*c*) followed by the faster process (*a*). (*c*) is thus the rate-determining step, and the vibrational energy of the upper modes is transferred to translational energy with a relaxation time β_{12}. The vibrational energy of the lowest mode, 1, relaxes independently by process (*a*) with the shorter relaxation time β_1. This behaviour has been observed for only three gases, SO_2[43], CH_2Cl_2[44] and C_2H_6[30]. The experimental data are summarized in Table 5. Two relaxation times are observed, β_1 and β_{12}, corresponding to processes (*a*) and (*c*) respectively. For all these molecules $\nu_2 > 2\nu_1$, and theoretical considerations show that the complex step (*c*) involves a transfer of energy between one quantum of mode 2 and two or three quanta of mode 1. (The frequency gap between 2 and the remaining upper modes is small in all cases, and transfer between these is very rapid.)

5.2 INTERMOLECULAR TRANSFER OF VIBRATIONAL ENERGY

5.2.1 *Ultrasonic dispersion in mixtures*

Experimental observation of relaxation phenomena in binary mixtures of polyatomic gases affords much more information about vibration–vibration transfer. The nature of the vibrational relaxation process for a mixture of a relaxing gas, A, with a non-relaxing gas, B, has been discussed in Section 4.3. It involves two collision processes

(1) $A^* + A \rightarrow A + A$ (vib → trans)

(2) $A^* + B \rightarrow A + B$ (vib → trans)

and the overall relaxation is governed by equation (20), giving a linear plot of reciprocal relaxation time, $1/\beta_A$, against the mole fraction of B in the mixture, *x*.

If both A and B are polyatomic relaxing gases, there will also be two collision processes, corresponding to (1) and (2), for vibration–translation energy transfer from B* in homomolecular and heteromolecular collisions. In addition there can be a vibration–vibration transfer between A* and B*, making five transfer processes in all

(1) $A^* + A \rightarrow A + A$ (vib → trans)

(2) $A^* + B \rightarrow A + B$ (vib → trans)

(3) $A^* + B \rightarrow A + B^*$ (vib → vib ± trans)

(4) $B^* + B \rightarrow B + B$ (vib → trans)

(5) $B^* + A \rightarrow B + A$ (vib → trans)

If process (3), vibration–vibration transfer, does not occur, the mixture will show a double relaxation phenomenon, characterized by two relaxation times, β_A and β_B, which will both be related to molar composition by equations of type (20), each giving a linear plot of $1/\beta$ against composition. If vibration–vibration transfer does occur, the picture is completely altered. Supposing, for convenience, that pure A relaxes slowly, and pure B rapidly, so that processes (4) and (5) are both much faster than processes (1) and (2), there are now two alternative possibilities for the overall relaxation process.

When vibration–vibration transfer is faster than all the other processes, (3) will maintain the vibrational energy of the whole system $(A^* + B^*)$ in continuous equilibrium, and the total vibrational heat content of both components will relax *via* the faster of processes (4) or (5). If (4) plays the predominant role, this will give rise to a near *quadratic* dependence of $1/\beta$ on mole fraction of B, since the rate of process (4) is proportional to x^2, and equation (20) no longer applies. This mechanism is analogous to the relaxation behaviour discussed in Section 5.1 for pure polyatomic gases giving single dispersion. The *near-resonant* collision process, (3), involving transfer of vibrational energy from a mode of frequency v_x of molecule A to a mode of frequency v_y of molecule B, plays exactly the same role as the *complex* collision process, involving transfer of vibrational energy from mode x to mode y of a single molecular species.

Alternatively, when process (3) is slower than (4) or (5), but faster than (1) or (2), A will again relax by the route (3) followed by (4) or (5), but now (3) will be rate determining. This will give a *linear* variation of $1/\beta_A$ with x. B will relax independently, and more rapidly, *via* (4) and (5), with linear dependence of $1/\beta_B$ on x. There will thus be a double relaxation phenomenon with two relaxation times, β_A involving only the vibrational heat capacity of A, and β_B only that of B, both showing linear concentration dependence. This mechanism is analogous to the relaxation behaviour discussed in Section 5.1 for pure polyatomic gases, which show double dispersion because vibration–vibration transfer between modes is *slower* than vibration–translation transfer from the lowest mode.

The nature of the overall relaxation process (whether single or double) for mixtures, and the concentration dependence of the relaxation times, are thus determined by the relative rates of processes (1) to (5). Observations of ultrasonic dispersion in the two pure gases and in a series of mixtures extending over the whole concentration range enable a diagnosis to be made of which type of mechanism is operating. The rates of processes (1) and (4) are obtained from measurements on the pure components, and the rates of processes (3) and (5) which give the best fit to the experimental observations may be estimated by trial and error. Results obtained by observations of this kind for a series of mixtures are given in the first two sections of Table 6[74]. Mixtures were chosen for which near-resonant vibration frequencies in the two components would be expected to give rise to rapid vibration–vibration transfer (small value of Z_{AB}). It will be seen that,

TABLE 6

EXPERIMENTAL COLLISION NUMBERS FOR INTERMOLECULAR VIBRATIONAL ENERGY TRANSFER AT 300 °K

A	B	ν_A (cm^{-1})	ν_B (cm^{-1})	i	$\Delta\nu$ (cm^{-1})	Z_{AB}	Z_{AA}	Z_{BB}
			Singly dispersing mixtures					
SF$_6$	CHClF$_2$	344	369	1	25	50	1005	122
C$_2$H$_4$	C$_2$H$_6$	810	821.5	1	11.5	40	970	74
			Doubly dispersing mixtures					
CCl$_2$F$_2$	CH$_3$OCH$_3$	260	250	1	10	5	73	<3
CH$_3$Cl	CH$_3$OCH$_3$	732	250	3	18	70	421	<3
SF$_6$	CH$_3$OCH$_3$	344	164	2	16	80	1005	<3
CHF$_3$	C$_2$F$_4$	507	507	1	0	50	1500	5.5
SF$_6$	C$_2$F$_4$	344	190	2	36	70	1005	5.5
CF$_4$	C$_2$F$_4$	435	220	2	5	110	2330	5.5
			Spectroscopic data					
NO($A^2\Sigma^+$)	N$_2$	2341	2330	1	11	7.9×10^2	—	—
NO($X^2\Pi$)	CO	1876	2143	1	267	1.0×10^4	—	—
NO($X^2\Pi$)	N$_2$	1876	2330	1	454	5.0×10^5	—	—
CO	O$_2$	2143	1556	1	587	4.5×10^6	—	—
NO($X^2\Pi$)	CH$_4$	1876	1534	1	342	1.1×10^3	—	—
CO	CH$_4$	2143	1534	1	609	3.3×10^4	—	—
NO($X^2\Pi$)	D$_2$S	1876	1892	1	16	94	—	—
NO($X^2\Pi$)	H$_2$O	1876	1595	1	281	160	—	—
NO($X^2\Pi$)	H$_2$S	1876	1290	1	586	310	—	—
NO($X^2\Pi$)	D$_2$O	1876	1179	1	697	1000	—	—

Z_{AB} is the collision number for vibration–vibration transfer between one quantum of mode of frequency ν_A of molecule A and i quanta of mode of frequency ν_B of molecule B. (Frequencies are for 0–1 vibrational excitation.)

where Z_{AB} lies *below* the collision numbers for vibrational relaxation of the pure components, Z_{AA} and Z_{BB}, single dispersion is observed; where Z_{AB} lies *between* Z_{AA} and Z_{BB}, double dispersion is observed.

The actual experimental data for one singly dispersing mixture, SF$_6$+CHClF$_2$, are shown in Fig. 14. The lowest vibrational modes of the two molecules lie close enough for rapid vibration–vibration transfer, and single dispersion is observed, with near-quadratic concentration dependence of $1/\beta$. This indicates that the homomolecular relaxation of CHClF$_2$ (B), process (4), is the rate-controlling step, and is faster than the heteromolecular relaxation process (5). The curve calculated for concentration dependence of $1/\beta$ was obtained by setting up the detailed energy and temperature relaxation equations developed by Tanczos[35], and solving over

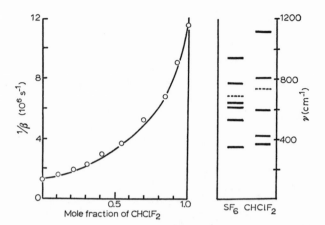

Fig. 14. Reciprocal relaxation times and energy level diagram for $SF_6+CHClF_2$ mixtures: ○ observed points; (———) curve calculated from theory. (Reproduced from Lambert *et al.*[74] by kind permission of the Council of the Royal Society)

the whole concentration range using an electronic computer. The value of Z_{AB} in Table 6 was estimated to give the best fit, and may be taken as 50 ± 15. Similar behaviour is shown by the mixture, $C_2H_4+C_2H_6$, investigated experimentally by Valley and Legvold[75]. This case is complicated by the double dispersion shown by pure ethane (see Section 5.1 above). The data are shown in Fig. 15. The torsional mode of ethane (290 cm^{-1}) relaxes independently in both pure gas and mixtures. There is near resonance between the second mode of ethane (825 cm^{-1}) and the lowest mode of ethylene (810 cm^{-1}); Z_{AB} is the collision number for vibration–vibration transfer between these modes, Z_{BB} for relaxation of the 821.5 cm^{-1}

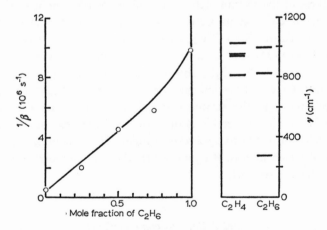

Fig. 15. Reciprocal relaxation times and energy level diagram for $C_2H_4+C_2H_6$ mixtures: ○ points recalculated from experimental data of Valley and Legvold[75]; (———) curve calculated from theory. (Reproduced from Lambert *et al.*[74] by kind permission of the Council of the Royal Society)

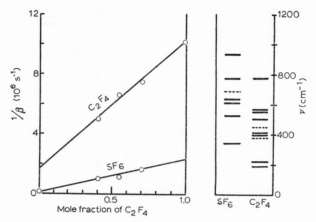

Fig. 16. Reciprocal relaxation times and energy level diagram for $SF_6 + C_2F_4$ mixtures: O observed points; (———) curve calculated from theory. (Reproduced from Lambert et al.[77] by kind permission of the Council of the Faraday Society)

mode of ethane, and Z_{AA} for the 810 cm^{-1} mode of ethylene. In this case the efficiencies of homomolecular and heteromolecular vibration–translation relaxation of ethane are roughly the same, so that processes (4) and (5) both play a rate-determining role, and an almost linear concentration dependence of $1/\beta$ is observed. Relaxation behaviour of the same type has also been reported by Rao and Srinivasachari[76] for mixtures of CO_2 with C_2H_4O (ethylene oxide). Here rapid vibration–vibration exchange is assumed to occur between the 667.3 cm^{-1} mode of CO_2 and the 704 cm^{-1} mode of C_2H_4O, and all mixtures show single dispersion with near-quadratic concentration dependence of $1/\beta$.

The mixtures of the second section in Table 6, which were investigated earlier (when erroneous conclusions were drawn)[77], all show double dispersion. The details for one mixture, $SF_6 + C_2F_4$, are shown in Fig. 16. There is near-resonance between the lowest (344 cm^{-1}) mode of SF_6 and the first harmonic of the lowest (190 cm^{-1}) mode of C_2F_4. C_2F_4 shows very efficient homomolecular vibration–translation transfer, and the estimated vibration–vibration transfer rate ($Z_{AB} = 70$) falls between this and the slower vibration–translation transfer rate of SF_6 ($Z_{AA} = 1005$). Double dispersion is observed, and the predicted linear variation with concentration of the two relaxation times. The remaining mixtures in this section, all of which involve B components whose homomolecular relaxation is very rapid, behave similarly.

5.2.2 Spectroscopic evidence

Direct information about the rate of vibration–vibration transfer between molecules with comparatively high vibrational frequencies can be obtained by techniques described in Section 2.3. Vibrationally excited NO in the ground electronic state, $NO(X^2\Pi)$, may be produced by flashing NO mixtures with ultraviolet light[78]. The initial flash forms electronically excited $NO(A^2\Sigma^+)$ which is quenched by collision with ground state $NO(X^2\Pi)$ $(v = 0)$ to produce vibrationally excited ground state molecules. The process may be written

$$NO(A^2\Sigma^+) + NO(X^2\Pi)(v = 0) \to\to NO(X^2\Pi)(v = 1) + NO(X^2\Pi)\ (v = 0)$$

The rate of relaxation of the vibrationally excited $NO(X^2\Pi)$ by vibration–vibration transfer to CO and N_2 in suitable mixtures can be measured spectroscopically by plate photometry. $NO(X^2\Pi)(v = 1)$ may also be produced directly by microwave-pulse excitation[19b], and this method has been used to investigate vibration–vibration transfer in collisions with D_2S, H_2O, H_2S, D_2O and CH_4. Optical fluorescence quenching in $NO + N_2$ mixtures has given information about the more rapid near-resonant transfer between electronically excited $NO(A^2\Sigma^+)$ and N_2[21b]. Infrared fluorescence quenching has been used to measure the rates of vibration–vibration transfer between CO and O_2[20a], and between CO and CH_4[20c]. The values of Z_{AB} for all these mixtures are included in Table 6, and will be discussed in Section 5.3.

In the optical fluorescence quenching experiments on mixtures of $NO(A^2\Sigma^+) + N_2$, the former was produced in vibrational levels, $v = 3, 2$ and 1, and transfers involving each of these levels were followed. It was found that more than 85 % of the transfers observed involved exchange of a *single* quantum, e.g.

$$NO(A^2\Sigma^+)(v = 2) + N_2(v = 0) \to NO(A^2\Sigma^+)(v = 1) + N_2(v = 1)$$

thus confirming that the selection rule $\Delta v = \pm 1$ applies to vibration–vibration transfer as it does to vibration–translation transfer (see Sections 3 and 4.1). There are thus three allowed transfer processes for deactivation of the excited $NO(A^2\Sigma^+)(v = 1, 2$ or $3)$ by $N_2(v = 0)$. Their collision numbers were found to be: $Z_{3,2:0,1} = 200$; $Z_{2,1:0,1} = 440$; $Z_{1,0:0,1} = 790$. This shows that the efficiency of transfers involving $\Delta v = \pm 1$ increases almost proportionately with increase in vibrational quantum number, which is in accord with the theory for a harmonic oscillator[79].

5.3 THEORETICAL DISCUSSION

The theoretical treatment of vibration–vibration transfer was outlined in Section 3. Sufficient data for *a priori* theoretical calculations are only available for the simpler molecules. It is interesting first to discuss the general pattern revealed by the collision numbers in Tables 5 and 6 in terms of equations (18) and (19).

For the nine mixtures listed first in Table 6 the exchanging vibrational frequencies lie close to exact resonance, 36 cm^{-1} being the largest energy discrepancy. The theoretical equation for resonant exchange, (19), applies here. The first striking conclusion which emerges is that exactly resonant vibration–vibration exchanges do not, as has often been wrongly assumed, have unit efficiency. This is illustrated experimentally by the mixture $CHF_3 + C_2F_4$, where both components have vibrational modes of identical frequency, 507 cm^{-1}, and the estimated value for Z_{AB} is 50. The efficiency of near-resonant collisions decreases with rising frequency of the exchanging modes. Were all other factors equal, Z_{AB} should be proportional to the square of v. This is illustrated by the increase in Z_{AB} from 5 for interchange between CCl_2F_2 ($v_A = 260$ cm^{-1})$+ CH_3OCH_3$ ($v_B = 250$ cm^{-1}), to 790 for the mixture $NO(A^2\Sigma^+)(v_A = 2371$ cm^{-1})$+N_2(v_B = 2359$ cm^{-1}), where the vibration frequencies increase by a factor of approximately 10, and Z_{AB} by a factor of rather more than 100. Z_{AB} values for the other single quantum near-resonant exchanges

$$CHF_3 \ (507 \ cm^{-1}) + C_2F_4 \ (507 \ cm^{-1}),$$

$$SF_6 \ (344 \ cm^{-1}) + CHClF_2 \ (369 \ cm^{-1}),$$

$$C_2H_4 \ (810 \ cm^{-1}) + C_2H_6 \ (821.5 \ cm^{-1}),$$

with molecules which have intermediate frequencies, all lie in the neighbourhood of 50. The other factor which is important is the relative inefficiency of multiple quantum transfers. This is illustrated by the remaining near-resonant mixtures in Table 6, all of which involve 2- or 3-quantum transitions and show values of Z_{AB} ranging from 70 to 110. The intramolecular 3-quantum transfer between the 290 cm^{-1} and 820 cm^{-1} modes of ethane (Table 5), with a value of $Z_{1,2} = 74$, also falls into this class. For all the mixtures of this group there is no apparent correlation between the size of the energy discrepancy, Δv, and Z_{AB}, which justifies the approximation of treating all these energy exchanges as resonant.

The following five mixtures in Table 6, all involving NO or CO as one component, show energy discrepancies ranging from 267 to 609 cm^{-1} and energy transfer can no longer be regarded as near-resonant: the exchange should follow the theoretical equation (18), modified for vibration–vibration transfer as described in Section 3. The four mixtures involving NO, CO, N_2 and O_2 (including the near-resonant pair, $NO(A^2\Sigma^+) + N_2$), form a 'uniform set'. All the molecules are diatomic and of similar molecular weight; the atoms involved have closely similar masses and the molecules would be expected to show similar gas-kinetic collision parameters. Equation (18) would thus predict a roughly linear relation between the energy discrepancy, Δv, and log P, and this is confirmed experimentally[19b,78]. The data are also quantitatively consistent with the more elaborate theoretical treatment of Rapp and Englander-Golden[80]. The higher transfer efficiency of CH_4 as a collision partner for CO, in spite of the large energy discrepancy, $\Delta v = 617 \text{ cm}^{-1}$, is mainly due to its lower molecular weight, combined with the 'hydrogen effect': *a priori* calculations by the SSH theory give a value of $Z_{AB} \sim 17000$, which is in fair agreement with experiment as compared with other similar calculations[81]. Similar considerations apply to the mixture $NO(X^2\Pi) + CH_4$[19b]. The four mixtures of $NO(X^2\Pi)$ with D_2S, H_2O, H_2S and D_2O, with energy discrepancies ranging from 16 to 697 cm^{-1} again, rather surprisingly, form a "uniform set" giving a linear variation of log Z with Δv[19b]; the absolute collision numbers are considerably lower than are predicted by the SSH theory (which, of course, includes the 'hydrogen effect'). Since NO has only a very small dipole moment, dipole–dipole interaction cannot be responsible, and it has been suggested that hydrogen bonding between NO and the hydrides and deuterides may be responsible for the high transfer probability[19b]. This explanation is similar to that discussed in Section 4.3 for the high efficiency of $CO_2 + H_2O$ collisions for vibration–translation transfer.

A number of experimental measurements have been made on the vibrational relaxation of oxygen, which demonstrate the powerful 'catalytic' effect of small quantities of various additives. The experimental data do not extend to sufficiently high concentrations of additive to make detailed interpretation possible in terms of vibration–vibration and vibration–translation transfers. The molecules involved are all simple enough to enable SSH calculations to be made with reasonable prospect of success. The results of *a priori* calculations by the procedures described by Stretton[33] are presented in Table 7[82]. They show clearly the striking efficiency

TABLE 7

CALCULATED COLLISION NUMBERS FOR INTERMOLECULAR VIBRATIONAL ENERGY TRANSFER IN OXYGEN MIXTURES AT 300 °K

Additive (B)	v_B (cm^{-1})	i	Δv (cm^{-1})	Z_{AB}	Z_{BB}
CH$_4$	1534	1	20	170	11600
CD$_4$	1092	1	462	16000	4960
C$_2$H$_4$	1444	1	110	490	4050
C$_2$H$_2$	729	2	96	1800	490
H$_2$O	1596	1	42	80	~1
D$_2$O	1178	1	376	1160	~1
HDO	1402	1	152	140	~1

Z_{AB} is the collision number for vibration–vibration transfer between one quantum of the fundamental mode of O_2, $v_A = 1554 \text{ cm}^{-1}$, and i quanta of mode of frequency v_B of molecule B. For pure O_2, $Z_{AA} = 8.31 \times 10^7$.

(This Table is reproduced from Lambert[86] by kind permission of the Council of the Chemical Society)

heteromolecular vibration–vibration transfers can have, and reveal the same kind of pattern as discussed above.

Bradley Moore and co-workers[167] have recently developed one of the most fruitful methods of measuring the rates of V–V transfer between simple molecules. A Q-switched CO_2 laser source was employed to populate $CO_2(0, 0°, 1)$ in an experimental gas and relaxation was monitored from the decay of fluorescence following termination of the laser pulse. Not only have the relaxation processes been quantitatively analysed, but direct evidence for V–V transfer was revealed by recording the infrared fluorescence of a number of acceptor molecules. The resonant exchange between $CO_2(0, 0°, 1)$ and N_2 was shown to occur at a similar rate to that observed for the NO $A^2\Sigma^+(v = 1)/N_2$ system. The effect of increasing the energy discrepancy was somewhat less marked than that found in the experiments with nitric oxide. A very striking result of the laser experiments has been the discovery that in a mixture of different isotopes of CO_2, V–V transfer occurs at practically every collision. According to Mahan[173], this extremely fast transfer is due to dipole–dipole coupling and represents a totally different type of process from that described in Section 3.

5.4 MECHANISMS OF VIBRATIONAL EXCITATION

Comparison of the collision numbers given above for vibration–vibration transfer with those for vibration–translation, given in Section 4, shows that in many cases vibration–vibration transfer between two resonant or near-resonant modes is much more efficient than vibration–translation transfer from either. This applies equally to homomolecular and heteromolecular collisions, and carries the interesting consequence that the quickest route for vibrational excitation of upper levels from the ground level by homomolecular collisions is an initial vibration–translation excitation to the $v = 1$ level, followed by successive vibration–vibration transfers to higher levels. Because of the selection rule, $\Delta v = \pm 1$, this will be a stepwise process[83]

$$2A(v = 0) + \text{K.E.} \rightleftharpoons A(v = 1) + A(v = 0) \qquad (\text{trans} \rightleftharpoons \text{vib})$$

$$2A(v = 1) \rightleftharpoons A(v = 2) + A(v = 0) \qquad (\text{vib} \rightleftharpoons \text{vib})$$

$$A(v = 2) + A(v = 1) \rightleftharpoons A(v = 3) + A(v = 0) \qquad (\text{vib} \rightleftharpoons \text{vib})$$

$$2A(v = 2) \rightleftharpoons A(v = 3) + A(v = 1) \qquad (\text{vib} \rightleftharpoons \text{vib}), \textit{etc.}$$

The slowest process will be the vibration–translation activation to the $(v = 1)$ level, which will be rate-determining, and the subsequent vibration–vibration transfers will occur at increasingly fast rates with increasing vibrational quantum number. (For harmonic oscillators $P_{m, m+1}^{n, n-1} = n(m+1)P_{0, 1}^{1, 0}$.) Shock-tube experiments have been conducted on mixtures of 10 % HI with 90 % Ar, in which the populations of HI molecules in levels, $v = 1, 2$ and 3 were followed spectroscopically[84]. The conditions of experiment were equivalent to the HI, initially at 300 °K, being suddenly plunged into a heat bath of inert gas at 2000 °K, and it was found that all three vibrational levels were produced at about the same rate, *i.e.* that of vibration–translation activation to the $(v = 1)$ level, which is clearly the rate-determining step. This step is faster than predicted by SSH theory, for reasons discussed already in Section 4.1, but is still much slower than the vibration–vibration transfers. The reverse situation would be expected in deactivation[85].

It has been suggested[84,85] that owing to anharmonicity, truly resonant collisions will only occur between two molecules in the same vibrational level, so that the fastest process will be

$$2A(v = n) \rightleftharpoons A(v = n+1) + A(v = n-1)$$

and this will predominate. This view is not supported by evidence given in Section 5.2.2, which shows that for $NO(A\,^2\Sigma^+) + N_2$ collisions, $Z_{3,2:0,1}$ is *smaller* than $Z_{1,0:0,1}$, while the former would be expected to show a larger energy discrepancy, due to anharmonicity, than the former. In fact only an extreme degree of anharmonicity would be expected to have any effect, as it has been shown above that collisions with energy discrepancy as large as 50 cm^{-1} still behave as near-resonant, and show no significant difference in efficiency from true resonant collisions.

It must be stressed that the rate-determining role of the vibration–translation activation from $v = 0$ to $v = 1$ only applies to conditions such as shock-wave heating from cold, where all the molecules are in the ground vibrational state initially. Under the equilibrium temperature conditions which usually apply in thermally activated chemical reactions, an appreciable fraction of molecules will be at least in the $v = 1$ level initially, so that the rate of chemical activation or deactivation will depend entirely on the faster vibration–vibration transfer. Heteromolecular collisions with polyatomic additives possessing suitable vibrational frequencies for near-resonant transfer may show the same, or slightly higher, efficiency than homomolecular collisions. Complex organic molecules, which contain hydrogen atoms and have a wide spectrum of frequencies, are very efficient energy transfer catalysts[86].

6. Rotation–translation transfer

Rotational quanta are much smaller than vibrational quanta, and rotational energy is therefore much more easily degraded to translational energy. For most molecules the collision number for rotation–translation transfer is less than 10, corresponding to relaxation times smaller than 10^{-9} sec at one atmosphere pressure.

It is thus more difficult to investigate experimentally by acoustic methods, which will require very high f/p values at which *classical* dispersion and absorption must also be taken into account[87]. Theoretical treatment is also much less satisfactory; molecules are usually distributed in a variety of rotational energy levels, so that a number of different transition probabilities have to be considered, and observed relaxation times usually represent averages over sets of transitions involving a range of J-states.

The only molecules for which quantum mechanical treatment by the distorted wave method is satisfactory are H_2 and D_2. Their small moments of inertia and correspondingly large rotational quanta lead to low transfer probabilities[88]. The transitions which need to be considered at 300 °K are $0 \rightleftharpoons 2$ transitions for p-H_2 and $1 \rightleftharpoons 3$ transitions for o-H_2. Brout[89] made the first quantum mechanical calculations, and found $Z_{2,0} = 329$ for p-H_2, and $Z_{3,1} = 338$ for o-H_2. These values compare well with the experimental collision number, 350, obtained from ultrasonic measurements on n-H_2 at 273 °K by Stewart and Stewart[90], who also found the value, 200, for n-D_2. Similar calculations by Takayanagi[91] also give a fair approximation to the experimental results, both for H_2 and D_2, and predict a higher transfer probability for HD, where one-quantum transitions are allowed. More recent measurements on p-H_2 by Geide[92] have shown definite evidence of a double relaxation process, giving $Z_{0,2} = 173$ and $Z_{2,4} = 301$ for $0 \rightleftharpoons 2$ and $2 \rightleftharpoons 4$ transitions respectively.

The diatomic and triatomic hydrides, which also have very small moments of inertia, would be expected to show relaxation efficiencies of the same order as hydrogen. But these are all polar, and dipole–dipole interaction clearly has a profound effect on transfer efficiency, so that acoustic measurements give for HCl[93], $Z_{rot} = 7$; for H_2O[61], $Z_{rot} = 4$; for H_2S[92], $Z_{rot} = 31$. The high efficiency of equilibration of rotational energy among HCl molecules is confirmed by shock-tube measurements[94]. The OH radical, which should resemble HCl, is shown by spectroscopic examination of detonation waves[95] to have $Z_{rot} = 10$.

Heavier homonuclear diatomic molecules have been treated by an approximate quantum mechanical method by Brout[96]. This gives the very simple result that the mean transition probability is $\frac{1}{2}(d_0/r_0)^2$, where d_0 is the internuclear distance in the molecule, and r_0 the kinetic collision diameter. The reasons for the simplicity of this result are interesting. Temperature disappears from the relation because the increased spread of occupied rotational levels at higher temperatures causes a lowering of the probability, which cancels out the favourable effect of increased velocity of approach. Mass disappears because a larger moment of inertia brings the rotational levels closer and increases the probability, cancelling out the adverse effect of the decreased velocity of approach of the heavier molecules. The inter-molecular potential is unimportant, as the collision time is always short with respect to the frequency of the rotation. The resulting values of Z_{rot} are 17 for O_2 and 23 for N_2. These are higher than experimental values obtained from ultrasonic dispersion[97], 5.3 for N_2 and 4.1 for O_2, and from ultrasonic absorption[98], 4.7 for N_2 and 4.1 for O_2. Shock-wave measurements[94] give 5.5 for N_2. Dipole–dipole interaction is also effective in promoting easy rotational relaxation in these heavier molecules, and NO, which on other grounds might be expected to resemble O_2 and N_2, gives an experimental value of $Z_{rot} \sim 1$[99]. Rotational relaxation of electronically excited $NO(A^2\Sigma^+)$ has been shown by fluorescence measurements[100] to be equally efficient, and to involve multiple quantum transitions up to at least $\Delta J = \pm 5$. In view of the general closeness of rotational levels such multiple transfer is likely to occur for most non-hydrides in contrast to vibrational relaxation, for which the selection rule $\Delta v = \pm 1$ usually seems to hold in systems which are not reacting chemically. Shock-tube measurements on the heavier triatomic molecules give values of Z_{rot} between 1 and 2 collisions[94], and it may be inferred that for the majority of polyatomic molecules rotational energy degrades at approximately every collision.

The occurrence of multiple quantum rotational transitions is also proved conclusively in the detailed studies of Klemperer et al.[22] on resonance fluorescence of $I_2 (B^3\Pi_{0^+u})$. Employing either sodium emission (D lines) to populate predominantly $J' = 44$ and 37 in $v' = 15$, or mercury emission (green line) to populate $J' = 34$ in $v' = 25$, they investigated rotational and vibrational energy transfer, and induced predissociation, in He, Ne, Ar, Kr, Xe, H_2, D_2, N_2, O_2, NO, CO_2, SO_2 and CH_3Cl. Several very interesting features were identified. The efficiency of R–T transfer increases smoothly with increase of reduced mass, except for the polyatomic species (CO_2, SO_2 and CH_3Cl) which are comparatively more efficient with cross-sections up to 70 A^2. For all added gases, $\Delta J \leqslant 40$ transitions occurred in single collision events. The high efficiency of the polyatomic molecules may arise because of the comparative ease of R–R transfer, resulting from the abundance of available rotational states and the consequent facility of conserving angular momentum for such processes. However, the interpretation is complicated because of the simultaneous occurrence of induced predissociation.

The NO and I_2 fluorescence demonstrates that although the optical rules do not hold for rotational energy transfer, complete thermalisation of rotation does not occur. The results suggest that the smaller the rotational spacing, the greater is the probability of a large ΔJ transition. As ΔJ increases, the decreasing translational overlap (equivalent to equation (14) with a three dimensional scattering model) will be restrictive at ambient temperature when the energy to be converted to translation exceeds 50 cm^{-1}; transitions with $\Delta E > 200$ cm^{-1} should have low probability. Three other novel features were demonstrated from the $I_2(B^3\Pi_{0^+u})$ results:

(a) Some persistence of the initial rotational states was observed, following $\Delta v' = \pm 1$ transitions in He, H_2, HD and D_2, but not in the other gases.

(b) No persistence of the initial rotational states was observed following $\Delta v' = \pm 2$ transitions, which have 1/4 of the probability of $\Delta v' = \pm 1$ transitions.

This was attributed to the stronger coupling required for the double quantum vibrational transition.

(c) The HD($J = 0 \rightarrow 1$) rotational spacing (91 cm^{-1}) is very nearly identical with the energy of the vibrational transition $v' = 15 \rightarrow 14$. HD is not abnormally effective at inducing vibrational relaxation.

In the previous paragraph it was suggested that if the energy to be converted to translation is greater than about 50 cm^{-1}, translational overlap, or tunnelling between the potential surfaces, will become restrictive. Supporting evidence is found in studies of the rotational relaxation of electronically-excited diatomic hydrides. Brenner and Carrington[174] have shown that rotational relaxation of the CH radical by the inert gases, is dominated by the $\Delta J = \pm 1$ rule. A similar conclusion was drawn by Kley and Welge[175] from observations of the effect of inert gases on the OH fluorescence. Presumably the optical rule holds in these systems because of the tunnelling restriction, due to the large differences in energy of adjacent J states especially with J large.

The rotational relaxation of polyatomic spherical top molecules can be treated approximately on the classical 'rough sphere' model. This has been done for homo-molecular collisions by Wang Chang and Uhlenbeck[101]. They find a simple expression resembling that obtained by Brout for diatomic molecules

$$Z_{rot} = \tfrac{3}{8}(1+2b)^2/b$$

where $b = I/Ma^2$, I is the moment of inertia, M the reduced mass of the collision, and a the sum of the molecular radii. This gives for CH$_4$, $Z_{rot} = 18$, which is in good agreement with the experimental values, 14 to 17, obtained ultrasonically by Kelley[103]. Widom[102] has applied a similar treatment to heteromolecular collisions between a spherical top and an inert gas molecule. He finds a theoretical expression

$$Z_{rot} = \tfrac{3}{8}(1+b^2)/b$$

giving values of Z_{rot} ranging from 2 for CCl$_4$+Ar to 22 for CH$_4$+Ar. No experimental data are available for comparison. Holmes et al.[98] have measured rotational relaxation in mixtures of N$_2$+He and O$_2$+He, and find for N$_2$, $Z_{rot} = 6.0$, and for O$_2$, $Z_{rot} = 7.5$. These are only slightly larger than for self-collisions. They are much larger than predicted by a modification of Widom's treatment, and much smaller than predicted by a classical theory due to Parker[104a].

It may be noted that all the above formulae for heavier molecules, which take into account only the repulsive part of the intermolecular potential, give Z_{rot} as independent of temperature. This is in striking contrast to vibrational relaxation; rotational relaxation *times*, which depend also on the gas-kinetic collision frequency, would thus be expected to show weak temperature dependence varying inversely as $T^{\frac{1}{2}}$. More elaborate classical theories (Parker[104a] for homonuclear diatomic molecules; Sather and Dahler[104b] for rough spherical molecules), which take into account the attractive part of the intermolecular potential as well as the repulsive, predict an *increase* in Z_{rot} with rising temperature. This has been confirmed by acoustic measurements on diatomic gases[105a,b], which show approximate doubling of Z_{rot} between 300 and 1300 °K.

An interesting development in molecular rotational relaxation has been the microwave double-resonance method[176–178]. The technique permits the exploration of the fine detail of the processes which occur in collisions of polyatomic molecules, and results for a number of symmetric tops have been reported. For example, Oka has described experiments on NH_3 in which inversion doublets for selected J values were pumped by high microwave power. Pumping disturbs the population of the inversion doublet, and also that of other doublets which are populated from the original pair by collision processes. By absorption measurements of 'other' inversion doublets with steady state irradiation, Oka has shown that in NH_3/NH_3 collisions, transitions which are allowed by the electric dipole selection rules ($\Delta J = 0, \pm 1, + \leftrightarrow -$) are preferred. Oka's analysis indicates that relaxation is most favourable in collision with molecules having similar J values, which are termed rotational resonances (R–R transfer). For example the process

$$(K = 4, J = 5) + (K = 3, J = 4) \rightarrow (K = 4, J = 4) + (K = 3, J = 5)$$

might be expected to have a large cross-section since the energy discrepancy is extremely small and angular momentum is conserved. The possibility of modulating the power signal has been discussed by Oka[176] and Gordon[178], which would allow cross-sections to be determined and perhaps lead to the identification of specific processes. Obviously such a method would have great promise.

7. Rotation–vibration transfer

Whether rotation–vibration transfer occurs, and how important it is, are questions of considerable dispute. The experimental observation by Millikan[106,107], that vibrational deactivation of CO in collision with $p\text{-}H_2$ is more than twice as efficient as in collision with $o\text{-}H_2$, seems to provide some evidence that rotational energy participates in vibrational relaxation. The only significant difference between o- and $p\text{-}H_2$ in the context of this experiment would appear to be the difference in rotational energy states, as illustrated by the fact that at 288 °K (the temperature of the experiment) the rotational specific heat of $o\text{-}H_2$ is 2.22, while that of $p\text{-}H_2$ is 1.80 cal.mole^{-1}.deg^{-1}. Cottrell *et al.*[108–110] have measured the vibrational relaxation times of a number of hydrides and the corresponding deuterides. On the basis of SSH theory for vibration–translation transfer the relaxation times of the deuterides should be systematically *shorter* than those of the hydrides. The

TABLE 8

RATIOS OF REDUCED RELAXATION TIMES FOR LOWEST MODES FOR
DEUTERIDES AND HYDRIDES

Molecules ($v$cm^{-1})	Theory β_D/β_H		Expt. β_D/β_H
	vib–rot	vib–trans	
CD$_4$ (996); CH$_4$ (1306)	1.7	0.58	1.7
SiD$_4$ (681); SiH$_4$ (914)	1.5	0.25	1.5
PD$_3$ (730); PH$_3$ (992)	1.5	0.13	1.5
AsD$_3$ (660); AsH$_3$ (906)	1.2	0.04	0.6

(This Table is reproduced from Cottrell et al.[110] by kind permission of the Council of the Faraday Society)

small increase in mass in passing from hydride to deuteride should be more than compensated by the lower vibrational frequency of the deuteride. Experimentally the opposite is found, as shown in Table 8. Cottrell et al. suggest that rotation-vibration transfer is responsible, and develop a theory of energy transfer between a classical rotator and a quantum-mechanical oscillator. Ease of transfer depends on interaction between the vibrator and rapidly moving peripheral atoms of the rotator. Since hydrides will clearly be rotating faster than the corresponding deuterides, they should show higher transfer efficiency, in accordance with the experimental findings. Table 8 lists rough quantitative estimates of the ratio of efficiencies calculated theoretically for vibration–rotation and for vibration–translation transfer. Except for AsD$_3$/AsH$_3$, the former alone are in satisfactory agreement with experiment. If, however, this theory is applied to the relative effects of o- and p-H$_2$ in deactivating CO, it would predict the ortho-form to be the more efficient, in contrast to experimental observation[107]. Millikan has put forward the alternative possibility that the higher efficiency of p-H$_2$ is due to near-resonance between the $J = 2$ to $J = 6$ rotational transition in p-H$_2$ and the vibrational quantum of CO. At 288 °K more than 30 % p-H$_2$ is in the $J = 2$ state, and this seems a plausible explanation. But preliminary experiments at lower temperatures, where the $J = 2$ level is largely depopulated, show the effect to persist almost unchanged.

The further suggestion has been made, by Cottrell[108] and by Bradley Moore[111], that the high vibration transfer efficiency shown by hydrogen-containing molecules (see Section 4.2, the Lambert–Salter plot) is due to their small moments of inertia leading to easy rotation–vibration transfer, which becomes entirely responsible for the vibrational relaxation. While this gives a plausible semi-quantitative explanation for many cases, it breaks down seriously over hydrogen-containing molecules, which do not possess low moments of inertia, but nevertheless fall firmly on the "hydrogen line" in the Lambert–Salter plot: see cyclo-propane, ethylene oxide and acetylene. The rotation–vibration theory also fails to give the correct temperature dependence[111]. The theoretical model for rotation–vibration transfer presented by both Cottrell and Bradley Moore depends on the rotational velocity of peripheral atoms being greater than the translational velocity of the molecules. A suitably oriented collision will thus produce a time-dependent perturbation of the vibrator, which will be more effective for producing transition than that due to the translational motion of the molecules. This presents the physical picture of a peripheral hydrogen atom as an independent sphere, attached by a string to the centre of gravity of the molecule. A more sophisticated view of molecular structure and shape would picture a peripheral hydrogen atom as a pimple on the surface of a "rough sphere", whose motion would produce a much less marked effect on the vibrator.

This concept of a rotation–vibration transfer is different from the concept of *simultaneously* occurring vibration–translation and rotation–translation transfers discussed in Section 4.2 in connection with the vibrational relaxation of polar molecules. This almost certainly occurs, but presents a very difficult theoretical problem to which no satisfactory solution is yet available[112]. Bauer and Liska[113] have made ultrasonic absorption measurements on mixtures of CO_2 and He, which show that $CO_2 + $He collisions are some 22 times more effective in promoting vibrational relaxation than are $CO_2 + CO_2$ collisions. They also show that the relaxing part of the total specific heat in helium-rich mixtures includes some of the rotational heat content of the CO_2 in addition to the vibrational heat content. They attribute this to the effect of rotational transitions simultaneous with, and *opposite* in direction to the vibrational transitions. These diminish the amount of energy degraded to translational in the collision process, thereby increasing its transfer efficiency. Rotational transitions in the *same* direction as simultaneous vibrational transitions will have the opposite effect, and occur less frequently. The same phenomenon was observed by Winter[73] for CO_2/H_2 and CO_2/D_2 collisions. These are some 200 times more efficient for vibrational relaxation of CO_2 than are self-collisions, and the relaxing specific heat contribution is 5 to 8 % higher than the vibrational specific heat of the CO_2. Here the large rotational quanta of the *additive*, H_2 or D_2, may play an important role (D_2 is less efficient than H_2 as would, in any case, be expected). The reversed temperature dependence shown by these two mixtures is curious.

8. Electronic–translation and electronic–vibration relaxation with $\Delta E < 1$ eV

8.1 GENERAL CONSIDERATIONS

At the present time it is difficult to find order or correlations in experimental data on any aspect of electronic energy transfer. Consideration of the factors which may influence the rates of such processes provides pointers which indicate where systematic correlations may lie, to give some guidance in an experimental study.

Considering for the moment the theory of vibrational relaxation (discussed in Section 3), the transition probability is determined by the product of two quantities, first a matrix element determined by the manner in which the internal energy couples with translation, and secondly a term which was described as the "overlap" of the translational wave functions. In the case of parallel potential surfaces without a substantial potential minimum, the translational term accounts for the uniform behaviour shown by the Lambert–Salter plot (Fig. 8) for V–T relaxation, and a similar plot for V–V transfer with the same slope[78]. The manner whereby translation interacts with vibration is obvious in classical or wave mechanics, because the potential is a simple function of the coordinates x and X. The squared vibrational matrix element is inversely proportional to the energy to be converted to translation (ΔE), and since the translational term contains ΔE in the exponential, it is the latter which dominates the form of the variation of transition probability with ΔE. The mechanism of coupling of translational energy and internal electronic energy is much more complex, and also usually includes a change in internal angular momentum. It is obvious that translation can interact with the energy of an electron since the energy of an orbital will be influenced by the proximity of a colliding molecule. However, evaluation of the off-diagonal matrix elements is complex, for example in the theory of Thorson[128], which deals with spin–orbit re-orientation in atomic sodium and potassium.

The absence of any simple and general equation to predict the efficiency of translational–electronic coupling limits the extent to which the controlling factors can be recognised from the magnitudes of the experimental cross-sections.

From consideration of a few fragmentary results it has been suggested that the translational term may be the same, or similar, in all energy transfer processes[79]. This serves as a rough guide in understanding the overall rates of electronic energy transfer processes, provided the translational term can be predicted. The latter restriction is rather severe, because it requires a knowledge of the interaction potential. We later examine evidence which indicates that, when an atom is excited by changing the orbital angular momentum or principal quantum number (usually requiring an energy of a few eV), there is a strong attractive interaction with all but the inert gases. Therefore, to find systems in which the simple theory given earlier may be applicable with respect to the translational part, we should investigate relaxation of atoms in low lying electronic states. There should exist numerous cases where relaxation of atoms amongst the spin–orbit components of the lowest term occurs between approximately parallel potential surfaces, at least for collision with inert gases and some of the stable diatomic molecules. The determination of the potential minima is accessible experimentally from the temperature dependence of the three-body combination of atoms[114]. For example, the potential minimum for interaction of $I(5\,^2P_{\frac{1}{2}})$ with N_2 is ~ 2 kcal.mole^{-1}; this would enhance the translational term by about a factor of 10 only.

Following these general comments, we consider two extreme models for electronic energy transfer:

(*a*) The first to be considered is a process in which the potential curves are parallel. In these cases the probability of energy transfer will decrease sharply with increasing ΔE, because of decreasing overlap of the translational wave functions near the classical turning point. Vibrational transitions will be controlled by the vibrational matrix elements given in Section 2, because the vibrational part is essentially the same as for V–T transfer. The energy which is transferred in this type of process is generally $\ll 1$ eV.

(*b*) The second involves a process in which energy transfer occurs because of a convergence or crossing of the potential energy surfaces (see Fig. 2 and the related discussion). The efficiency of quenching of highly electronically excited species, *e.g.* Hg($6\,^3P$), Na($3\,^2P$), appears to correlate roughly with the ionisation potential of the quenching gas, which suggests that charge transfer occurs in the transition complex. The strong attractive interaction with the excited state would have precisely this effect of converging the potential surfaces. The optical rules for vibrational transitions need not apply in this case where a chemical complex is involved. The formation and fragmentation of the transition complex will produce a low yield of energy in vibration, tending towards a random distribution in the available degrees of freedom. The exact distribution depends on the precise disposition of the potential curves. It is not possible to account for the behaviour of highly excited species, unless potential curve-crossing is postulated.

Into this conceptual framework, we shall attempt to fit the experimental results. Examples of (*a*) are discussed in this Section, and examples of (*b*) are deferred to the next Section.

8.2 CROSS SECTIONS FOR ENERGY TRANSFER

A 'classical' molecular system, which until recently was supposed to show systematic dependence on the change in internal energy, is the collisional spin–orbit relaxation of Hg($6\,^3P$)

$$\text{Hg}(6\,^3P_1) + \text{AB}(v = 0) \rightarrow \text{Hg}(6\,^3P_0) + \text{AB}(v = 1)$$

The problem has recently been studied by flash spectroscopy[115], which is proving to be capable of general application to the investigation of electronic energy transfer processes. The experiments are interesting even though the results may be difficult to understand from first principles. Fig. 17 shows the formation and decay of Hg($6\,^3P_0$) in a flashed mixture of N_2 and Hg vapour. The atoms can be detected in absorption *via* any member of the $n\,^3S_1$–$6\,^3P_0$ or $n\,^3D_1$–$6\,^3P_0$ Rydberg series. The spin–orbit relaxation is probably accompanied by excitation of the N_2

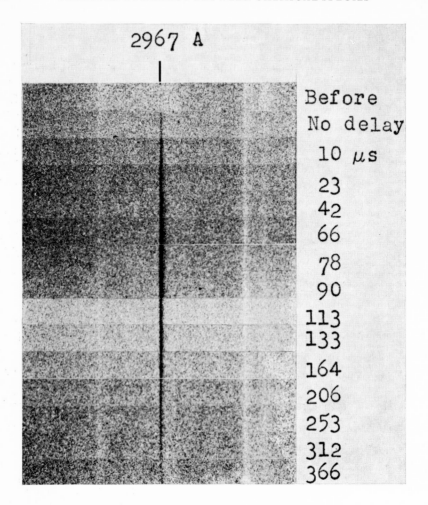

Fig. 17. Absorption spectrum of $Hg(6^3P_0)$.

to the first vibrational level, because Matland[116] has shown that the activation energy for quenching is about 560 cm^{-1}, which corresponds to the difference between the spin–orbit splitting (1767 cm^{-1}) and the N_2 fundamental (2330 cm^{-1}). The $Hg(6^3P_0)$ state is metastable because J cannot remain zero in an optical transition, and the main process which causes its decay over $\sim 10^{-3}$ sec is collision with ground state atoms

$$Hg(6^3P_0) + Hg(6^1S_0) \rightarrow 2Hg(6^1S_0)$$

(This efficient conversion of substantial electronic energy to translation requires crossed potential curves. The process has a large cross-section and therefore cannot be a radiative or three body transition.) Although most simple molecules had been assumed to cause the spin–orbit relaxation, the flash-photolysis experiments narrowed the candidates down to N_2, CO, H_2O and D_2O. It is interesting to note that only a minute fraction of the available light from the photoflash is usefully employed at 2537 A to excite $Hg(6\,^3P_1)$ and subsequently $Hg(6^3P_0)$. However, the transition 6^3D_1–6^3P_0 at 2967 A is sufficiently intense for detection of metastable atoms at concentrations of about 10^{10} atoms. cc^{-1}.

According to Scheer and Fine[117], who detected metastable atoms by means of their property of ejecting electrons from a metal electrode, CO quenches $^3P_1$ atoms both to the $^3P_0$ and the ground $^1S_0$ states, in the ratio of about 1 : 7. The fundamental of CO at 2143 cm^{-1} is 376 cm^{-1} in excess of the separation between the spin–orbit multiplets; taking the total cross-section for quenching of $^3P_1$ by CO to be 4 A^2, it follows that the cross-section for spin–orbit relaxation is 3.5 A^2 in collisions where the energy deficit is available from kinetic energy. (This assumes that the 1 : 7 ratio is independent of the relative velocity of the collision.) Similarly the effective cross-section for N_2 is increased from 0.42 A^2 to 6.7 A^2 when the energy deficit is available. The activation or Boltzmann factors have to be taken out if we expect to make a sensible comparison of processes with varying energy discrepancies. The derived cross-sections may be considered to represent the efficiencies of the reverse process, per unit weight of the $^3P_1$ state.

Fig. 18 is a plot of cross-section *vs.* energy discrepancy for the four molecules which quench efficiently to the $^3P_0$ state. There is no systematic variation with the minimum energy which has to be converted to translation. The random behaviour must arise because of individual peculiarities in the manner in which the potential curves converge. The quenching processes have been discussed along similar lines by Bykhovskii and Nikitin[118]. Following the general comments which

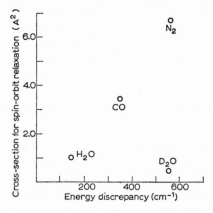

Fig. 18. The effect of energy discrepancy on the relaxation rate. The ordinate is the cross-section for the transition $^3P_1 \to {}^3P_0$ for collisions which have sufficient energy to excite vibration in the quenching molecule.

introduce this section, the failure to find systematic behaviour in the $Hg(6^3P_0)$ relaxation prompted a study of ground state atoms.

The magnitude of the cross-sections is not the only enigma in the mercury experiments. The second curiosity concerns the reason why the four molecules N_2, CO, H_2O and D_2O should efficiently induce spin–orbit relaxation, whereas most other simple molecules (e.g. CO_2, NH_3) are rather inefficient[119,179].

TABLE 9

CROSS-SECTIONS FOR DEACTIVATION OF $Hg(6^3P_1)$ AND $Hg(6^3P_0)$

Gas	$Q(A^2)$, $^3P_1$	$Q(A^2)$, $^3P_0$
N_2	0.42	9×10^{-6}
H_2	6.01	0.018
O_2	13.9	0.093
CO	4.07	0.028
CO_2	2.48	0.0014
H_2O	1.0	0.0066
D_2O	0.46	0.0048
N_2O	12.6	0.51
NO	24.7	0.34
CH_4	~ 0.10	0.007
C_2H_6	0.11	0.011
C_3H_8	1.6	0.16
C_2H_4	22	0.6
NH_3	2.9	0.0033

The third point concerns the data of Table 9, which compares the cross-sections for $^3P_1$ quenching to those for $^3P_0$, both to the $^1S_0$ ground state. The latter cross-sections are all small in comparison to the former. Although in particular cases the decrease of the available energy may effectively forbid electronic energy transfer, e.g. $NO(X^2\Pi) \rightarrow NO(a^4\Pi)$, the occurrence of a general disparity is not yet understood. The mechanism of quenching of $^3P_1$ and $^3P_0$ by the paraffins has been discussed by Kang Yang[180] and differences have been accounted for on the basis of symmetry rules for radiationless transitions.

8.3 SPIN–ORBIT RELAXATION IN SELENIUM

Atomic selenium was the first ground state atom for which spin–orbit relaxation was measured[120]. The deposition of the sub-levels is shown in Table 10.
Excited atoms can be conveniently produced by flashing CSe_2 in a large excess of inert gas which prevents any significant temperature rise. Relaxation processes can be identified, and their rates measured, by kinetic spectroscopy and plate density measurement. The atoms are produced by direct photolysis into the 4^3P_J states

$$CSe_2 + h\nu(\sim 2300 \text{ A}) \rightarrow CSe(X^1\Sigma\ v \leqslant 3) + Se(4^3P_J)$$

The CSe formed is vibrationally excited. The overall spin change can be accounted for by postulating that the upper singlet is predissociated by a triplet state which correlates with the ground electronic-state fragments.

In mixtures containing initially 0.05 torr CSe_2 with 20–60 torr of argon, the rate of decay of $Se(4^3P_0)$ is proportional to the pressure of argon, which indicates that the only important relaxation step is

$$Se(4^3P_0) + Ar \rightarrow Se(4^3P_1) + Ar$$

TABLE 10

ENERGY LEVELS OF $Se(4^3P)$

State	Energy (cm^{-1})
$Se(4^3P_0)$	2534
$Se(4^3P_1)$	1990
$Se(4^3P_2)$	0

TABLE 11

DERIVED RATE COEFFICIENTS AND CROSS-SECTIONS

Point in Fig. 24, p. 250	Process	Rate coefficient $(cm^3.sec^{-1})$	Cross-section (A^2)	ΔE (cm^{-1})
1	$Se(4^3P_0) + Ar \rightarrow Se(4^3P_1) + Ar$	2.4×10^{-14}	4.7×10^{-3}	544
2	$Se(4^3P_0) + N_2O(v_2 = 0) \rightarrow Se(4^3P_1) + N_2O(v_2 = 1)$	1.2×10^{-10}	31	45
3	$Se(4^3P_0) + CO_2(v_2 = 0) \rightarrow Se(4^3P_1) + CO_2(v_2 = 1)$	1.4×10^{-10}	50	128
4	$Se(4^3P_0) + H_2 \rightarrow Se(4^3P_1) + H_2$	3.5×10^{-10}	19	544
5	$Se(4^3P_0) + CO \rightarrow Se(4^3P_2) + CO(v = 1)$	1.1×10^{-12}	0.2	391
6	$Se(4^3P_0) + N_2 \rightarrow Se(4^3P_2) + N_2(v = 1)$	3×10^{-12}	0.54	203
7	$Se(4^3P_0) + O_2 \rightarrow Se(4^3P_1) + O_2$	1.5×10^{-12}	0.29	544
8	$Cs(6^2P_{3/2}) + Ar \rightarrow Cs(6^2P_{\frac{1}{2}}) + Ar$	—	5.2×10^{-4}	554
9	$NO(X^2\Pi_{\frac{3}{2}}) + NO \rightarrow NO(X^2\Pi_{\frac{1}{2}}) + NO$	—	2.4	121
10	$Fe(a^5D_3) + Ar \rightarrow Fe(a^5D_4) + Ar$	2.1×10^{-15}	1.3×10^{-4}	416
11	$Fe(a^5D_3) + H_2 \rightarrow Fe(a^5D_4) + H_2$	7.3×10^{-12}	0.13	416
12	$Fe(a^5D_3) + D_2 \rightarrow Fe(a^5D_4) + D_2$	6.0×10^{-12}	0.15	416
13	$Fe(a^5D_3) + He \rightarrow Fe(a^5D_4) + He$	6.0×10^{-14}	0.0015	416
14	$Fe(a^5D_3) + N_2 \rightarrow Fe(a^5D_4) + N_2$	1.9×10^{-13}	0.011	416
15	$Fe(a^5D_3) + CO \rightarrow Fe(a^5D_4) + CO$	2.7×10^{-12}	0.16	416
16	$Fe(a^5D_3) + Fe \rightarrow Fe(a^5D_4) + Fe$	1.07×10^{-10}	7.3	416
17	$Rb(5^2P_{\frac{3}{2}}) + Ar \rightarrow Rb(5^2P_{\frac{1}{2}}) + Ar$	—	1.6×10^{-3}	238
18	$Ne(3^3P_0) + Ne \rightarrow Ne(3^3P_1) + Ne$	—	6×10^{-4}	359
19	$Ne(3^3P_0) + Ne \rightarrow Ne(3^3P_2) + Ne$	—	6×10^{-4}	777
20	$Ne(3^3P_1) + Ne \rightarrow Ne(3^3P_2) + Ne$	—	5×10^{-3}	418
21	$K(4^2P_{\frac{3}{2}}) + Ar \rightarrow K(4^2P_{\frac{1}{2}}) + Ar$	—	22	58

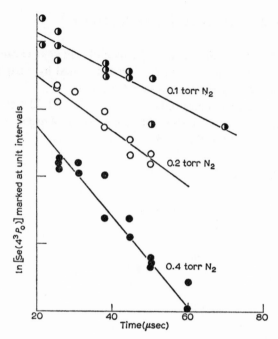

Fig. 20. Semi-logarithmic plots showing the decay of $Se(4^3P_0)$ in 25 torr Ar with added nitrogen (298 °K).

Unfortunately, with the available equipment it was not possible to measure the rate of the decay of the (4^3P_1) state, because the transition at 2040 A $(5^3S_1 \leftarrow 4^3P_1)$ is comparatively weak. The formation and decay of the excited atoms is illustrated in Fig. 19. The increased rate of decay of $Se(4^3P_0)$ caused by addition of various gases was shown using plate photometry, and some semi-log plots for N_2 are given in Fig. 20. Table 11 lists some derived cross-sections (Q) and rate coefficients. The former are derived from the latter by dividing by the mean velocities. Relaxation was effectively complete by 100 μsec in all experiments, during which time removal of atoms by three body combination, or other chemical reactions, was shown to be negligible, either by observation of the slow formation of Se_2, or by photometry of the 1960 A resonance line $(5^3S_1 \leftarrow 4^3P_2)$. The recorded cross-section for deactivation by Ar could be influenced by trace impurities such as N_2, and should be regarded as an upper limit. The processes of Table 11 that are postulated to occur are those in which the minimum energy is converted to translation.

8.4 RELAXATION OF ATOMIC IRON

The selenium atom experiments demonstrated the possibility of a general study of the relaxation of ground state atoms. The second system, involving atomic iron,

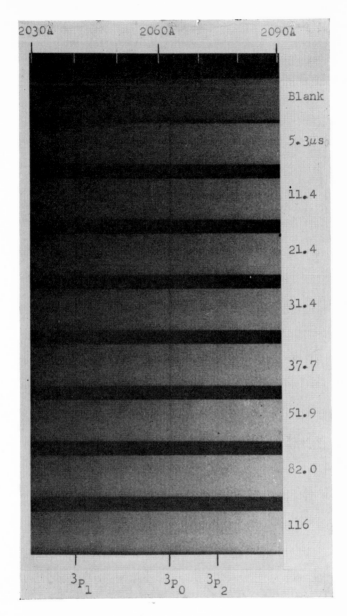

Fig. 19. Formation and decay of hot selenium atoms in argon. 0.05 torr CSe_2 + 50 torr Ar, 2025 J flash energy.

was selected by a more experienced appraisal of the possibilities offered by the periodic table of elements, considering the need for results with particular changes of internal energy, and with electronic absorption at longer wavelengths than in the atomic selenium transitions. Ideally the parent compound should be an intense

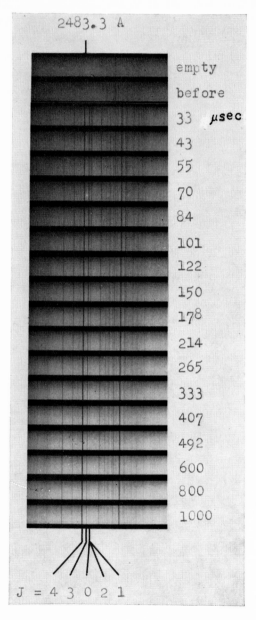

Fig. 21. Production and relaxation of Fe(a^5D_J). 50 torr Ar, 0.001 torr Fe(CO)$_5$, 1670 J flash energy.

absorber which can be entirely destroyed early in the photolysis flash. Otherwise the residue of polyatomic gas, with a range of vibrational modes, would induce rapid relaxation. Metal carbonyls are particularly suitable because the effect of traces of liberated CO can be investigated in controlled experiments.

Flash photolysis of iron carbonyl[121] ($\sim 10^{-3}$ torr) in an excess of Ar (~ 50 torr) destroys the carbonyl in < 30 μsec to produce atomic iron (and iron ions) in a large number of different electronic states; the translational temperature of the system is fixed by the huge excess of inert gas. Photolysis requires consecutive absorption of several light quanta. The highly excited states of iron, and the iron ions, decay rapidly and at 50 μsec delay, only the sub-levels of the ground $a\,^5D_J$ state are significantly populated. The energy levels are given in Table 12.

Fig. 21 shows the formation and relaxation of atomic iron, which attains a Boltzmann distribution after about 300 μsec. Depletion of the total atomic concentration is not significant at times shorter than 1 msec in argon at pressures less than 100 torr. However, Fe*–Fe collisions do appear to play an important role in the overall relaxation process.

Fig. 22 shows the results of photometry of plates similar to that illustrated in Fig. 21. The relative intensities of suitable transitions were determined from the asymptotic limit at long time delays when the system attains equilibrium. (These resemble, but are not identical to, the relative f values because of the usual instrumental effects which depend on line width.) The time variation of the relative concentrations is shown in Fig. 23; the upper four levels attain Boltzmann equilibrium amongst themselves after ~ 100 μsec, to form a "coupled" (by collision) system overpopulated with respect to the $^5D_4$ state. The equilibration of the upper four levels causes the initial rise (Fig. 22) in the population of Fe($a\,^5D_3$). Thus relaxation amongst the sub-levels is formally similar to vibrational relaxation in most polyatomic molecules, in which excitation to the first vibrational level is the rate determining step. In both cases, this result is due to the translational overlap term, for example, in the simple form of equation (14) of Section 3.

By assuming a linear dependence of log. cross-section for deactivation on the change in internal energy (discussed in detail below), and determining k_{34} from

TABLE 12

DISPOSITION OF THE SUB-LEVELS OF Fe(a^5D_J)

State	Energy (cm^{-1})
Fe(a^5D_0)	978
Fe(a^5D_1)	888
Fe(a^5D_2)	704
Fe(a^5D_3)	416
Fe(a^5D_4)	0

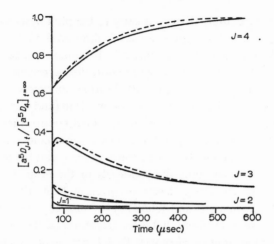

Fig. 22. Comparison of experimental decay of $Fe(a^5D_J)$ with a theoretical cascade model. (——) theory; (– – –) experiment. 50 torr Ar, 0.001 torr $Fe(CO)_5$.

the last phase of the atomic relaxation, a complete set of rate coefficients was deduced. A numerical solution of the resulting five first-order and linear differential equations is illustrated in Fig. 22, fitted to the experimental results at 70 μsec delay. (At shorter delay times the problem is complicated by processes involving higher levels.) The population ratios are shown as 'dotted' curves in Fig. 23. Evidently, from the general agreement of theory and experiment, it follows that collisional relaxation on the ^5D_J cascade occurs largely by jumps between adjacent levels. The $\Delta J = \pm 1$ rule is dominant in this particular system because of the

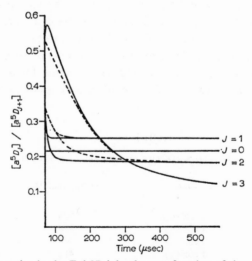

Fig. 23. Population ratios in the $Fe(a^5D_J)$ levels, as a function of time, illustrating the quasi-equilibrium in the upper levels. (——) theory; (– – –) experiment.

restriction of translational overlap, and is probably not influenced by the electronic matrix element.

From the final phase of the atomic relaxation the first-order rate coefficients for the $J = 3 \rightarrow 4$ transition were determined at different argon pressures, to separate out the individual rate coefficients for the processes

$$Fe(a\,^5D_3) + M \rightarrow Fe(a\,^5D_4) + M$$

where M = Fe or Ar. Other rate coefficients were determined by adding additional gases, and determining the enhancement of the rate of the $J = 3 \rightarrow 4$ transition. Results are included in Table 11.

8.5 SPIN–ORBIT RELAXATION OF HIGHLY EXCITED SPECIES

The rate of spin–orbit relaxation of $NO(X^2\Pi)$ has been measured using ultrasonic absorption by Bauer et al.[46], and is included in Table 11.

We now briefly mention measurements of spin–orbit relaxation of highly excited species by the inert gases. Phelps[122] has investigated collisional transitions between the spin–orbit components of $Ne(3\,^3P)$ produced in a long cylindrical vessel with a pulsed DC (15 kV) discharge, duration 10 μsec. The metastable species were detected photoelectrically by observing reversal of suitably isolated lines from a discharge lamp. Relaxation processes amongst the ^3P states and the establishment of a quasi-equilibrium occur in a manner similar to that in the $Fe(a\,^5D)$ system. The ultimate decay of the coupled system occurs via the $^3P_1$ radiating state, which has some singlet character due to mixing with the $3\,^1P_1$ state.

Krause et al.[123–125] have recently reported a series of measurements of the spin–orbit relaxation of the alkali metals in their first excited states (^2P). The technique, for example for atomic caesium with $\Delta E = 554\ \text{cm}^{-1}$, consists of irradiating the metal vapour with light from a monochromator to excite only one of the ^2P states. The vapour pressure of the metal is controlled at $\sim 10^{-6}$ torr to avoid imprisonment of the resonance radiation. The components of the fluorescence light are measured with a photomultiplier by isolating the $^2P \rightarrow {}^2S$ lines with interference filters. In the presence of added gases which cause the transitions

$$^2P_{\frac{1}{2}} + M \rightleftharpoons {}^2P_{\frac{3}{2}} + M$$

the ratio of the intensities of the two lines yields the cross-section for the process because the radiative lifetimes are known. Relaxation of K, Rb and Cs in He, Ne, Ar, Kr and Xe was investigated. Only results with Ar are included in Table 11. With all the systems investigated, helium is more efficient than neon due to the reduced mass effect, as would be predicted by equation (14), p. 199. However, the spin–orbit relaxation rates for xenon collisions are larger than those for krypton; this effect may be due either to a dependence of the electronic matrix element on the polarisability of the colliding atom, or to a slight increase in the long-range attractive interaction.

Spin–orbit relaxation of $Rb(5^2P)$ appears to be abnormally slow in all the inert gases except helium (cross-section 10^{-17} cm^2). In fact, according to Pitre et al.[123], the relaxation rate in Kr and Xe is slower than for the equivalent transitions of atomic Cs, which correspond to a three-fold larger change in internal energy. Pitre et al. discuss complications in the rubidium experiments, including the formation of van der Waals complexes with the inert gases, in order to account for the apparently abnormal relaxation rates. Efficient removal of $Rb(5^2P)$ by radiationless processes could upset the derived rate coefficients. The results were discussed in relation to Zener's semi-classical equivalent of equation (14).

8.6 VARIATION OF CROSS-SECTION WITH CHANGE IN INTERNAL ENERGY

Fig. 24 is a plot of log. cross-section against the change in the internal energy for the data of Table 11. The Fe–He result has been omitted because, except for those involving H_2 and D_2 which appear to illustrate a special point, the molecular systems are of roughly the same reduced mass. The theoretical equations for the translational part would predict a reduced mass effect of 10^4 in relaxation of $Fe(a^5D_3)$ by He compared to Ar. The theory may exaggerate this effect even in

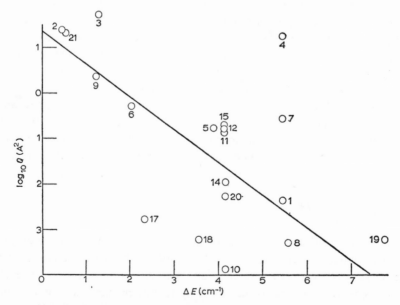

Fig. 24. Diagram for spin–orbit relaxation (for data see Table 11, p. 243).

vibrational relaxation. However, it is reasonable to suppose that in electronic energy transfer the mass effect is to some extent offset by a decrease in the efficiency of coupling between the electronic and translational motion. The ratio of the cross-sections for relaxation by He and Ar is similar for the iron and caesium experiments (~ 10), but is much larger for atomic rubidium (~ 100).

The line in Fig. 24 has the Lambert–Salter slope, and is arbitrarily drawn through the point $Q = 30$ A^2 at $\Delta E = 0$, corresponding to relaxation at every gas kinetic collision for the limiting case of zero splitting. Some 12 out of the 19 experimental points correlate reasonably well with this line, and although this may eventually be shown to be fortuitous in a number of cases, there does seem to be a significant indication that the same translational term as in V–T and V–V transfer is again dominant. Three experimental systems involving atom–atom collisions give very small inelastic cross-sections. The Fe–Ar measurement was made with considerable care over many experiments and cannot be seriously in error. The small cross-sections presumably imply that the translational–electronic coupling is abnormally weak in these systems.

It is not unexpected that O_2 causes abnormally fast relaxation of $Se(4\,^3P_0)$ since the two species must have appreciable chemical affinity. This causes the potential surfaces to converge and may also substantially split the degenerate levels of the free atom.

The fast relaxation of $Se(4\,^3P_0)$ by H_2 is possibly due to E–R transfer. The $J = 0 \rightarrow 2$ transition in p-H_2 corresponds to 365 cm^{-1}, supplying all but 179 cm^{-1} of the $^3P_0$–$^3P_1$ energy separation without significantly affecting the change in angular momentum required due to translation away from the centre of gravity. To test the view that the effect of hydrides may be a general phenomenon in this type of energy transfer, the Fe–H_2 and Fe–D_2 systems were explored. The results appear to confirm the E–R theory, and it is particularly interesting to note the disparity of 10^2 between the efficiencies of D_2 and He for relaxation of $Fe(a\,^5D_3)$, where the reduced mass is constant. (Comparing vibrational relaxation, the absence of any V–R transfer in the relaxation of $CO(v = 1)$ was demonstrated by Millikan[107], who showed that He and D_2 are equally effective collision partners.) The hydrogens are especially efficient for E–R transfer because changing the angular momentum of the rotator requires a comparatively large energy.

8.7 RELAXATION OF ATOMIC IODINE

Finally, we discuss briefly the results of Donovan and Husain[126] on the relaxation of $I(5\,^2P_{\frac{1}{2} \rightarrow \frac{3}{2}})$, for which $\Delta E = 7603$ cm^{-1}. The problem is even more complex than the foregoing, because the splitting is large and the extent of vibrational excitation accompanying electronic deactivation is not known. Atomic iodine in the $5\,^2P_{\frac{1}{2}}$ state can be produced by flash photolysis of various iodides,

CF_3I being particularly suitable because the residual parent compound is only moderately effective at inducing relaxation. Decay rates are conveniently measured by photometry of the atomic lines in the far ultraviolet. Some rate coefficients are listed in Table 13. The inert gases are not effective for this process because the splitting is too large to permit pure E–T relaxation to occur at a measurable rate.

Deactivation by I_2 is very fast and corresponds to $Z \sim 15$; the efficiency of relaxation may be comparable with the rate of the chemical exchange

$$I(5^2P_{\frac{1}{2}})+I_2 \rightarrow I(5^2P_{\frac{3}{2}})+I_2$$

It has recently been demonstrated[127] that $I(5^2P_{\frac{1}{2}})$ reacts with propane

$$I(5^2P_{\frac{1}{2}})+C_3H_8 \rightarrow HI+C_3H_7$$

with an activation energy of ~ 5 kcal.mole^{-1}. Therefore the potential curves should converge with all the hydrides, and also the chemical interaction provides direct coupling between the electronic and vibrational energy. Relaxation of $I(5^2P_{\frac{1}{2}})$ in N_2 does occur at a measurable rate; there would be no justification for postulating converging potential curves in this case, because the N_2 bond dissociation energy is much larger than that of NI. The theoretical rate for excitation of N_2 to $v = 1$ would correspond to $Z \sim 10^{40}$. The reaction

$$I(5^2P_{\frac{1}{2}})+N_2(v = 0) \rightarrow I(5^2P_{\frac{3}{2}})+N_2(v = 3)$$

for which $\Delta E = -697$ cm^{-1}, is a possibility even though it is restricted by the small $V_{0,3}(V_{0,3}^2/V_{0,1}^2 \sim 10^{-4})$.

Clearly a great deal of experimental and theoretical work is required to clarify some of the points which have been raised in this section.

TABLE 13

DEACTIVATION OF $I(5^2P_{\frac{1}{2}})$ TO THE $I(5^2P_{\frac{3}{2}})$ GROUND STATE

Collision partner	Rate coefficient (cm^3.sec^{-1})
I_2	5×10^{-12}
H_2	9×10^{-14}
D_2	1.1×10^{-13}
HI	1.5×10^{-13}
C_3H_8	6×10^{-14}
N_2	2×10^{-16}

9. Electronic–vibration and electronic–translation energy transfer with $\Delta E \gg 1$ eV

It is of great interest, especially in photochemistry, to discover how much energy is converted to vibration accompanying transfer of a substantial quantity of electronic energy, corresponding either to a change in the orbital angular momentum, or to a change of principal quantum number of an atom. Such highly energetic species are usually deactivated very efficiently by any polyatomic gas, and the quenching cross-sections show some approximate correlation (especially for members of homologous series), with polarisability and ionisation potential. Energy transfer has a high probability because of chemical complex formation, and consequent crossing of potential surfaces.

9.1 QUENCHING AND EXCITATION OF ATOMIC SODIUM

The quenching and excitation (2.09 eV) of $Na(3^2P)$ has been superficially investigated

$$AB + Na(3^2P) \rightarrow AB^\dagger + Na(3^2S)$$

Inert gases have very small cross-sections for quenching $Na(3^2P)$, whereas for all the polyatomic molecules which have been studied, the quenching cross-sections are of the same order of magnitude as gas kinetic cross-sections. It is suggested later that this is not because a polyatomic molecule can accept most of the energy internally. Table 14 lists quenching cross-sections for a number of diatomic mol-

TABLE 14

CROSS-SECTIONS FOR DEACTIVATION OF $Na(3^2P)$

Quenching gas	N_2	NO	O_2	CO	H_2
Quenching cross-section (A^2)	24.9	31.6	52.2	13.4	15.7
Vibrational level	7	9	11	8	4
Energy discrepancy (eV)	0.15	0.09	0.12	0.07	0.21

ecules, and has been modified from that given by Pringsheim[5]. In each case, for the given vibrational level the quenching is exothermic, the cross-sections being too large to correspond to vibrational levels of higher energy than the electronic energy of the sodium. Oxygen is exceptional in having electronically excited states of lower energy than $Na(3^2P)$. There is no relationship between quenching cross-section and the minimum energy that cannot be converted to vibration. Recently Jenkins has reported accurate cross-sections for quenching $Na(3^2P)$, derived from flame studies[181].

Experimental work of Gaydon *et al.*[129] shows conclusively that vibrationally excited CO and N_2 will excite $Na(3\,^2P)$, and conversely that vibrationally excited species must be produced in the quenching process. The temperature of a flame can be measured by focusing a black-body source through it onto the slit of a spectrograph. If the light source and flame subtend the same angle to the detector, when they have the same temperature the flame does not affect the light flux falling from the black-body onto the detector. If reversal of a strong atomic resonance line is observed, for example the sodium D lines, the appropriate temperature is defined by the populations in the $(3\,^2S)$ and $(3\,^2P)$ states. The electronic temperature of the sodium may be higher or lower than the translational temperature. If the excited state is populated either by chemical reaction or by absorption of radiation, in addition to excitation by collision (thermal), the electronic temperature is higher than the translational temperature[130,131]. Clouston *et al.*[129] discovered that in shock-wave heated N_2, the electronic temperature of sodium is below the translational temperature immediately behind the shock front, and from the relaxation time it was evident that the electronic temperature is related to the vibrational temperature of the N_2. The same phenomenon was later observed with CO, and also with both gases and chromium atom reversal. Evidently the process

$$Na(3\,^2S) + N_2(v = n) \rightarrow Na(3\,^2P) + N_2(v = 0)$$

occurs rapidly, though the shock-wave experiments have not given any information about the magnitude of n.

9.2 DETECTION OF VIBRATIONAL EXCITATION BY INFRARED EMISSION

Karl and Polanyi[132] have described an experiment whereby the yield of vibrational energy in a quenching molecule was observed directly. Highly vibrationally excited CO was detected by infrared emission, in a mixture of mercury vapour and CO on irradiation with mercury resonance radiation. Evidently the processes

$$Hg(6\,^3P_{1\text{ and }0}) + CO(v = 0) \rightarrow Hg(6\,^1S_0) + CO(v = n)$$

occur to populate a range of vibrational states. Detailed results[133] show that about 25 % of the electronic energy is converted to vibration, with a fairly smooth distribution into the lower vibrational levels up to $v = 8$, a small yield into $v = 9$ and negligible population into $v = 10$ or higher levels. Excitation to $v = 10$ corresponds to about one half at the total electronic energy. Similar experiments and results were later reported for NO, though higher vibrational levels are populated[172].

9.3 THEORY OF QUENCHING

Dickens, Linnett and Sovers[134] have given a distorted wave calculation of the cross-section for electronic quenching of an atom by a diatomic molecule. An *ultra-simplified* model was chosen in which the electronic wave-functions are spherically symmetrical in both electronic states. Despite the crudity of this model, the conclusions are important, and probably have some general validity. The quenching cross-sections increase with decreasing energy discrepancy but are extremely small because of the vanishingly small vibrational matrix elements for multiple quantum vibrational transitions. Therefore "resonance quenching" can only result in excitation to the few lowest vibrational levels in the quenching molecule; all the processes would have extremely small cross-sections because of the small translational term and the small matrix elements for vibrational transitions. (The intermediate case of relaxation by N_2 of $I(5\,^2P_{\frac{1}{2}})$, excitation energy 0.9 eV, provides an example where the translational and vibrational terms maximise for the $v = 0 \rightarrow 3$ transition. However, even in this system where the total energy is small compared, for example, to that with $Na(3\,^2P)$, the theoretical probability of relaxation per collision would be about 10^{-7}.) Dickens *et al.* concluded that for efficient quenching there must occur either a crossing or near-crossing of potential curves. They suggest that the diatomic species takes up only a single quantum of vibration. It is unlikely that this always will be true, and of course, multiple quantum vibrational excitation has now been demonstrated by Polanyi's method. Chemical interaction generally will be required for efficient quenching in order that the potential curves of the initial and final states may approach. Because of the change in electronic configuration, in the transition complex the separated species lose their identity and selection rules for these species will not be important. Excitation of vibration occurs by the usual coupling process with translation, and also directly because the electronic interaction may change the intermolecular separation. Excitation to a range of vibrational states occurs, depending on the bond length of the quenching molecule in the transition complex and the form of the potential surfaces. An explanation somewhat along these lines was given some time ago by Laidler[135], and more recently by Polanyi[136]. Two classes of electronic–vibrational energy transfer were defined at the beginning of this section. According to the views expressed here, collisional quenching, when several eV of electronic energy is converted to other forms, belongs to the second type. Polyatomic molecules quench electronically excited atoms more efficiently than do the inert gases, because the former are chemically more reactive than the latter. Excitation of vibration is an incidental consequence of the dynamics of the interaction.

It has been pointed out that the shock tube reversal experiments are perfectly consistent with inefficient conversion of electronic to vibrational energy[79].

10. Electronic–electronic energy transfer

10.1 INTRODUCTION

The exchange of electronic excitation between two atoms frequently results in sensitised fluorescence and one of the earlier examples was the discovery of emission of the fluorescence of atomic sodium, which occurs when a mixture of sodium and mercury vapour is irradiated with mercury resonance radiation at 2537 A

$$Hg(6^1S_0) + h\nu \rightarrow Hg(6^3P_1)$$

$$Hg(6^3P_1) + Na(3^2S_{\frac{1}{2}}) \rightarrow Hg(6^1S_0) + Na(9^2S_{\frac{1}{2}})$$

$$Na(9^2S_{\frac{1}{2}}) \rightarrow h\nu + Na(n^2P)$$

Some aspects of the theory of the exchange of electronic energy between atoms at near resonance has been reviewed by Massey and Burhop[137], and Mott and Massey[138]. If the change in internal energy is very small and the transitions are optically allowed for both atoms, long-range resonance interaction should result in a large cross-section for excitation transfer, which may approach 5×10^{-14} cm^2. The theory predicts a sharp decrease of cross-section with increasing energy discrepancy. If the transitions of each atom are associated with a quadrupole, the cross-section for $\Delta E = 0$ falls to 10^{-15} cm^2, which is approximately the gas kinetic cross-section. Qualitatively, cross-sections for electronic energy transfer should be less than for gas kinetic collisions if $l|\Delta E|/h\nu \gg 1$, where v is the relative speed, l is a length characteristic of the interaction potential, and ΔE is the change in internal energy. It may be supposed that this is the ratio of the duration of the collision (l/v) to a time interval characteristic of the electronic motion $(h/|\Delta E|)$ (a generalised version of the Landau–Teller condition). At ambient temperature, the mean molecular speed is about 5×10^4 cm.sec^{-1}, and with $l = 10^{-8}$ cm, the above ratio is equal to unity when $\Delta E = 133$ cm^{-1}. Therefore the probability of exchange of electronic energy per gas kinetic collision should fall below unity if $\Delta E > \sim 200$ cm^{-1}. An essentially similar conclusion is implicit in equation (14), p. 199. For the few experimental systems which have been investigated quantitatively, the cross-sections for near resonance show some agreement with the theory, and certainly, as seen later in Fig. 26, the limiting cross-section at $\Delta E = 0$ does approach 5×10^{-14} cm^2.

In several of the known examples of sensitized fluorescence, the acceptor species can be excited to any one of a large number of near-resonant states. Processes with $\Delta E \sim 0$ predominate, though there appear to be several examples where a transition to a state slightly off resonance is preferred to another at almost exact resonance[172].

10.2 MERCURY SENSITISED FLUORESCENCE

Pringsheim[5] has reviewed some of the early qualitative experiments. Beutler and Josephy[139] examined the fluorescence of sodium sensitized by mercury, and concluded that the process with the largest cross-section is

$$Hg(6\,^3P_1) + Na(3\,^2S_{\frac{1}{2}}) \rightarrow Hg(6\,^1S_0) + Na(9\,^2S_{\frac{1}{2}})$$

for which $\Delta E = +162 \text{ cm}^{-1}$. The absolute cross-section is not known. The intensities of the sharp and diffuse series of lines, measured by plate photometry, indicated that the states $8S$ ($\Delta E = -444 \text{ cm}^{-1}$), $7D$ ($\Delta E = +212 \text{ cm}^{-1}$) and $8D$ ($\Delta E = +317 \text{ cm}^{-1}$) are also populated directly by excitation transfer from the mercury, the intensities of the various lines in emission being about one-fifth that of the $9\,^2S \rightarrow 3\,^2P$ line. (The intensity of a single line is not, of course, a direct measure of the rate of population of a state. To measure the latter, it is necessary to consider the rate of emission of radiation at this wavelength, and the rates of all other radiative and radiationless transitions.) Beutler and Josephy did not record emission from the P states of sodium. The transitions to $Na(3\,^2S_{\frac{1}{2}})$ (principal series) are reversed, and presumably the 7000–8000 A region, for detection of $n\,^2P \rightarrow 4\,^2S$ transitions, was inaccessible. Rautian and Sobelman[140] postulated that the sharing between the $9\,^2S$ and $8\,^2P$ states should be according to the scheme

$$Hg(6\,^3P_1) + Na(3\,^2S) \begin{cases} \nearrow \quad \frac{1}{6}\, Na(9\,^2S) + Hg(6\,^1S_0) \\ \searrow \quad \frac{5}{6}\, Na(8\,^2P) + Hg(6\,^1S_0) \end{cases}$$

This apparently is based on the assumption that the cross-section depends on $\exp(-\Delta E/kT)$. However, Frish and Bochkova[141] claim to have demonstrated that the $8P$ state is produced only in very low yield. They observed the emission from excited sodium atoms in an electric discharge in He, Hg and Na mixtures at very low total pressures. Their data appear to show that the cross-section for excitation of the $8P$ state, the process with the smallest energy discrepancy and for which optical transitions of both atoms are allowed, is smaller than the cross-section into the $9S$ state, which is optically forbidden for the sodium.

Excitation transfer from $Hg(6\,^3P_1)$ to indium is again consistent with the rule that the exchange is accompanied by a very small change in internal energy. Photographic examination of the sensitized fluorescence indicates that at 900 °C the processes

$$Hg(6\,^3P_1) + In(5\,^2P) \rightarrow Hg(6\,^1S_0) + In(7\,^2P) \dots \Delta E(\text{cm}^{-1})$$
$$= -551 \quad \text{and} \quad -440,$$

$$Hg(6\,^3P_1) + In(5\,^2P) \rightarrow Hg(6\,^1S_0) + In(6\,^2D) \dots \Delta E(\text{cm}^{-1})$$
$$= -364 \quad \text{and} \quad -314$$

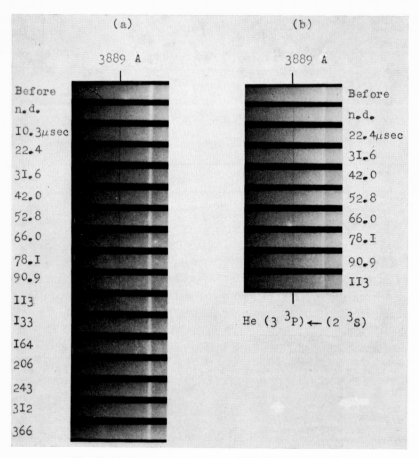

Fig. 25. (a) Formation and decay of He(2^3S_1) in 5 torr He.
(b) Decay of He(2^3S_1) in 5 torr He$+1.2\times10^{-2}$ torr Ne.

have approximately equal cross-sections. The first is optically forbidden and the second allowed.

The mercury photosensitized fluorescence of thallium vapour appears to provide an exception to the rule, because the reaction with the largest cross-section corresponds to a substantial change in internal energy, notwithstanding the opportunity of electronic energy transfer to a state of almost identical internal energy. At 900 °C the reaction

$$\mathrm{Hg}(6\ ^3P_1)+\mathrm{Tl}(5\ ^2P_{\frac{1}{2}}) \rightarrow \mathrm{Hg}(6\ ^1S_0)+\mathrm{Tl}(6\ ^2D)\ .\ .\ \Delta E(\mathrm{cm}^{-1})$$

$$= -3218 \quad \text{and} \quad -3294$$

is apparently preferred to

$$Hg(6\,^3P_1)+Tl(5\,^2P_{\frac{1}{2}}) \rightarrow Hg(6\,^1S_0)+Tl(8\,^2S)\,..\,\Delta E\,=\,-666$$

Pringsheim[5] discusses a number of complications, and suggests that excitation of $Tl(6\,^2D)$ may occur by species other than $Hg(6\,^3P_1)$. The violation of the $\Delta E \simeq 0$ rule has been confirmed in recent experiments[142].

10.3 ELECTRONIC EXCITATION TRANSFER BETWEEN INERT GAS ATOMS

Research on laser systems provided the first quantitative information about electronic excitation transfer between atoms. In an electric discharge in helium containing a trace of neon, energy transfer occurs from metastable helium atoms, $He(2\,^3S_1)$, to neon and excites $Ne(2s)$, to produce population inversion with respect to the $(2p)$ and $(1s)$ states. With pure helium, Phelps and Brown[143] have measured the rate of decay of $He(2\,^3S_1)$, $He(2\,^1S_0)$ and $He_2\,(a\,^3\Sigma_u^+)$ following a pulsed DC discharge. The metastable species were detected photoelectrically, by observing reversal of suitably isolated lines from a helium discharge lamp. The production and decay of $He(2\,^3S_1)$ in an electric discharge, obtained by the microwave pulse method, is illustrated in Fig. 25(a). Javan et al.[144] employed the method of Phelps and Brown to observe the increased rate of decay of $He(2\,^3S_1)$ in the presence of neon and recorded a cross-section of $3.7\pm0.5\times10^{-17}$ cm^2. This value was confirmed by the microwave pulse measurements, Fig. 25(b). Javan et al. proved that transfer to the neon occurred, by observing the $2s \rightarrow 2p$ emission lines during the afterglow; the decay of the emission was simultaneous with the $He(2\,^3S_1)$ decay. Apparently all four of the $2s$ levels are populated directly, though the individual cross-sections have not yet been reported. The internal energy changes for the process

$$He(2\,^3S_1)+Ne(2\,^1S_0) \rightarrow He(1\,^1S_0)+Ne(2s_{2,\,3,\,4\,or\,5})$$

are, respectively, -314, -469, -1053 and -1247 cm^{-1}. The transitions are optically forbidden for the helium, and the second and fourth are forbidden for the neon. Benton et al.[145] employed the same technique to measure excitation transfer from $He(2\,^1S_0)$ to neon. Transfer occurs predominantly to the $3s_2$ state, viz.

$$He(2\,^1S_0)+Ne(2\,^1S_0) \rightarrow He(1\,^1S_0)+Ne(3s_2)$$

The changes in internal energy for transfer to $3s_2$, $3s_3$, $3s_4$ and $3s_5$ are, respectively, $+387$, $+337$, -357 and -441. The transition is optically forbidden for the helium, and allowed for the neon. The cross-section was recorded as 4×10^{-16} cm^2. If this value is correct, it represents very efficient energy transfer because at 300 °K, on average only one collision in 6.8 has sufficient kinetic energy to supply the internal energy increase. $He(2\,^1S_0)$ is short-lived even in pure helium, and it is difficult to measure its decay rate accurately.

10.4 CROSS-SECTIONS FOR ELECTRONIC ENERGY TRANSFER

Krause and co-workers[123-125] have recently accomplished a quantitative experimental survey of the rates of E–E transfer between the alkali metal atoms K, Rb and Cs. The experimental method was discussed in the first section on E–T transfer. Their results are very interesting, and for E–E transfer provide the only evidence for a fall in cross-section with increasing energy discrepancy. Table 15 lists all the E–E results, and the quantitative data are plotted in Fig. 26 as a $\log_{10} Q$ versus ΔE diagram. Also included is the experimental rate of spin–orbit relaxation of $Fe(a^5D_3)$ by $Fe(a^5D_4)$; although this is not formally an E–E process, it probably involves long-range chemical interaction which is common to all these systems. It is of interest to note that, notwithstanding the unknown complexities of the interacting potential surfaces, the processes still 'remember' the overall change in internal energy. The slope of the line is higher than that of the Lambert–Salter diagram for non-hydrides by a factor of 4. Czajkowski et al.[124] point out that the cross-sections are roughly proportional to ΔE^{-2}. At exact resonance, the extrapolated line seems correctly to predict the theoretical limit. These systems must still show dependence

TABLE 15

SUMMARY OF RESULTS ON EXCITATION TRANSFER BETWEEN ATOMS

Donor		Acceptor		ΔE (cm^{-1})	Cross-section (Å^2)
Initial	Final	Initial	Final		
$Hg(6^3P_1)$	$(^1S_0)$	$Na(3\,^2S_{\frac{1}{2}})$	$(9\,^2S_{\frac{1}{2}})$	$+$ 162	—
$Hg(6^3P_1)$	$(^1S_0)$	$In(5\,^2P)$	$(7\,^2P)$	$-$ 440	⎰about equally
$Hg(6^3P_1)$	$(^1S_0)$	$In(5\,^2P)$	$(6\,^2D)$	$-$ 314	⎱probable
$Hg(6^3P_1)$	$(^1S_0)$	$Tl(5\,^2P)$	$(6\,^2D)$	-3218	
$He(2\,^3S_1)$	$(1\,^1S_0)$	$Ne(2\,^1S_0)$	$(2s_{2,3,4,5})$	$-$ 314, $-$ 469, -1053, -1247	$\Sigma = 0.37$ about equally probable
$He(2\,^1S_0)$	$(1\,^1S_0)$	$Ne(2\,^1S_0)$	$(3s_2)$	$+$ 387	4
$Rb(5\,^2P_{\frac{1}{2}})$	$(5\,^2S_{\frac{1}{2}})$	$Cs(6\,^2S_{\frac{1}{2}})$	$(6\,^2P_{\frac{1}{2}})$	847	1.5
$Rb(5\,^2P_{\frac{1}{2}})$	$(5\,^2S_{\frac{1}{2}})$	$Cs(6\,^2S_{\frac{1}{2}})$	$(6\,^2P_{\frac{3}{2}})$	1401	0.5
$Rb(5\,^2P_{\frac{3}{2}})$	$(5\,^2S_{\frac{1}{2}})$	$Cs(6\,^2S_{\frac{1}{2}})$	$(6\,^2P_{\frac{1}{2}})$	1084	0.9
$Rb(5\,^2P_{\frac{3}{2}})$	$(5\,^2S_{\frac{1}{2}})$	$Cs(6\,^2S_{\frac{1}{2}})$	$(6\,^2P_{\frac{3}{2}})$	1638	0.3
$K(4\,^2P_{\frac{3}{2}})$	$(4\,^2S_{\frac{1}{2}})$	$K(4\,^2S_{\frac{1}{2}})$	$(4\,^2P_{\frac{1}{2}})$	258	250
$Rb(5\,^2P_{\frac{3}{2}})$	$(5\,^2S_{\frac{1}{2}})$	$Rb(5\,^2S_{\frac{1}{2}})$	$(5\,^2P_{\frac{1}{2}})$	238	70
$Cs(6\,^2P_{\frac{3}{2}})$	$(6\,^2S_{\frac{1}{2}})$	$Cs(6\,^2S_{\frac{1}{2}})$	$(6\,^2P_{\frac{1}{2}})$	554	31
$K(4\,^2P_{\frac{3}{2}})$	$(4\,^2S_{\frac{1}{2}})$	$Rb(5\,^2S_{\frac{1}{2}})$	$Rb(5\,^2P_{\frac{1}{2}})$	464	3.2

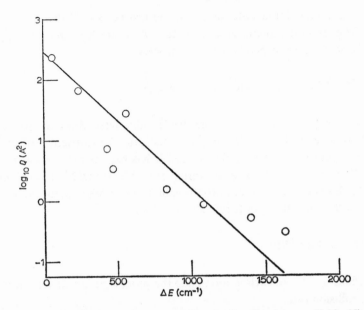

Fig. 26. Diagram for electronic energy transfer (for data see Table 15).

on ΔE near exact resonance, because the large cross-sections correspond to internuclear distances where the attractive interaction is still weak. Fig. 26 emphasises that the largest cross-sections correspond to $\Delta E \to 0$, and that the scatter is due to peculiarities in the interaction potentials. The effect of the latter is manifest in cases where, given a choice, a process slightly off resonance is preferred to one at almost exact resonance.

Insufficient experimental data is available to demonstrate either the occurrence or lack of selection rules. In Table 15, some of the fastest reactions do correspond to transitions allowed for both atoms. Singlet helium transfers to Ne at a rate consistent with the data in Fig. 26, whereas triplet helium transfers comparatively slowly; neither of the He transitions is allowed. However, there appear to be other cases, discussed earlier, where a forbidden transition is preferred to an allowed transition. Stepp and Anderson[146] have suggested that there is partial conservation of electronic angular momentum accompanying energy transfer between atoms, and interpreted experiments on mercury fluorescence by means of the steps

$$Hg(6\,{}^3P_1) + Hg(6\,{}^3P_0) \to Hg(6\,{}^1S_0) + Hg(6\,{}^3D_1)$$

$$Hg(6\,{}^3P_1) + Hg(6\,{}^3P_2) \to Hg(6\,{}^1S_0) + Hg(6\,{}^3D_3)$$

The large cross-sections for spin–orbit relaxation of Na and K in the ^2P states, however, seem to show that angular momentum change is not generally restrictive. The observations of Stepp and Anderson perhaps have their origin in the peculiarities of the interaction potentials. All the processes of Table 15 occur according to Wigner's rule, to conserve the vector of the electron spin. An apparent exception is production of $Ne(2s_2)$ by the process

$$He(2\,^3S_1) + Ne(2\,^1S_0) \rightarrow He(1\,^1S_0) + Ne(2s_2)$$

However, the $Ne(2s_2)$ state has some triplet character due to mixing with $Ne(2s_4)$. It is of considerable interest to investigate the restrictive effect of an overall spin-change in any electronic process. It has been noted that, of the inert gases, only Xe induces collisional deactivation of $O(2\,^1D)$ to $O(2\,^3P)$, which may be due either to chemical interaction, or to the spin–orbit coupling in the heavy atom. Most gases will induce the spin forbidden transition

$$As(4\,^2D) \rightarrow As(4\,^4S)$$

with high efficiency[147]. In carbon monoxide the process occurs at very nearly the gas kinetic collision rate.

10.5 ELECTRONIC EXCITATION TRANSFER IN COMPLEX SYSTEMS

This section is concluded with a brief discussion of electronic excitation transfer in more complex systems, where the subject expands into photochemistry[148]. There are several examples known where the excitation energy of $Hg(6\,^3P_1)$ is transferred to a collision partner, and the spin vector appears to be conserved if the quenching cross-section is large, in accordance with Wigner's rule. Electronic excitation of diatomic or polyatomic molecules is rather more complex than in atomic systems, because in the former the change of electronic state alters the equilibrium internuclear separations. This Franck–Condon restriction can forbid excitation transfer. In principle, there should again be two quite different cases, first a resonance process with parallel surfaces in which the additional energy for the "nuclear displacement" (from the equilibrium configuration) is exactly available, and secondly the intermediate chemical complex mechanism with random distribution of the excess. As yet, there is insufficient experimental and theoretical data available with which to test this point of view.

Considering the cross-sections given in Table 9 (p. 242), transfer from $Hg(6\,^3P_1)$ to NO, O_2 and C_2H_4 produces metastable electronic states which undergo secondary reactions. Transfer to N_2O, NH_3 and saturated hydrocarbons produces unstable triplets which decompose to free atoms or radicals. Further details are described in text-books on photochemistry. An example of transfer between an atom and a diatomic molecule has been described by Brennen and Kistiakowsky[149], who observed emission of atomic iron excited by $N_2(A^3\Sigma)$. The transfer appears to occur indiscriminately to a large number of the states of atomic iron in the region of close resonance; interpretation of the experiments is complicated because cascade processes, both collisional and radiative, complicate the problem of determining the initial distribution from observations which necessarily require finite delay with respect to the initial act of transfer. Transfer is accompanied by a substantial contraction (20 %) of the internuclear separation for the transition $A^3\Sigma \rightarrow X^1\Sigma$, and the process of transferring all the electronic energy to the atomic ion is perhaps more easily understood if it is assumed to occur *via* a chemical intermediate, rather than by resonance between parallel potential curves. Transfer between $N_2(A^3\Sigma)$ and NO has recently been reported by Callear and Smith[150] and Sagert and Thrush[151]. The process

$$NO(C^2\Pi) + N_2(X^1\Sigma) \rightarrow NO(X^2\Pi) + N_2(A^3\Sigma)$$

occurs at practically every collision, and can only be understood if the N–N internuclear separation is directly coupled to translation because of intermediate chemical complex formation. In the transfer step

$$N_2(A^3\Sigma) + NO(X^1\Sigma) \rightarrow N_2(X^1\Sigma) + NO(A^2\Sigma)$$

the energy distribution may show some dependence on the extent to which each of the squares of the internuclear separations have to change in the overall process (sharing according to Hooke's Law). Both of these processes are forbidden from a simple consideration of the Franck–Condon principle. Cundall and co-workers[152,153] have reported a series of investigations of electronic energy transfer from excited triplet states of simple organic molecules, to *cis*- or *trans*-butene-2. The latter are excited to the triplet state, and the process can be investigated experimentally by using chemical analysis to determine the extent of *cis–trans* isomerisation. Transfer from excited NH_3 to atomic Tl has been reported by Andreeva *et al.*[154]. Provided the spin rule is not violated, transfer between two complex molecules seems to occur at practically every collision. Wilkinson[155] has reviewed the energy transfer processes which can occur between complex organic molecules in solution.

11. Application to reaction kinetics and photochemistry

11.1 UNIMOLECULAR REACTIONS

The mechanism of unimolecular decomposition can be described semi-qualitatively by the simple theory of Kassel[156] which is adequate for this brief review. Molecules exchange energy in bimolecular collisions, and if a molecule acquires sufficient energy to decompose, it may do so between collisions because of intra-molecular rearrangement of the energy to acquire a favourable configuration for decomposition. For a given energy in excess of the critical amount required for decomposition, the mean lifetime of an excited molecule increases with increasing complexity of the molecule, because of the increase in the number of ways the energy can be arranged in the available oscillators. Except at enormous pressures, a diatomic molecule will always decompose if it has sufficient energy of vibration, since the period of oscillation is $\sim 10^{-13}$ sec. At 1 atm pressure, the collision frequency is $\sim 10^{10}$ per second. The decomposition of more complex molecules frequently shows asymptotic first-order kinetics as the pressure tends to infinity; the mean lifetime of the excited molecules is then much greater than the duration between collisions, and the population of excited species corresponds closely to thermodynamic equilibrium. At limiting zero pressure, the kinetics become second-order because the rate is collision controlled; if sufficient energy is transferred to a molecule during collision, it is almost certain to decompose. Molecules with insufficient energy to decompose will retain a normal Boltzmann distribution of energy (except at low pressure in shock-tube conditions), and the rate of reaction is given by the rate at which molecules are promoted to the high energy Boltzmann 'tail' above the critical energy.

There is very little direct experimental data on energy transfer in polyatomic gases in the typical temperature range of about 700–1000 °K for unimolecular decompositions. Obviously several different vibrational modes will be simultaneously involved and the low frequency vibrations would be excited to high quantum states. Systematic measurements have as yet only been successful in the comparatively simple problem of low quantum states. It is obvious that V–V transfer can be a very efficient means of redistributing energy between the molecules, and if for example polyatomic hydrogen-containing molecules were added to the system as an energy transfer catalyst, the reaction should be greatly accelerated. On the other hand, monatomic gases (with which energy equilibration can occur solely because of V–T transfer) should have much less effect. Such behaviour is found experimentally, though the difference between polyatomic and monatomic gases is only about 20-fold[157].

Experiments on the fluorescence of β-naphthylamine provide some information concerning the behaviour of highly vibrationally excited species[158]. When electronically and vibrationally excited by absorption of monochromatic radiation, β-naphthylamine suffers either unimolecular decomposition or spontaneous radiation. Shifting the wavelength of the exciting radiation to higher energies has the effect of increasing the decomposition rate at the expense of spontaneous radiation. This is qualitatively consistent with the theory of Kassel, which predicts that the probability that the dissociation energy D is concentrated in the critical bond (a single oscillator) when the whole set has energy E, is

$$\left(\frac{E-D}{E}\right)^{s-1},$$

where s is the total number of classical oscillators. Addition of inert gas to the system enhances the fluorescence by removing energy from vibrationally excited β-naphthylamine by collision processes. Boudart and Dubois[159] have interpreted Neporent's experimental data[158] to determine the accommodation coefficients (α) for collision with various gases. The latter quantities are normally employed for collisions with surfaces and are defined by the equation

$$\alpha = (T_1 - T_2)/(T_2^{vib} - T_1^{vib})$$

where T_1 and T_2 are the initial and final "temperatures" of the colliding foreign gas, and T_2^{vib} and T_1^{vib} are the initial and final vibrational temperatures of the β-naphthylamine. Some results are listed in Table 16[160].

TABLE 16

ACCOMMODATION COEFFICIENTS FOR ENERGY TRANSFER FROM EXCITED
β-NAPHTHYLAMINE

Added gas	α
NH_3	0.85
CO_2	0.4
$CHCl_3$	0.5
H_2	0.1
He	0.2
N_2	0.3

The results show that collisional transfer is very efficient from a molecule in the highly excited and closely spaced quantum states, which behave as a classical system with little restriction due to quantisation. The monatomic and diatomic gases appear to be rather less effective than the polyatomic molecules, which may emphasise the importance of V–V transfer in this type of process. However, it is desirable to establish experimentally how the energy is distributed in both molecules following collision.

We conclude that collisional activation, preceding unimolecular decomposition of complex molecules, seems to occur with very high efficiency, but precisely how efficiently is not known in any one case[161]. A detailed examination of energy distributions may be found in recent papers by Rabinovitch and coworkers[183].

11.2 ENERGY DISTRIBUTION IN CHEMICAL REACTIONS

During the last few years an increasing interest has been shown in the chemistry of individual quantum states. The main experimental techniques have been flash photolysis, afterglows in discharge tubes (especially examination of infrared emission), and crossed molecular beams. One problem is to determine how the heat of a chemical reaction is transferred to and distributed amongst the various degrees of freedom in the product species. A similar problem in photochemistry is to determine the distribution of energy accompanying photolytic dissociation. This chapter is devoted almost entirely to non-reactive energy transfer processes, which comprise some of the simplest of all kinetic systems. The subject aids our understanding of reactive energy transfer in two ways. Examples in Sections 9 and 10 have theoretical and empirical features in common with "ordinary" chemical reactions and photochemical processes. From the theoretical discussion, it is to be anticipated on classical and wave-mechanical grounds that a reaction in which there is a contraction in molecular dimensions will tend to retain the chemical energy as internal energy, usually as vibrational energy; *i.e.* there will be no recoil forces to separate the product fragments. As well as providing an introduction to reactive energy transfer, an understanding of non-reactive processes is required for the analysis of experimental data from reacting systems. Under what conditions can the initial distributions of energy be observed, and what are the main processes which establish equilibrium? The feasibility of observing an initial distribution, with a particular experimental technique, depends on what the distribution is. If a large number of quantum states are populated, because relaxation of the higher levels is very rapid, it is difficult to achieve adequate time resolution. An initial distribution cannot be derived from a "stationary" distribution unless all energy transfer processes have been independently established and measured. It is comparatively easy to establish an initial distribution, if only a few low vibrational states are populated, especially if the vibrational frequency is high and relaxation is slow. A good example is the reaction of $S(3\,^2P)$ with O_2, which produces SO in the lowest few vibrational states[58]. A frequent complication seems to be relaxation by collision with transient species. Thus far, very few experimental systems have been adequately analysed, the most sophisticated attempts being those of Polanyi's group, who studied infrared emission during reaction.

Much of the pioneering work on energy transfer in chemical reactions was initiated by Norrish with various co-workers. Norrish *et al.*[55] showed that vibrationally excited oxygen produced by the reaction

$$O+NO_2 \rightarrow O_2^{\ddagger}+NO$$

can be observed by kinetic absorption spectroscopy. However, the nitric oxide is not highly vibrationally excited, and these experiments and similar studies with the $O + ClO_2$, $O + O_3$, and $O + RH$ reactions, indicated that the nascent bond is rich in vibrational energy, which is what one would expect intuitively. Recent semi-quantitative measurements show that quite a small fraction of the total exothermicity appears as vibration in each of these reactions. It would be of considerable interest to obtain precise rate coefficients for the formation of the excited product in each quantum state. Norrish and Oldershaw[162] have detected vibrationally excited $SO(v \lesssim 3)$, produced by photolysis of SO_3, $P_2 (v \lesssim 7)$ from phosphine, and $PN(v \lesssim 1)$ from PH_3 and NH_3 mixtures. The last two provide rare (as yet) examples of vibrationally excited products of radical–radical reactions, e.g.

$$2PH \rightarrow P_2^{\dagger} + H_2$$

A problem which is even more difficult is to investigate the yield of vibrational energy accompanying three-body combination of atoms. The combination rate coefficients tend to be small compared to the fast (and second order) vibrational relaxation. Thus the stationary concentration of the diatomic product usually corresponds very closely to an ambient Boltzmann distribution. Callear[58] observed $S_2^{\dagger}$ from S-atom combination, but concluded that the relaxation was too fast for any meaningful quantitative measurements to be made.

Polanyi et al.[163,164] have described some detailed investigations of the reaction of H with Cl_2 in a flow system

$$H + Cl_2 \rightarrow HCl^{\dagger} + Cl$$

From the infrared emission, a non-Boltzmann vibrational distribution was observed in the HCl product and by analysing the various relaxation processes, the absolute rates into each quantum state were investigated. Charters and Polanyi[163] employed total pressures of $\sim 10^{-2}$ torr, with HCl partial pressures of 10^{-4} torr, in order to avoid "Boltzmannisation" by V–V transfer. According to the equations of Section 3, $Z_{0,1}^{1,0}$ is 9 at 400 °K, corresponding to a relaxation time of $\sim 3 \times 10^{-2}$ sec for the process

$$HCl(v = 2) + HCl(v = 0) \rightarrow 2HCl(v = 1)$$

under their conditions. However, according to Findlay and Polanyi[164] Z for the exchange

$$HCl(v = 6) + HCl(v = 0) \rightarrow HCl(v = 5) + HCl(v = 1)$$

is ca. 5×10^{-3}. The process is apparently slowed down because there is an internal energy change of 620 cm^{-1} due to anharmonicity. This is an advantageous situation which protects the initial distribution in the higher vibrational levels. The time resolution should be determined not only by the rate of spontaneous radiation ($\sim 10^{-2}$ sec) and pumping rate, but also by wall deactivation. Under the conditions of Charters and Polanyi[163], the mean time for diffusion to the wall is $\sim 10^{-4}$ sec. At 10^{-2} torr total pressure, before returning to the bulk phase, a molecule will make $\sim 10^4$ collisions with the wall and therefore wall removal should be diffusion-controlled, providing an efficient trap for vibrationally excited species. From the $H + Cl_2$ reaction, the yield of vibrational energy is less than 10 % of the heat of reaction.

Formation of vibrationally excited species by photochemical methods also has been pioneered by means of the flash photolysis techniques. Photochemical dissociation of molecules invariably produces excited fragments though this is not always easy to demonstrate by kinetic spectroscopy. Polyatomic fragments relax extremely rapidly through low-frequency bending modes, and even vibrationally excited diatomic hydrides are difficult to detect because of rapid V–T transfer, and V–V transfer to the parent hydride. One of the most spectacular of the photochemical reactions is the formation of NO with $v \leqslant 11$ from dissociation of $NOCl^{56}$. Some V–V exchanges were investigated. Production of vibrationally excited CN radicals from photolysis of $(CN)_2$ and CNBr also has been studied in some detail[165]. Photolysis of CS_2 and CSe_2 produces excited CS and CSe respectively[59].

The most attractive and direct technique of studying reactive energy transfer is the method of crossed molecular beams (Volume 1, p. 172) in which reactants and products are completely isolated from destructive collision processes[166]. However, it does involve rather complex engineering and very low pressures. The concentration of products in the region of the crossing point is $\sim 10^{-9}$ torr, and the total yield of product would correspond to about one monolayer per month. Suitable sensitivity has so far been achieved only for alkali metal atoms and alkali metal compounds, by means of surface ionisation detectors. However, detectors may soon be available for halogens. If the yield of translational energy is small, the products are thrown forward in the direction of the initial centre of gravity vector, and considerations of angular momentum conservation show that the product beam must peak at certain specific angles, which has been verified experimentally. This technique has been confined largely to reactions of alkali metal atoms with alkyl halides, and almost the entire heat of reaction is usually retained as internal energy. Presumably there is little recoil or repulsive force at the point of the broken bond in this type of reaction.

12. Conclusion

As yet only V–T and V–V energy transfer have been investigated experimentally in any detail. Except for certain hydrides, the theory interprets the orders of magnitude of the observed data. Other types of energy transfer have been only super-ficially investigated, both experimentally and theoretically. The initial studies indicate that for some simple processes in which the mechanism is established with reasonable certainty, there is an approximate general law for the transfer of energy, such that for given masses and temperature

$$\log Z = A\Delta E + B$$

where ΔE is the energy which is converted to translation, and A is independent of the type of process. Such a law appears to hold because the "overlap" of the trans-lational wave functions is restrictive in all energy transfer processes.

REFERENCES

1 G. W. PIERCE, *Proc. Am. Acad. Sci.*, 60 (1925) 271.
2 O. OLDENBERG, *Phys. Rev.*, 37 (1931) 194.
3 L. LANDAU AND E. TELLER, *Physik. Z. Sowjetunion*, 10 (1936) 34.
4 C. ZENER, *Phys. Rev.*, 37 (1931) 556.
5 P. PRINGSHEIM, *Fluorescence and Phosphorescence*, Interscience, New York, 1949.
6 A. C. G. MITCHELL AND M. W. ZEMANSKY, *Resonance Radiation and Excited Atoms*, Oxford University Press, Oxford, 1949.
7 M. POLANYI, *Atomic Reactions*, Williams and Norgate, London, 1932.
8 W. R. BENNETT, JR., *Appl. Opt. Suppl.*, 1 (1962) 24.
9 A. B. CALLEAR AND R. G. W. NORRISH, *Proc. Roy. Soc. (London)*, A266 (1962) 299.
10 S. W. BENSON, *The Foundations of Reaction Kinetics*, McGraw-Hill, New York, 1960.
11 J. D. LAMBERT AND J. S. ROWLINSON, *Proc. Roy. Soc. (London)*, A204 (1950) 424.
12 T. L. COTTRELL AND J. C. McCOUBREY, *Molecular Energy Transfer in Gases*, Butterworths, London, 1961.
13 E. MEYER AND G. SESSLER, *Z. Physik*, 149 (1957) 15;
    W. TEMPEST AND H. D. PARBROOK, *Acustica*, 7 (1957) 354;
    R. HOLMES, H. D. PARBROOK AND W. TEMPEST, *Acustica*, 10 (1960) 155.
14 P. D. EDMONDS AND J. LAMB, *Proc. Phys. Soc. (London)*, 72 (1958) 940.
15 S. J. LUKASIK AND J. E. YOUNG, *J. Chem. Phys.*, 27 (1957) 1149.
16 H. A. BETHE AND E. TELLER, *Aberdeen Proving Ground Rept.*, X (1941) 117.
17 E. F. GREEN AND D. F. HORNIG, *J. Chem. Phys.*, 21 (1953) 617;
    E. F. SMILEY AND E. H. WINKLER, *J. Chem. Phys.*, 22 (1954) 2018;
    M. W. WINDSOR, N. DAVIDSON AND R. TAYLOR, *J. Chem. Phys.*, 27 (1957) 315;
    J. G. CLOUSTON, A. G. GAYDON AND I. R. HURLE, *Proc. Roy. Soc. (London)*, A252 (1959) 143.
18 A. KANTROWITZ, *J. Chem. Phys.*, 14 (1946) 150.
19 (a) A. B. CALLEAR, J. A. GREEN AND G. J. WILLIAMS, *Trans. Faraday Soc.*, 61 (1965) 1831.
    (b) A. B. CALLEAR AND G. J. WILLIAMS, *Trans. Faraday. Soc.*, 62 (1966) 2030.
20 R. C. MILLIKAN, *J. Chem. Phys.*, (a) 38 (1963) 2855; (b) 40 (1964) 2594; (c) 42(1965) 1439.
21 (a) B. STEVENS AND M. BOUDART, *Ann. N.Y. Acad. Sci.*, 67 (1957) 570.
    (b) A. B. CALLEAR AND I. W. M. SMITH, *Trans. Faraday Soc.*, 59 (1963) 1720, 1735.
22 R. L. BROWN AND W. KLEMPERER, *J. Chem. Phys.*, 41 (1964) 3072;
    J. F. STEINFELD AND W. KLEMPERER, *J. Chem. Phys.*, 42 (1965) 3475.
23 P. V. SLOBODSKAYA, *Izv. Akad. Nauk SSSR, Ser. Fiz.*, 12 (1948) 656.
24 M. E. JACOX AND S. H. BAUER, *J. Phys. Chem.*, 61 (1957) 833;
    W. E. WOODMANSEE AND J. C. DECIUS, *J. Chem. Phys.*, 36 (1962) 1831;
    M. G. FERGUSON AND A. W. READ, *Trans. Faraday Soc.*, 61 (1965) 1559.
25 J. M. JACKSON AND N. F. MOTT, *Proc. Roy. Soc. (London)*, A137 (1932) 703.

26 R. N. Schwartz, Z. I. Slawsky and K. F. Herzfeld, *J. Chem. Phys.*, 20 (1952) 159.
27 R. N. Schwartz and K. F. Herzfeld, *J. Chem. Phys.*, 22 (1954) 767.
28 F. H. Mies, *J. Chem. Phys.*, 40 (1964) 523.
29 E. E. Nikitin, *Opt.i Spektroskopiya*, 9 (1960) 16.
30 J. D. Lambert and R. Salter, *Proc. Roy. Soc. (London)*, A253 (1959) 277.
31 J. O. Hirschfelder, C. F. Curtis and R. B. Bird, *Molecular Theory of Gases and Liquids*, Wiley, New York, 1955.
32 K. F. Herzfeld and T. A. Litovitz, *Absorption and Dispersion of Ultrasonic Waves*, Academic Press, New York, 1959.
32 a K. F. Herzfeld, *Dispersion and Absorption of Sound by Molecular Processes*, Proceedings of the International School of Physics "Enrico Fermi", Course 27, D. Sette (Ed.), Academic Press, New York, 1964, p. 311.
33 J. L. Stretton, *Trans. Faraday Soc.*, 61 (1965) 1053.
34 F. H. Mies, *J. Chem. Phys.*, 41 (1964) 903.
35 F. Tanczos, *J. Chem. Phys.*, 25 (1956) 439.
36 P. G. Dickens and A. Ripamonti, *Trans. Faraday Soc.*, 57 (1961) 735.
37 H. Knötzel and L. Knötzel, *Ann. Physik (Leipzig)*, (6)2 (1948) 393.
38 V. H. Blackman, *J. Fluid Mech.*, 1 (1956) 61.
39 M. W. Windsor, N. Davidson and R. Taylor, *Symp. Combustion*, 10th (1959) 80.
40 E. G. Richardson, *J. Acoust. Soc. Am.*, 31 (1959) 152.
41 F. D. Shields, *J. Acoust. Soc. Am.*, 32 (1960) 180.
42 P. W. Huber and A. Kantrowitz, *J. Chem. Phys.*, 15 (1947) 275.
43 J. D. Lambert and R. Salter, *Proc. Roy. Soc. (London)*, A243 (1957) 78.
44 D. Sette, A. Busala and J. C. Hubbard, *J. Chem. Phys.*, 23 (1955) 787.
45 A. Eucken and S. Aybar, *Z. Physik. Chem.*, B46 (1940) 195.
46 H. J. Bauer, H. O. Kneser and E. Sittig, *J. Chem. Phys.*, 30 (1959) 1119.
47 K. L. Wray, *J. Chem. Phys.*, 36 (1962) 2597; F. Robben, *J. Chem. Phys.*, 31 (1959) 420.
48 N. Basco, A. B. Callear and R. G. W. Norrish, *Proc. Roy. Soc. (London)*, A260 (1961) 459.
49 P. Borrell, *Molecular Relaxation Processes*, The Chemical Society, London, 1966, p. 263.
50 E. Bauer and F. W. Cummings, *J. Chem. Phys.*, 36 (1962) 618.
51 K. Takayanagi, *Sci. Repts. Saitama Univ.*, 3A (1959) 101.
52 W. J. Hooker and R. C. Millikan, *J. Chem. Phys.*, 38 (1963) 214.
53 J. C. Decius, *J. Chem. Phys.*, 32 (1960) 1262.
54 E. W. Montroll and K. E. Shuler, *J. Chem. Phys.*, 26 (1957) 454.
55 F. J. Lipscomb, R. G. W. Norrish and B. A. Thrush, *Proc. Roy. Soc. (London)*, A233 (1956) 455.
56 N. Basco and R. G. W. Norrish, *Proc. Roy. Soc. (London)*, A268 (1962) 291.
57 R. V. Fitzsimmonds and E. J. Bair, *J. Chem. Phys.*, 40 (1964) 451.
58 A. B. Callear, *Proc. Roy. Soc. (London)*, A276 (1963) 401.
59 A. B. Callear and W. J. R. Tyerman, *Trans. Faraday Soc.*, 61 (1965) 2395.
60 K. E. Shuler and R. Zwanzig, *J. Chem. Phys.*, 33 (1960) 1778.
61 H. Roesler and K. F. Sahm, *J. Acoust. Soc. Am.*, 37 (1965) 386.
62 R. Holmes, G. R. Jones and R. Lawrence, *Trans. Faraday Soc.*, 62 (1966) 46.
63 T. A. Litovitz, *J. Chem. Phys.*, 26 (1957) 469;
   R. Bass and J. Lamb, *Proc. Roy. Soc. (London)*, A247 (1958) 168.
64 M. C. Henderson and L. Peselnick, *J. Acoust. Soc. Am.*, 29 (1957) 1074.
65 P. G. Corran, J. D. Lambert, R. Salter and B. Warburton, *Proc. Roy. Soc. (London)*, A244 (1958) 212.
66 Y. Fujii, R. B. Lindsay and K. Urushihara, *J. Acoust. Soc. Am.*, 35 (1963) 961.
67 J. R. Olson and S. Legvold, *J. Chem. Phys.*, 39 (1963) 2902.
68 S. D. Hamann and J. A. Lambert, *Australian J. Chem.*, 7(1954) 1.
69 A. Eucken and E. Nümann, *Z. Physik. Chem.*, B36 (1937) 163.
70 A. Eucken and L. Küchler, *Z. Tech. Physik.*, 19 (12) (1938) 517.
71 B. Widom and S. H. Bauer, *J. Chem. Phys.*, 21 (1953) 1670.
72 J. W. L. Lewis and K. P. Lee, *J. Acoust. Soc. Am.*, 38 (1965) 813.
73 T. G. Winter, *J. Chem. Phys.*, 38 (1963) 2761.

74 J. D. LAMBERT, D. G. PARKS-SMITH AND J. L. STRETTON, *Proc. Roy. Soc. (London)*, A282 (1964) 380.
75 L. M. VALLEY AND S. LEGVOLD, *J. Chem. Phys.*, 36 (1962) 481.
76 T. SESHAGIRI RAO AND E. SRINIVASACHARI, *Nature*, 206 (1965) 926.
77 J. D. LAMBERT, A. J. EDWARDS, D. PEMBERTON AND J. L. STRETTON, *Discussions Faraday Soc.*, 33 (1962) 61.
78 A. B. CALLEAR, *Discussions Faraday Soc.*, 33 (1962) 28.
79 A. B. CALLEAR, *J. Appl. Optics*, Supplement on Chemical Lasers, (1965) 145.
80 D. RAPP AND P. ENGLANDER-GOLDEN, *J. Chem. Phys.*, 40 (1964) 573; 3120.
81 D. G. JONES, J. D. LAMBERT AND J. L. STRETTON, *J. Chem. Phys.*, 43 (1965) 4541.
82 D. G. JONES, J. D. LAMBERT AND J. L. STRETTON, *Proc. Phys. Soc. (London)*, 86 (1965) 857.
83 J. C. POLANYI, *Chemistry in Britain*, 2 (1966) 151.
84 C. C. CHOW AND E. F. GREENE, *J. Chem. Phys.*, 43 (1965) 324.
85 R. G. W. NORRISH, *Discussions Faraday Soc.*, 33 (1962) 87.
86 J. D. LAMBERT, *Quart. Rev. (London)*, 20 (1967) 67.
87 R. HOLMES, G. R. JONES AND N. PUSAT, *Trans. Faraday Soc.*, 60 (1964) 1220; *J. Chem. Phys.*, 41 (1964) 2512.
88 K. TAKAYANAGI, in *Advances in Atomic and Molecular Physics*, D. R. BATES AND I. ESTERMANN (Eds.), Academic Press, New York, Vol. 1, 1965, p. 149.
89 R. BROUT, *J. Chem. Phys.*, 22 (1954) 934.
90 J. L. STEWART AND E. S. STEWART, *J. Acoust. Soc. Am.*, 24 (1952) 194.
91 K. TAKAYANAGI, *Proc. Phys. Soc. (London)*, A70 (1957) 348.
92 K. GEIDE, *Acustica*, 13 (1963) 31.
93 M. A. BREAZEALE AND H. O. KNESER, *J. Acoust. Soc. Am.*, 32 (1960) 885.
94 W. H. ANDERSEN AND D. F. HORNIG, *Mol. Phys.*, 2 (1959) 49.
95 G. B. KISTIAKOWSKY AND F. D. TABBUTT, *J. Chem. Phys.*, 30 (1959) 577.
96 R. BROUT, *J. Chem. Phys.*, 22 (1954) 1189.
97 M. GREENSPAN, *J. Acoust. Soc. Am.*, 31 (1959) 155.
98 R. HOLMES, G. R. JONES, N. PUSAT AND W. TEMPEST, *Trans. Faraday Soc.*, 58 (1962) 2342.
99 H. J. BAUER AND K. F. SAHM, *J. Chem. Phys.*, 42 (1965) 3400.
100 H. P. BROIDA AND T. CARRINGTON, *J. Chem. Phys.*, 38 (1963) 136.
101 C. S. WANG CHANG AND G. E. UHLENBECK, *University of Michigan Report, CM-681* (1951), Project N-Ord-7294.
102 B. WIDOM, *J. Chem. Phys.*, 32 (1960) 913.
103 B. T. KELLEY, *J. Acoust. Soc. Am.*, 29 (1957) 1005.
104 a J. G. PARKER, *Phys. Fluids*, 2 (1959) 449.
104 b N. F. SATHER AND J. S. DAHLER, *J. Chem. Phys.*, 37 (1962) 1947.
105 a E. H. CARNEVALE, C. CAREY AND G. LARSON, *J. Chem. Phys.*, 47 (1967) 2829.
105 b T. G. WINTER AND G. L. HILL, *J. Acoust. Soc. Am.*, 42 (1967) 848.
105 c C. H. TOWNES AND A. L. SCHAWLOW, *Microwave Spectroscopy*, McGraw-Hill, New York, 1955.
106 R. C. MILLIKAN AND L. A. OSBURG, *J. Chem. Phys.*, 41 (1964) 2196.
107 R. C. MILLIKAN, *Molecular Relaxation Processes*, The Chemical Society, London, 1966, p. 219.
108 T. L. COTTRELL AND A. J. MATHESON, *Trans. Faraday Soc.*, 58 (1962) 2336.
109 T. L. COTTRELL AND A. J. MATHESON, *Trans. Faraday Soc.*, 59 (1963) 824.
110 T. L. COTTRELL, R. C. DOBBIE, J. MCLAIN AND A. E. READ, *Trans. Faraday Soc.*, 60 (1964) 241.
111 C. BRADLEY MOORE, *J. Chem. Phys.*, 43 (1965) 2979.
112 K. TAKAYANAGI, *Progr. Theoret. Phys. (Kyoto)*, Suppl. No. 25 (1963) 1.
113 H.-J. BAUER AND E. LISKA, *Z. Physik*, 181 (1964) 356.
114 G. PORTER, *Discussions Faraday Soc.*, 33 (1962) 198.
115 A. B. CALLEAR AND G. J. WILLIAMS, *Trans. Faraday Soc.*, 60 (1964) 2158.
116 C. G. MATLAND, *Phys. Rev.*, 92 (1953) 637.
117 M. D. SCHEER AND J. FINE, *J. Chem. Phys.*, 36 (1962) 1264.
118 V. K. BYKHOVSKII AND E. E. NIKITIN, *Opt. i Spektroskopiya*, 16 (1964) 201.
119 S. PENZES, A. J. YARWOOD, O. P. STRAUSZ AND H. E. GUNNING, *J. Chem. Phys.*, 43 (1965) 4524.

120 A. B. CALLEAR AND W. J. R. TYERMAN, *Trans. Faraday Soc.*, 62 (1966) 2313.
121 A. B. CALLEAR AND R. OLDMAN, *Trans. Faraday Soc.*, 63 (1967) 2888.
122 A. V. PHELPS, *Phys. Rev.*, 114 (1959) 1011.
123 B. PITRE, A. G. A. RAE AND L. KRAUSE, *Can. J. Phys.*, 44 (1966) 731.
124 M. CZAJKOWSKI, D. A. McGILLIS AND L. KRAUSE, *Can. J. Phys.*, 44 (1966) 741.
125 G. D. CHAPMAN AND L. KRAUSE, *Can. J. Phys.*, 44 (1966) 753.
126 R. J. DONOVAN AND D. HUSAIN, *Trans. Faraday Soc.*, 62 (1966) 11.
127 A. B. CALLEAR AND J. F. WILSON, *Nature*, 211 (1966) 517.
128 W. R. THORSON, *J. Chem. Phys.*, 34 (1961) 1744;
    W. R. THORSON AND J. W. MOSKOWITZ, *J. Chem. Phys.*, 38 (1963) 1848.
129 J. G. CLOUSTON, A. G. GAYDON AND I. I. GLASS, *Proc. Roy. Soc. (London)*, A248 (1958) 429;
    A. G. GAYDON AND I. R. HURLE, *The Shock Tube in High-Temperature Chemical Physics*,
    Chapman and Hall, London, 1963.
130 P. J. PADLEY AND T. M. SUGDEN, *Proc. Roy. Soc. (London)*, A248 (1958) 248;
    H. P. HOOYMAYERS, *Dissertation*, Bronder-Offset, Rotterdam, 1966.
131 T. CARRINGTON, *J. Chem. Phys.*, 31 (1959) 1243.
132 G. KARL AND J. C. POLANYI, *J. Chem. Phys.*, 38 (1963) 271.
133 G. KARL, P. KRUUS AND J. C. POLANYI, *J. Chem. Phys.*, 46 (1967) 224.
134 P. G. DICKENS, J. W. LINNETT AND O. SOVERS, *Discussions Faraday Soc.*, 33 (1962) 52.
135 K. J. LAIDLER, *J. Chem. Phys.*, 10 (1942) 43.
136 J. C. POLANYI, *J. Quant. Spectry. Radiative Transfer*, 3 (1963) 471.
137 H. S. W. MASSEY AND E. H. S. BURHOP, *Electronic and Ionic Impact Phenomena*, Clarendon
    Press, Oxford, 1952.
138 N. F. MOTT AND H. S. W. MASSEY, *The Theory of Atomic Collisions*, Oxford University Press,
    Oxford, 1949.
139 B. BEUTLER AND H. JOSEPHY, *Z. Physik*, 53 (1929) 747.
140 S. G. RAUTIAN AND I. I. SOBELMAN, *Zh. Eksperim. i Teor. Fiz.*, 39 (1960) 217.
141 S. E. FRISH AND O. P. BOCHKOVA, *Zh. Eksperim. i Teor. Fiz.*, 43 (1962) 1831.
142 R. A. ANDERSON AND R. H. McFARLAND, *Phys. Rev.*, 119 (1960) 693;
    B. C. HUDSON AND B. CURNUTTE, *Phys. Rev.*, 148 (1966) 60;
    K. E. KRAULINYA AND A. E. LEZDIN, *Opt. i Spektroskopiya*, 20 (1966) 539.
143 A. V. PHELPS AND S. C. BROWN, *Phys. Rev.*, 86 (1952) 102.
144 A. JAVAN, W. R. BENNETT AND D. R. HERRIOTT, *Phys. Rev. Letters*, 6 (1961) 106. See also
    I. M. BETEROV AND V. P. CHEBOTAEV, *Opt. i Spektroskopiya*, 20 (1966) 1078.
145 E. E. BENTON, F. A. MATSON, E. E. FERGUSON AND W. W. ROBERTS, *Phys. Rev.*, 128 (1962)
    206;
    V. P. CHEBOTAEV AND L. S. VASILENKO, *Opt. i. Spektroskopiya*, 20 (1966) 505;
    L. COLOMBO, B. MARKOVIĆ, Z. PAVLOVIĆ AND A. PERSIN, *J. Opt. Soc. Am.*, 56 (1966) 890.
146 E. E. STEPP AND R. A. ANDERSON, *J. Opt. Soc. Am.*, 55 (1965) 31.
147 A. B. CALLEAR AND R. J. OLDMAN, *Trans. Faraday Soc.*, 64 (1968) 840.
148 J. G. CALVERT AND J. N. PITTS, JR., *Photochemistry*, Wiley, New York, 1966.
149 W. R. BRENNEN AND G. B. KISTIAKOWSKY, *J. Chem. Phys.*, 44 (1966) 2695.
150 A. B. CALLEAR AND I. W. M. SMITH, *Trans. Faraday Soc.*, 61 (1965) 2383.
151 N. H. SAGERT AND B. A. THRUSH, *Discussions Faraday Soc.*, 37 (1964) 223.
152 R. B. CUNDALL AND A. S. DAVIES, *J. Am. Chem. Soc.*, 88 (1966) 1329.
153 W. A. NOYES, JR. AND I. UNGER, in *Advances in Photochemistry*, W. A. NOYES, JR., G. S.
    HAMMOND AND J. N. PITTS (Eds.), Interscience, New York, Vol. 4, 1966, p. 49.
154 T. L. ANDREEVA, V. A. DUDKIN, V. I. MALYSHEV AND V. N. SOROKIN, *Opt. i Spektroskopiya*,
    20 (1966) 333.
155 F. WILKINSON, in *Advances in Photochemistry*, W. A. NOYES, JR., G. S. HAMMOND AND
    J. N. PITTS (Eds.), Interscience, New York, Vol. 3, 1964, p. 241.
156 L. S. KASSEL, *J. Phys. Chem.*, 32 (1928) 225.
157 A. F. TROTMAN-DICKENSON, *Gas Kinetics*, Butterworths, London, 1955.
158 B. S. NEPORENT, *Dokl. Akad. Nauk SSSR*, 72 (1950) 35.
159 M. BOUDART AND J. T. DUBOIS, *J. Chem. Phys.*, 23 (1955) 223.
160 B. STEVENS AND M. BOUDART, *Ann. N.Y. Acad. Sci.*, 67 (1957) 447.

161 B. S. RABINOVITCH AND M. C. FLOWERS, *Quart. Rev. (London)*, 18 (1964) 122.
162 R. G. W. NORRISH AND G. A. OLDERSHAW, *Proc. Roy. Soc. (London)*, A262 (1961) 1.
163 P. E. CHARTERS AND J. C. POLANYI, *Discussions Faraday Soc.*, 33 (1962) 107.
164 F. D. FINDLAY AND J. C. POLANYI, *Can. J. Chem.*, 42 (1964) 2176.
165 N. BASCO, J. E. NICHOLAS, R. G. W. NORRISH AND W. H. J. VICKERS, *Proc. Roy. Soc. (London)*, A272 (1963) 147.
166 D. R. HERSCHBACH, *Appl. Opt. Suppl.*, 2 (1965) 128.
167 J. T. YARDLEY AND C. BRADLEY MOORE, *J. Chem. Phys.*, 45 (1966) 1066; *ibid.*, 46 (1967) 4491; C. BRADLEY MOORE, R. E. WOOD, BEI-LOK AND J. T. YARDLEY, *J. Chem. Phys.*, 46 (1967) 4222; H.-L. CHEN AND C. BRADLEY MOORE, *Chem. Phys. Lett.*, 2 (1968) 542.
168 H. K. SHIN, *J. Am. Chem. Soc.*, 90 (1968) 3029; *Chem. Phys. Lett.*, 1 (1968) 635; 2 (1968) 83; *J. Quantum Chem.*, 11 (1968) 265.
169 S. L. THOMPSON, *J. Chem. Phys.*, 49 (1968) 3400; H. K. SHIN, *J. Chem. Phys.*, 48 (1968) 3644.
170 J. BILLINGSLEY AND A. B. CALLEAR, *Nature*, 221 (1969) 1136.
171 D. SECREST AND B. R. JOHNSON, *J. Chem. Phys.*, 45 (1966) 4556.
172 G. KARL, P. KRUUS, J. C. POLANYI AND I. W. M. SMITH, *J. Chem. Phys.*, 46 (1967) 2944.
173 B. H. MAHAN, *J. Chem. Phys.*, 46 (1967) 98.
174 W. BRENNEN AND T. CARRINGTON, *J. Chem. Phys.*, 46 (1967) 7.
175 D. KLEY AND K. H. WELGE, *J. Chem. Phys.*, 49 (1968) 2870.
176 T. OKA, *J. Chem. Phys.*, 45 (1966) 754; *ibid.* 47 (1967) 13, 4852; *ibid.*, 48 (1968) 4919; *ibid.*, 49 (1968) 3135.
177 A. M. RONN AND E. B. WILSON, *J. Chem. Phys.*, 46 (1967) 3262; A. M. RONN AND D. R. LIDE, *J. Chem. Phys.*, 47 (1967) 3669.
178 R. G. GORDON, *J. Chem. Phys.*, 46 (1967) 4399.
179 A. B. CALLEAR AND R. E. M. HEDGES, *Nature*, 218 (1968) 163.
180 KANG YANG, *J. Am. Chem. Soc.*, 99 (1967) 5344.
181 D. R. JENKINS, *Proc. Roy. Soc. (London)*, A293 (1966) 493.
182 R. E. ROBERTS, *J. Chem. Phys.*, 49 (1968) 2880.
183 D. C. TARDY AND B. S. RABINOVITCH, *J. Chem. Phys.*, 45 (1966) 3720; B. S. RABINOVITCH, Y. N. LIN, SIU C. CHAN AND K. W. WATKINS, *J. Phys. Chem.*, 71 (1967) 3715.

Index

26 - XII - 69